Topics in Applied Physics
Volume 101

Topics in Applied Physics

Topics in Applied Physics is a well-established series of review books, each of which presents a comprehensive survey of a selected topic within the broad area of applied physics. Edited and written by leading research scientists in the field concerned, each volume contains review contributions covering the various aspects of the topic. Together these provide an overview of the state of the art in the respective field, extending from an introduction to the subject right up to the frontiers of contemporary research.

Topics in Applied Physics is addressed to all scientists at universities and in industry who wish to obtain an overview and to keep abreast of advances in applied physics. The series also provides easy but comprehensive access to the fields for newcomers starting research.

Contributions are specially commissioned. The Managing Editors are open to any suggestions for topics coming from the community of applied physicists no matter what the field and encourage prospective editors to approach them with ideas.

Burkard Hillebrands   André Thiaville   (Eds.)

Spin Dynamics in Confined Magnetic Structures III

With 164 Figures, 42 in Color

Springer

Burkard Hillebrands

Fachbereich Physik
Technische Universität Kaiserslautern
Erwin-Schrödinger-Str. 56
67663 Kaiserslautern, Germany

hilleb@physik.uni-kl.de

André Thiaville

Laboratoire de Physique
Université Paris XI Orsay
91405 Orsay Cedex, France

thiaville@lps.u-psud.fr

Library of Congress Control Number: 2006921165

Physics and Astronomy Classification Scheme (PACS):
67.57; 33.15, 75.30, 85.70, 75.60

ISSN print edition: 0303-4216
ISSN electronic edition: 1437-0859
ISBN-10 3-540-20108-4 Springer Berlin Heidelberg New York
ISBN-13 978-3-540-20108-3 Springer Berlin Heidelberg New York

Springer is a part of Springer Science+Business Media

springer.com

Typesetting: DA-TEX · Gerd Blumenstein · www.da-tex.de
Production: LE-TEX Jelonek, Schmidt & Vöckler GbR, Leipzig
Cover design: WMXDesign GmbH, Heidelberg

Printed on acid-free paper 57/3100/YL 5 4 3 2 1 0

Preface

This is the third volume of the book series on "Spin Dynamics in Confined Magnetic Structures". The first two volumes (Topics in Applied Physics **83** (2002) and **87** (2003), hereafter referred to as SDI and SDII, respectively) introduced the bases of the subject, on the conceptual as well as on the experimental side.

Three years later, it appears worthwhile to present the results of the efforts of this active community in a consolidated way. Indeed, new experimental setups have been developed that have allowed investigation of samples of smaller dimensions down to smaller time scales. The results of these experiments have been compared in depth to theoretical predictions and, when necessary, new models better adapted to the experimental situation have been developed. Thus, we have now a better understanding of magnetization dynamics, be it of a macrospin or of more complex objects like a vortex or a domain wall, as will be described in this book. Moreover, we would like also to continue the trend established in the first two volumes, by including the new experimental techniques and application concepts that have appeared in recent years. Prominent among these is the concept of the spin-transfer torque, evoked earlier only in Chap. 6 of SDII, to which a large chapter is devoted here.

In terms of the three time scales mentioned in SDII, most of the content of this volume belongs to the micromagnetic precessional regime (100 ps–10 ns), with the notable exception of the contribution by Torsten Ohms et al. on hot electrons in GaAs.

The first chapter by Thibaut Devolder, Hans-Werner Schumacher and Claude Chappert describes one of the first experimental demonstrations of precessional switching in small elements patterned out of thin metallic films. It is an all-electrical setup with coplanar waveguides and detection of the magnetization state by the GMR effect. In order to understand finely this process, without having to rely on numerical simulation for each case, the authors introduce an analytic approximate solution of the Landau–Lifshitz and Gilbert equation in the macrospin regime. This work, which builds on Chap. 1 in SDI, is emblematic of the research conducted presently in the domain of fast magnetization dynamics, where damping is introduced as a perturbation. The same technique will also be employed in Chap. 7 of this volume addressing spin-transfer torque and dynamics. The analytic approach has, on top of its advantages in terms of physical clarity, an operational power. The authors can thus propose a detailed evaluation of the potential of precessional switching in a realistic model of a magnetic random access memory (MRAM).

In the second chapter, Christian Bayer, Jörg Jorzick, Sergej O. Demokritov, Andrei N. Slavin, Konstantin Y. Guslienko, Dmitry V. Berkov, Natalia L. Gorn, Mikhail P. Kostylev and Burkard Hillebrands revisit the spin-wave spectra in patterned rectangular elements. The beginning of this research was described in Chap. 2 in SDI. Since that time, new experimental results have been obtained on more complex geometries, and these are presented here. Most notable, however, is the progress in the understanding of these modes. The results presented in SDI indeed did trigger much theoretical activity. Micromagnetic calculations were first performed, and could justify the initial analysis of the experimental data as quantized Damon–Eschbach modes. The cost of such simulations is, however, quite high so they cannot be repeated for all experimental cases. Therefore, the analytic modeling of the spin waves in such samples was extended, as explained in this contribution. Through an accurate calculation of the inhomogeneous magnetostatic field, it is shown that the mode profiles and frequencies can be extremely well reproduced.

Many of the magnetic microscopy techniques can be extended to the dynamic regime so as to map the local dynamic response of the magnetization: MOKE (Chap. 4 in SDI, Chap. 3 in SDII, and Chap. 4 of this volume), and XMCD (Chap. 5 in SDII). For atomic force microscopy, a new mechanical detection scheme based on magnetic resonance was developed in recent years (MRFM: magnetic resonance force microscopy). In chapter 3, Philip E. Wigen, Michael L. Roukes and P. Chris Hammel describe the application to ferromagnetic materials of that technique, called ferromagnetic resonance force microscopy (FRFM). They show how the magnetic modes (that are the objects studied in the preceeding and next contributions also) can be characterized in frequency and in space, and even influenced by the tip field. Although the present spatial resolution is still largely above a micrometer, the single-spin sensitivity that was demonstrated by MRFM shows that, by downsizing the probe, FRFM could provide the same information at scales well below one micrometer.

During recent years, the dynamics of well-defined magnetic structures within a confined sample has attracted a lot of attention. The best known example of such structures is the magnetic vortex in a thin film. In two dimensions, this is a point structure of size of the order of the exchange length, with an internal "angular momentum" that affects its dynamics. Chapter 4 by Christian Back, Danilo Pescia and Martin Buess focuses on this structure. Recent experimental achievements that have revealed the spatial profiles and frequencies of the modes in structures containing a vortex are reviewed. No static field is applied here in order to keep the domain structure. This chapter reveals also the results of advanced analytical modeling of the magnetization vibration modes in such samples. Such modeling, in which the magnetostatic interaction is central, is similar to that exposed in the preceding chapter.

An equally important magnetic structure whose dynamics was well studied is the domain wall. The sample geometry adapted to a domain wall involves confinement along two dimensions only, thus forming nanowires. Chapter 5 by André Thiaville and Yoshinobu Nakatani describes first the taxonomy of domain walls in nanowires, so as to understand in detail their dynamics under a field. This subject is old if one considers samples that are not confined or confined only in one dimension. The authors show how this previous knowledge can be used and has to be adapted to the nanowire geometry. Here again, an analytic model is introduced so as to understand physically the results, both from numerical simulations and from experiments reported in the literature, including those reported in Chap. 6.

Chapter 6 by Del Atkinson, Colm C. Faulkner, Dan A. Allwood and Russel P. Cowburn dwells also on domain-wall dynamics in nanowires, this time having in mind an application to magnetic logic. In this radically new concept, domain walls, as they mark the transition between two magnetic states of the wire, materialize by their motion the process flow in a calculation. Clever design of wire shapes and connections is shown to lead to logical operations. Important experimental results on domain-wall propagation at high speed in nanowires are also presented, as well as the experimental technique to obtain them (local magneto-optical Kerr effect).

The study of the spin-transfer effect has developed very rapidly in recent years, so that this area is now an extremely active field in magnetics. Spin transfer occurs when a large spin-polarized current flowing through a magnetic material affects its magnetization. Chapter 7 by Mark D. Stiles and Jacques Miltat provides first an overview of this blossoming subject, quoting many experimental results obtained using a variety of structures. The confinement of the magnetic sample is here essential, in order that the spin-transfer torque dominates over the torque linked to the field created by the current itself (the so-called Œrsted field). The chapter proposes a step-by-step understanding of the physics of this effect, starting from the quantum mechanical treatment of the electron spin and reaching the micromagnetic regime. It is restricted to metals, like the vast majority of available data. Spin transfer leads to static effects (switching by current) that initially attracted much activity. But most of the research now concentrates on the magnetization dynamics under that torque as spectacular effects, like the generation and control of GHz magnetization precession by a DC current, were predicted and observed. The contribution describes the already quite good understanding of this new effect, but also highlights the gaps that remain between experiments and theory.

The study of the fundamental processes governing the spin dynamics has become more and more important. For example, as discussed in the preceding contribution, this controls the spin transfer between a spin-polarized current and the magnetization. Semiconductors and, especially, magnetic semiconductors emerge now as promising materials for spintronics applications, where electrical currents and magnetization interact. Such materials have indeed the advantages of a good matching to semiconductor circuits, and of high spintronic effects because of their small magnetization. The magnetic semiconductor most studied to date is GaAs:Mn, with Curie temperatures still rising as the chemistry of the manganese ions in the lattice is better controlled. Chapter 8 by Torsten Ohms, Kevin Hiebbner, Hans-Christian Schneider and Martin Aeschlimann starts with a presentation of the various spin-relaxation processes in semiconductors. Experimental results at the GaAs/metal interface are then shown, and the technique that provided them (two-photon photoemission) is presented in detail. The relaxation time for hot electrons is found to be larger than in the bulk. An explanation of this surprising fact is proposed and tested experimentally. These results open interesting perspectives for the control of spin relaxation in magnetic heterostructures involving semiconductors.

This book is intended both for graduate students and for researchers in the field looking for an up-to-date overview of a subject they are not specialists in. Some chapters start from the basics, others offer a guided tour through the recent and abundant literature, and several also contain original results. We hope that this third volume of the series does justice to the field of magnetization dynamics in confined magnetic structures in witnessing to its sustained activity.

Kaiserslautern, Orsay
November 2005

Burkard Hillebrands
André Thiaville

Contents

Precessional Switching of Thin Nanomagnets with Uniaxial Anisotropy

Thibaut Devolder, Hans Werner Schumacher, and Claude Chappert

Institut d'Electronique Fondamentale, CNRS UMR 8622, Bât. 220, Université Paris-Sud, 91405 Orsay, France
thibaut.devolder@ief.u-psud.fr

Abstract. This review describes the evolution of the magnetization of uniaxial thin magnets when subjected to fast-rising magnetic-field pulses. We report detailed "all-electrical" experimental investigations of precessional switching on soft uniaxial micrometer-sized thin magnets, and we discuss them using a comprehensive, mostly analytical framework. General criteria are derived for the analytical assessment of the switching ability of any arbitrary set of experimental parameters. For this, we start from the Landau–Lifshitz equation and first consider the precessional switching in a much idealized macrospin, easy-plane loss-free system. We then test the main outputs of this model with time-resolved experiments on advanced Magnetic Random Access Memories (MRAM) cells. Using applied fields above the anisotropy field H_k, we prove the quasiperiodic nature of the magnetization trajectory and we demonstrate experimental conditions ensuring a sub-200 ps ballistic magnetization reversal. We then upgrade our model accuracy by taking into account the uniaxial anisotropy and the behavior in hard-axis fields of the order of H_k. We derive a simple though reliable estimate of the switching speed; its limiting factors highlight the experimental poor switching reproducibility when close to the minimal hard-axis reversal field $H_\mathrm{k}/2$. The latter field does not correspond to the minimal energy cost of the reversal, whose prospective evolution in the future generations of MRAM is predicted. Small departures from the macrospin state are discussed. The effect of damping is modeled using perturbation theory. Finite damping alters the precessional motion periodicity and puts some constraints on the field rise time. A special focus is dedicated to the relaxation-dominated precessional switching: the minimal hard-axis field triggering the switching is shown to be above $H_\mathrm{k}/2$ by an extra field cost linked to the damping constant times the square root of $M_\mathrm{S}H_\mathrm{k}$. Finally, the selective addressing and the direct-write of a magnetic cell with combined easy-axis and hard-axis fields are studied. We introduce the concept of bounce and revisit the dynamical astroid to derive the related characteristic reversal durations and their margins. We propose a field timing that is immune to the delay jitter between the combined addressing fields. We finish by investigating briefly the challenges and the promises of the "precessional" strategy for future MRAM generations.

1 Introduction

Switching the magnetization of ferromagnetic objects far in the subnanosecond regime has recently become one of the most challenged objectives in the

B. Hillebrands, A. Thiaville (Eds.): Spin Dynamics in Confined Magnetic Structures III, Topics Appl. Physics **101**, 1–56 (2006)

applied magnetism community [1]. The review articles gathered in this book are an illustration of the intense activity that has been carried out on this topic in the last few years on both the experimental front and the theoretical side. The first revolution came from the development of innovative magneto-optical and magnetotransport setups that have pushed time resolution far into the subnanosecond regime. These new opportunities have triggered a wealth of propositions. Among them, one of the most innovative has been certainly the concept of "precessional switching" [2], i.e., a switching strategy that takes advantage of the inherently high frequencies associated with the magnetic precession phenomenon. In a precessional-switching scenario, a sufficiently fast-rising field applied roughly perpendicularly to the magnetization can toggle the magnetization from one direction to its reverse in a quite fast (subnanosecond) manner with a reduced energy cost.

The main difference between "standard" quasistatic switching and precessional switching relies on the path that the magnetization follows to reverse. In a quasistatic reversal event, the magnetization follows the most direct energy-minimization path. In a simplistic picture of the precessional scenario, the magnetization rotates far from equilibrium conditions, using a path of only slightly decreasing energy.

The detailed concept of precessional switching was introduced in the contribution of *Miltat* et al. [2] in the first volume of this series. This Chapter focused on macrospin samples and on the additional effects that arise when the sample has a more complex micromagnetic state. The subsequent Chapter of *Fassbender* and coworkers [3, 4] has contributed a lot to the theoretical understanding of precessional switching and has discussed in particular the quantitative requirement regarding the rise time of the applied field. The first experimental evidence for the existence of precessional switching events came in 2002 [5–10], and was partly reported in the second volume of this book.

Much enlightening information has already arisen from these previous reviews. However, because of multiple-parameter intricacy (damping, field magnitude and orientation, rise time, anisotropy, saturation magnetization), only snapshots in the parameter space have been reported so far, obtained through cumbersome numerical computations or through a few experiments.

In this Chapter, we report very detailed "all-electrical" experimental investigations [5, 6] of the physics of precessional switching on soft uniaxial nanomagnets, and we discuss them using a comprehensive, mostly analytical framework [11–14]. Since our measurement method is convenient and versatile, a large portion of the parameter space could be explored experimentally. As a result, we can show the main characters of the precessional switching: its quasiperiodic nature together with the existence of near-ballistic magnetization trajectories with sub-200 ps reversal durations. The scope of this Chapter is also to crosslink these experimental data with a fully analytical analysis of the precessional switching, so as to derive the general behavior versus the full set of magnetic parameters. In so doing, we express very general criteria for the success of a precessional switching strategy. We try and keep a simple though rather accurate formalism, which we hope to be sufficiently transparent to underline the relevant physical parameters.

Let us begin with a quick historical overview of precessional switching.

2 Basic Facts About Precessional Dynamics in Transverse Field

2.1 From Stoner–Wohlfarth Energy Minimization to Precessional Switching

When aiming at reversing the magnetization of ferromagnetic body, the conventional strategy is to apply a field in the direction of the desired final magnetization state. This often means applying a field $\boldsymbol{H}$ antiparallel to the spatial average of the initial magnetization $\boldsymbol{M}$ as is, for instance, practised in hard-disk technology. However, in the case of a macrospin (i.e., a magnetic body with perfectly uniform magnetization), this antiparallel field orientation generates no torque on $\boldsymbol{M}$. The ferromagnetic body thus needs to wait for some thermal fluctuation of its magnetization to gain a finite susceptibility. As a result, applying $\boldsymbol{H}$ antiparallel to $\boldsymbol{M}$ is ineffective when the reversal is aimed to be faster than the typical thermal attempt frequency, i.e., some nanoseconds.

When the macrospin has a uniaxial anisotropy, a more effective reversal strategy is a compromise between the initial torque magnitude $\boldsymbol{H} \times \boldsymbol{M}$ and the unbalance of the energies $\boldsymbol{H} \cdot \boldsymbol{M}$ of the initial and reversed magnetization positions. The nonvanishing torque (i.e., $\boldsymbol{H}$ not antiparallel to $\boldsymbol{M}$) ensures that the system has a finite susceptibility, while the imbalance of the energies (i.e., $\boldsymbol{H}$ not perpendicular to $\boldsymbol{M}$) determines the final state the system will preferably end in.

In a pioneering paper [15] where the sole energy imbalance was considered, Stoner and Wohlfarth showed that the lowest switching field was obtained for an angle of 135° between $\boldsymbol{H}$ and $\boldsymbol{M}$. Their theory assumed coherent motion (i.e., macrospin behavior) and zero temperature, and derived the critical fields H_c from the sole criteria of local energy minimization. This gives the well-known "quasistatic astroid" curve (Fig. 1) when plotting the orientation dependence of H_c in the plane of the macrospin.

Despite the generality of this result, the quasistatic astroid does not necessarily give the smallest possible switching field of a macrospin at $T = 0\,\mathrm{K}$. The reason is that in the Stoner–Wohlfarth scenario the field is swept *infinitely slowly* so that energy dissipation is *always* fast enough to stabilize the magnetization in the nearest energy minimum. While the field is increased *slowly*, the system *always* maintains a vanishing torque, by *continuously* dissipating the excess energy to stay in the *continuously* changing equilibrium state [11]. Note that this continuously dissipated energy may be considered as wasted.

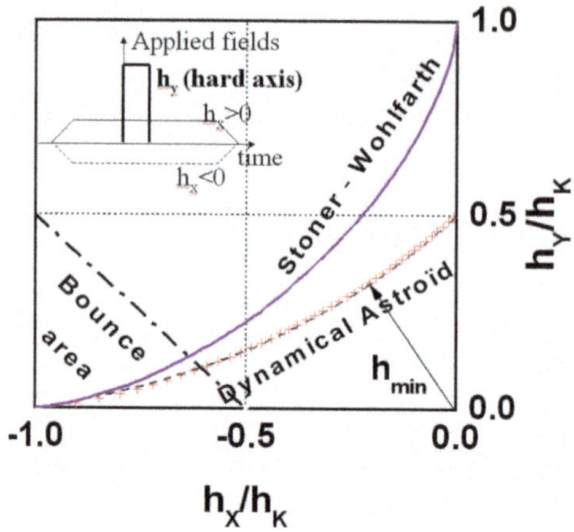

Fig. 1. Stoner–Wohlfarth astroid and its dynamical counterpart. Asymptotic analytical expression of the dynamical astroid near [$h_x = -h_k$ (44), *dotted line*], full interpolation [DynAst function (45), *dashed line*] and exact numerical value for $\alpha = 0.001$ (from [2], *cross symbols* almost superimposed to the interpolation). The bounce-occurrence criterion (42) is plotted in the *dashed-dotted line.* It intersects the dynamical astroid at $h_x = -0.605h_k$ and $h_y = 0.105h_k$. The smallest reversal field (*arrow* $h_{\min}$) obeys $h_x = -0.224h_k$ and $h_y = 0.315h_k$. Its magnitude and orientation are $\|h_{\min}\| = 0.387h_k$, $\{h_{\min}, x\} = 125°$. *Inset:* Timing of the easy- and hard-axis magnetic field pulses

However, if the field is swept very rapidly, i.e., much faster than the energy-loss rate, the magnetization undergoes a finite torque and then evolves without being in an energy local minimum for a while. Hence, the primary assumption of the Stoner–Wohlfarth model is no longer valid. The magnetization motion is mostly determined by the (precession) torques and marginally by the energy minimization. For that reason, these reversal processes are generally gathered in the terminology as "precessional switching". The corresponding switching fields and precessional switching trajectories are rather different from that of the quasistatic astroid: the magnetization escapes substantially out of the thin-film plane and the magnetization does not follow the most direct energy-minimization path.

In 1994, *He* et al. [16,17] were one of the first groups to realise that the precessional switching fields could be lower than that of the quasistatic astroid. Using numerical computations, they showed that due to the precessional nature of the magnetization motion, fast-rising applied fields could induce very fast switching events for fields even smaller than the Stoner–Wohlfarth ones. This was confirmed in more detail by several authors and reviewed in SDI by *Miltat* et al. [2]. The switching fields were predicted versus their orientations with respect to the easy axis; a so-called "dynamical astroid" (Fig. 1) was introduced as the precessional analog of the Stoner–Wohlfarth astroid. The effect of finite rise time and damping were studied by numerical integration of the Landau–Lifshitz–Gilbert equation [18, 19] (LLG) and then reviewed by *Fassbender* [3,4]. Deep subnanosecond precessional switching events, lasting half a precession period (typically 200 ps) were experimentally confirmed in late 2001 at the MMM conference and published shortly after [5,6]. Some of these results were discussed in the SDII by *Russek* et al. [9,10] and *Rasing* and coworkers [7,8]. Later, the first attempts to address the direct-write reliability of precessional switching were reported in a crosswire configuration mimicking a magnetic random access memory (MRAM) architecture [20,21].

In this Chapter, we report very detailed experimental studies of the precessional switching, using an "all-electrical" setup and advanced MRAM cells fabricated by R. C. Sousa and P. P. Freitas. Analytical descriptions of precessional switching are highly desirable to avoid time-consuming numerical integrations and to help highlight these when they cannot be avoided; we discuss our experimental results in that spirit.

As a starting point, we first detail (Sect. 2) an analytical study [12] of an idealized precessional switching event at high applied fields (Sect. 2.3).

We then (Sect. 3) report detailed experiments proving unambiguously the concept of precessional switching. On samples (Sect. 3.1) mimicking an advanced MRAM architecture, we demonstrate [5,6] the quasiperiodic nature of the magnetization trajectory (Sect. 3.2) and we study the conditions to be fulfilled for a ballistic, no-ringing magnetization reversal (Sect. 3.3).

We then (Sect. 4) model precessional switching at moderate transverse fields H_y of the order of the anisotropy field H_k. The values are taken to mimic representative standard MRAM cells [13]. Three cases can occur (Sect. 4.1). Low fields (i.e., $H_y < H_\mathrm{k}/2$) do not permit switching. At moderate fields $H_\mathrm{k} > H_y > H_\mathrm{k}/2$, the switching trajectory depends critically on the anisotropy, while for $H_y > H_\mathrm{k}$ the anisotropy plays no significant role. The switching speed (Sect. 4.2) can be modeled using physical arguments to divide a reversal event into three meaningful steps. These steps help to understand why the reproducibility of the switching is questionable when attempting to reach the ideal minimal field cost $H_\mathrm{k}/2$ (Sect. 4.3). Long-range fluctuations of the magnetization (i.e., small departures from the macrospin state) can be qualitatively dealt with using the same simplified formalism (Sect. 4.4). This helps to assess the technological potential of precessional switching on more realistic devices, and to evaluate 1. the minimal energy cost of the reversal (Sect 4.5), and 2. how it may scale in future generations of MRAM (Sect. 4.6).

In Sect. 5, the effect of damping is analyzed [13] in a perturbative manner. Damping both alters the periodicity of the precession and puts additional constraints on the field rise time (Sect. 5.1). The decay rate of the energy determines the effective number of precession cycles gone through (Sect. 5.2) before the system slows down towards rest. Criteria are derived for the analytical assessment of the switching ability of a given set of experimental parameters. A special focus is dedicated to the relaxation-dominated precessional switching, where the minimal cost in applied field is shown to increase linearly with the damping constant, the slope of this dependence scaling with the square root of the anisotropy. Also, for some relaxation conditions, the magnetization toggles only once whatever the pulse duration above a given threshold (Sect. 5.3).

In Sect. 6, addressing with combined easy-axis and hard-axis fields is reviewed [14]. We introduce the concept of bounce occurrence (Sect. 6.2) and revisit the dynamical astroid (Sect. 6.3) using a semianalytic formalism. This approach allows us to derive semiquantitative trends of the reversal duration for combined easy- and hard-axis fields (Sect. 6.4). Finally, we discuss the issue of selectivity and direct-write in a potential "precessional" MRAM (Sect. 6.5) before suggesting further readings for the technical optimizations [14, 22, 23].

2.2 Definitions: Geometry and Notations

In this Chapter we shall describe the magnetization dynamics using the Landau–Lifshitz–Gilbert equation [18, 19] discussed thoroughly in SDI [2]:

$$\frac{\mathrm{d}\boldsymbol{M}}{\mathrm{d}t} = \gamma_0 \boldsymbol{H}_{\mathrm{eff}} \times \boldsymbol{M} - \frac{\alpha}{M_{\mathrm{S}}} \left(\frac{\mathrm{d}\boldsymbol{M}}{\mathrm{d}t} \times \boldsymbol{M} \right), \qquad (1)$$

where $\gamma_0 = 221\,\mathrm{kHz} \cdot \mathrm{A}^{-1} \cdot \mathrm{m}$ is the gyromagnetic factor, chosen positive. The instantaneous effective field is the sum of the applied field $\boldsymbol{H}$, the anisotropy field $\boldsymbol{H}_{\mathrm{k}}$ and the demagnetizing field $\boldsymbol{H}_{\mathrm{demag}}$. We will restrict our discussion to the macrospin approximation, where the magnetization $\boldsymbol{M}$ is assumed uniform and the exchange field is zero. The relaxation towards equilibrium is described phenomenologically by the damping constant α. Throughout this Chapter, SI units are used: $\boldsymbol{H}$ and $\boldsymbol{M}$ are in Ampere per meter, $\mu_0 \boldsymbol{H}$ and $\mu_0 \boldsymbol{M}$ are in Tesla, and $\gamma_0 H$ and $\gamma_0 M_S$ are angular frequencies.

The system has an easy axis along x, i.e., in the main dimension of the (assumed thin and flat) sample. This easy axis is assumed to arise from both shape and magnetocrystalline uniaxial anisotropy. The macrospin will be treated as an ellipsoid with principal axes along x, y and z, and corresponding demagnetizing factors N_x, N_y and N_z. We are concerned with a thin-film macrospin, such that $N_x \le N_y \ll N_z \approx 1$.

The external field $\boldsymbol{H}$ will always be applied in the plane of the sample. Its projections are written H_x along the easy axis x and H_y along the hard axis y. Longitudinal (respectively, transverse) fields refer to fields along x (resp. along y). We define $h_{x,y} = H_{x,y}/M_{\mathrm{S}}$ their reduced strengths. We define $\boldsymbol{m} = \boldsymbol{M}/M_{\mathrm{S}}$ the reduced magnetization, with m_x, m_y, m_z being its projections. The initial magnetization will always be assumed at rest exactly along the $+x$-axis, except in Sect. 4.4 for which $\boldsymbol{m}_{(t=0)}$ is arbitrary.

The magnetocrystalline anisotropy field is along x and it is $H_{\mathrm{k}} = h_k M_{\mathrm{S}} m_x$. We consider that $h_k + (N_y - N_x) > 0$, consistent with an easy axis along x. For simplicity we write the equations with only the parameter h_k assuming that $N_z = 1$. The demagnetizing field is considered solely along z and is $H_{\text{demag}} = -M_S m_z$. To take into account the inplane components of the demagnetizing field our equations can be straightforwardly generalized by systematically replacing h_k by $h_k + (N_y - N_x)$. Under these assumptions, the total reduced energy $e(t) = \frac{E(t)}{\mu_0 M_S^2}$ of the system is:

$$e(t) = \tfrac{1}{2}(m_z^2 - h_k m_x^2) - h_y m_y - h_x m_x \,. \tag{2}$$

In the majority of this Chapter, we will consider that damping is negligible. It is thus convenient to rewrite the LLG equations in the frame of the sample:

$$\dot{m}_x = \gamma_0 M_S m_z (m_y + h_y)\,, \tag{3a}$$
$$\dot{m}_y = -\gamma_0 M_S m_z [(1 + h_k) m_x + h_x]\,, \tag{3b}$$
$$\dot{m}_z = -\gamma_0 M_S m_x (h_y - h_k m_y) + \gamma_0 M_S h_x m_y \,. \tag{3c}$$

Exact expressions where we consider h_y, $h_x \ll 1$ or $h_k \ll 1$ are displayed using the symbol "$\cong$". Approximate expressions implying other arguments use the symbol "$\approx$".

2.3 A Tutorial Example: The Bascule Precession of a Loss-Free Isotropic Thin Film

Before going into the details of precessional switching in complicated situations, we would like first to introduce the main specificities of precessional switching in a much simplified situation, where the relevant parameters come out straightforwardly from an analytical formalism. In this subsection [12], we make several major simplifications: we assume a macrospin behavior, an infinitely extended thin film ($N_z = 1$, $N_x = N_y = 0$), an isotropic easy plane ($h_k = 0$), a purely transverse field ($h_x = 0$) applied abruptly at $t > 0$ and energy conservation ($\alpha = 0$). We first search for the trajectory of the magnetization vector.

2.3.1 Magnetization Trajectory for a Loss-Free Isotropic Film Subjected to a Transverse Field

Using the initial conditions $m_x = 1$ and applying abruptly the field h_y at $t = 0$, the magnetization trajectory can be obtained from the direct integration of the LLG equations, or equivalently from a combination of the energy conservation and the magnetization norm conservation:

$$m_z^2 = 2 h_y m_y \,. \tag{4}$$

This equation means that $m_y \geq 0$ after $t > 0$. As is intuitive, the y-component of the magnetization always points towards the (transverse) applied field h_y.

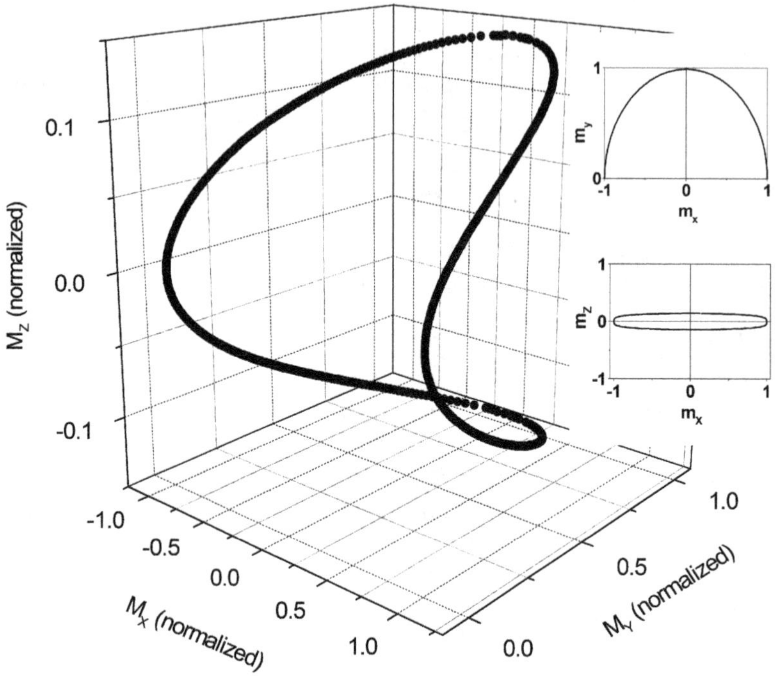

Fig. 2. Trajectory (4)–(6) of the magnetization vector of a loss-free macrospin isotropic thin film of initial magnetization along x when subjected to a transverse field $h_y = 0.01$ applied along y. The items "+1" and "−1" refer to the initial and reversed magnetizations. The items "+0" and "−0" refer to the positions when the magnetization is above or below the transverse axis y. Note the expanded scale of the m_z-axis, and the offcentered m_y-axis. *Upper inset:* quasi-half-circle trajectory in the xy-plane (easy plane). *Lower inset:* trajectory projected in the xz-plane (out of the easy plane incursion)

Also, since $h_y \ll 1$, the out-of-plane magnetization component m_z will remain small along the trajectory. The other projections of the trajectory are:

$$m_x^2 = 1 - m_z^2 - \frac{m_z^4}{4h_y^2}, \tag{5}$$

$$m_x^2 = 1 - 2h_y m_y - m_y^2 . \tag{6}$$

These trajectories are displayed in Fig. 2 for a typical value of $h_y = 0.01$, which would correspond to 10.8 mT for a Permalloy macrospin. The $\boldsymbol{M}$ trajectory is almost a semicircular back and forth rotation in the xy-plane and does not escape much out of the sample plane (Fig. 2).

The reversal is as follows. At the onset of the transverse field H_y, the torque $\boldsymbol{h} \times \boldsymbol{m}$ is along $-z$ and expels $\boldsymbol{m}$ below the film plane (Fig. 2, item "+1"). This m_z-component, in turn, generates a demagnetizing field $\boldsymbol{H}_{\mathrm{demag}}$ also oriented perpendicular to the plane but pointing in the direction opposite to m_z.

Under the torque $\boldsymbol{H}_{\text{demag}} \times \boldsymbol{M}$, the magnetization is set in a forward rotational motion slightly below the film plane. During this semicircular motion, $\boldsymbol{H}_{\text{demag}}$ and $\boldsymbol{M}$ stay oriented almost perpendicularly to each other allowing for fast quasi-inplane rotation of $\boldsymbol{M}$ (Fig. 2, item "+0"). When the magnetization finally comes back in the film plane, $\boldsymbol{m}$ points opposite to its initial state and the demagnetizing field has vanished (Fig. 2, item "−1"). The magnetization has switched once, in a bascule fashion. If the h_y pulse terminates, then a switched state is stabilized. If the field is let on, the sole remaining torque $\boldsymbol{h} \times \boldsymbol{m}$ is now along $+z$ and expels $\boldsymbol{m}$ above the film plane, thus resetting a finite demagnetizing field, and a backward inplane rotation starts until a full precession cycle has been gone through. The magnetization has then switched back and forth in a double-bascule fashion (Fig. 2, back to item "+1"), and the precession can in principle iterate again and again (remember that $\alpha = 0$).

Let us calculate the magnitude of the relevant torques. From (5) and (6), we get the maxima of m_z and m_y occurring for $m_x = 0$:

$$m_z^{\max} = \pm\sqrt{2h_y\left(\sqrt{1+h_y^2} - h_y\right)} \cong \pm\sqrt{2h_y}\,, \tag{7}$$

$$m_y^{\max} = \sqrt{1+h_y^2} - h_y \cong 1 - h_y\,. \tag{8}$$

In practical cases where $h_y \ll 1$, the maximum demagnetizing field occurring at $m_x = 0$ reduces to $H_{\text{demag}}^{\max} \approx -\sqrt{2H_y M_S}$. In this expression, the square root can be seen as an amplification factor: the maximum demagnetizing field is very much greater than the applied field H_y. For instance, if $\mu_0 H_y = 10.8\,\text{mT}$, then $\mu_0 H_{\text{demag}}$ can reach up to 153 mT in Permalloy. This corresponds to a maximum out-of-plane incursion angle of 8°. This high value of H_{demag} results in a high rotation speed of the magnetization. For instance at $m_x = 0$ (half-way of reversal), its rate of change is $\dot{m}_x \approx \gamma_0\sqrt{2H_y M_S}$.

Along the trajectory, the energy flows back and forth from the Zeeman contribution when $|m_x| = 1$ (Fig. 2, items "±1") to the demagnetizing energy when $m_x = 0$ (Fig. 2, items "±0"). As a result, the magnitude of the effective field varies a lot along the trajectory and cannot be used to derive a characteristic frequency, which would be though desirable. In the next section, we report a simple variational approach that derives the switching frequency using basic periodic trial functions of $\boldsymbol{m}(t)$, expanded in their main Fourier components.

2.3.2 Switching Speed for a Loss-Free Isotropic Film Subjected to a Transverse Field

Figure 3 displays the time evolution of the magnetization of a Permalloy macrospin subjected to $\mu_0 H_y = 10.8\,\text{mT}$, as calculated from direct numerical integration of the LLG equation. From this figure and the periodic character

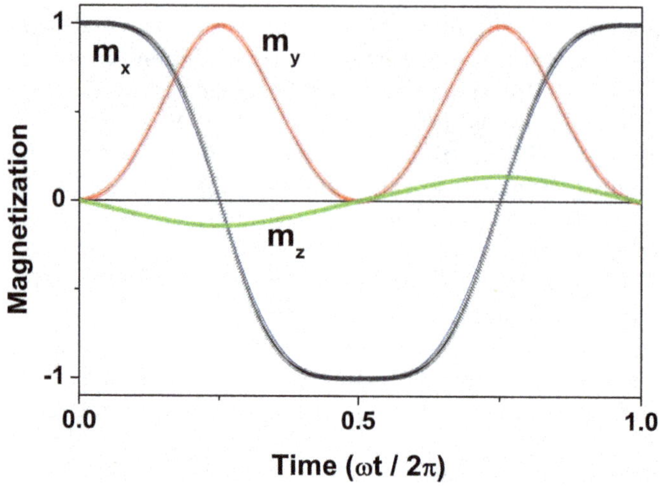

Fig. 3. Time evolution of the magnetization of a loss-free isotropic thin film subjected to a transverse field step $h_y = 0.01$: comparison between the exact time evolution of the magnetization obtained through numerical integration of the LLG equations (*cross symbols*), and that obtained through the Fourier expansion [up to 4ω, thin continuous lines, (14), (15)]. The time coordinate has been normalized to a single precession period

of $\boldsymbol{m}(t)$, intuitive trial functions for the time dependence of magnetization are the following simple trigonometric functions:

$$\begin{aligned} m_x(t) &= \cos(\omega_\mathrm{r} t)\,, \\ m_y(t) &= m_y^{\max}\frac{1-\cos 2\omega_\mathrm{r} t}{2}\,, \\ m_z(t) &= -m_z^{\max}\sin(\omega_\mathrm{r} t)\,. \end{aligned} \tag{9}$$

The $m_y^{\max}$ and $m_z^{\max}$ are those derived formerly (7), (8). The “r” subscript to the angular frequency ω_r is used to recall that these trial functions are rough estimates. They do not satisfy the magnetization norm conservation, since $\boldsymbol{m}^2$ oscillates typically between 0.8 and 1. Using the classical properties of the trigonometric functions, we can easily get that $\omega_\mathrm{r} \approx \gamma_0\sqrt{(1/2)H_y M_\mathrm{S}}$. From these rough trial functions, it is thus likely that the frequency of the magnetization evolution scales with $\gamma_0\sqrt{H_y M_\mathrm{S}}$. Note that this could also be derived by combining the LLG equations to achieve a second-order differential equation of m_z:

$$\ddot{m}_z + \gamma_0 M_\mathrm{S}^2 h_y (h_y + m_y) m_z = 0\,. \tag{10}$$

This equation is harmonic when $m_y = m_y^{\max} \approx 1$ because $\dot{m}_y = 0$. It also leads to a characteristic frequency scaling with $\gamma_0\sqrt{H_y M_\mathrm{S}}$.

Obtaining the exact scaling factor requires the use of more elaborate trial functions, for instance by developing $\boldsymbol{m}(t)$ in its main Fourier components as is done below. The variational approach can be carried further by considering improved trial functions, for which symmetry arguments were used [12] to eliminate most of the Fourier terms:

$$m_x(t) = X_1 \cos \omega_\mathrm{r} t + X_3 \cos 3\omega_\mathrm{r} t \tag{11}$$
$$m_y(t) = Y_0 - Y_2 \cos 2\omega_\mathrm{r} t - Y_4 \cos 4\omega_\mathrm{r} t \tag{12}$$
$$m_z(t) = -Z_1 \sin \omega_\mathrm{r} t - Z_3 \sin 3\omega_\mathrm{r} t\,. \tag{13}$$

These improved trial functions can be used to gain a better accuracy in estimating the solutions of the LLG equation. There are eight unknown parameters: the seven Fourier components and the frequency ω_r. Each of the three LLG equation projections contains two harmonics, which provides thus six equations. The initial conditions give the needed further two equations: $m_x(t=0) = 1$ provides $X_1 + X_3 = 1$ and $m_y(t=0) = 0$ provides $Y_0 - Y_2 - Y_4 = 0$. The solving procedure is reported in [12] and assumes that $Z_3 \ll Z_1$, $X_3 \ll 1$ (which were checked a posteriori) and that $h_y \ll 1$.

These improved trial functions lead to an improved estimate of the main frequency, and to Fourier components being:

$$\omega = \gamma_0 \sqrt{\tfrac{3}{4} H_y M_\mathrm{S}} = 0.866 \gamma_0 \sqrt{H_y M_s}\,, \tag{14}$$
$$X_1 \approx \tfrac{9}{8}\,, \qquad X_3 \approx -\tfrac{1}{8}\,, \qquad Y_0 \approx \tfrac{29}{64} m_y^{\max}\,, \qquad Y_2 = \tfrac{1}{2} m_y^{\max}\,,$$
$$Y_4 \approx -\tfrac{3}{64} m_y^{\max}\,, \quad Z_1 = \tfrac{23}{24} m_z^{\max}\,, \quad Z_3 = -\tfrac{1}{24} - m_z^{\max}\,. \tag{15}$$

This improved Fourier development is compared to the numerical integration of the LLG equation for $h_y = 0.01$ in Fig. 3. The time coordinate has been normalized to the respective precession frequencies to ease the comparison. The agreement is almost perfect. With no fitting parameter and the spectral weights of (15), the overall maximum absolute disagreement is 3 % in the magnetization components. The magnetization norm (not shown) fluctuates between 0.969 and 1.

The improved frequency estimate in (14) may be compared to the real periodicity that can be obtained by direct numerical integration of the LLG equation. Such a comparison [12] indicates that (14) overestimates the real frequency by typically 3 %. Equivalently, our improved frequency estimate can be compared to the (more general) analytical result obtained recently by *Serpico* et al. (equation 15 of [24]). Expanding their result when both h_k and h_y tend to zero leads to the following law:

$$\omega \approx 0.847 \gamma_0 \sqrt{H_y M_\mathrm{S}}\,. \tag{16}$$

This exact expression falls very near our estimate (14).

It is worth recalling that the small-angle magnetization eigenexcitations of the system are well known from the ferromagnetic resonance theory [25]. During a uniform precession (UP), the magnetization rotates around the applied field with a frequency $\omega_\mathrm{UP} = \gamma_0 \sqrt{H_y (H_y + M_\mathrm{S})} \approx \gamma_0 \sqrt{H_y M_\mathrm{S}}$. Although our trajectories very significantly depart from small-angle excitations, it is remarkable that the characteristic frequencies scale with the same quantity. The main difference comes from the existence of a substantial variation of m_y and from the spectral weight of the harmonics.

To summarize this section, we have first recalled the nature of precessional switching by reconsidering the validity of the Stoner–Wohlfarth assumptions. We have emphasized that in the quasistatic Stoner–Wohlfarth scenario, the field is swept infinitely slowly and so the magnetization always follows the most direct energy-minimization path. In contrast, in the precessional switching scenario the magnetization follows a (faster) path given by the precession torque, with the energy dissipation playing a marginal role. We then have described analytically and in some detail a very simplistic precessional switching event by making strong assumptions: macrospin behavior, no energy dissipation, infinitely extended isotropic thin film and purely transverse step-like applied field. In that approximation, we have described the main characters of a precessional switching trajectory of the magnetization. These are the buildup of a strong demagnetizing field and the back-and-forth periodic quasi-inplane motion of the magnetization (Fig. 2). Finally, we have shown that the precession frequency has a simple expression (14) with a reasonable scaling law and that the magnetization time response has a relatively simple spectral content (15).

Let us now study how this simple model compares with the experimental behavior at high applied fields.

3 Precessional Switching at High Hard-Axis Fields

In this section, we report detailed experimental investigations [5, 6] of the precessional switching of micrometer-sized uniaxial magnetic cells subjected to hard-axis magnetic fields. These measurements correlate at best with our former simplistic assertions in the regime of applied fields much above the anisotropy field (i.e., $H_y \gg H_\mathrm{k}$). For didactic purposes, we presently restrict our investigations to this regime where the finite anisotropy of the samples does not play the major role. The more complicated situation of smaller applied fields will be considered afterwards (Sect. 4).

We first detail the experimental aspects of our "all-electrical" setup and samples (Sect. 3.1), and then dwell on the two main experimental findings: the precessional motion of the magnetization is quasiperiodic (Sect. 3.2), and it can be ballistic if the field is applied for a duration that is a small multiple of the half-precession period (Sect. 3.3).

3.1 "All-Electrical" Experimental Methods

The reported experiments were carried out on exchange-biased spin valves (SV) consisting of Ta 65 Å / $Ni_{81}Fe_{19}$ 40 Å / $Mn_{78}Ir_{22}$ 80 Å / $Co_{88}Fe_{12}$ 43 Å /

Cu 24 Å / $Co_{88}Fe_{12}$ 20 Å / $Ni_{81}Fe_{19}$ 30 Å / Ta 8 Å. The films are structured into either $4 \times 2\,\mu m^2$ or $5 \times 2.3\,\mu m^2$ SV cells, with electrical contacts measuring the sample's current-in-plane giant magnetoresistance (GMR). Due to the overlap of the contacts and the SV the GMR signal mainly probes the magnetization within the center of the cell, away from the flux closure domains (if any). The magnetization of the pinned layer $Co_{88}Fe_{12}$ 43 Å is aligned along the long SV dimension x by the exchange bias field. The minor GMR loops of the free layer $Co_{88}Fe_{12}$ 20 Å / $Ni_{81}Fe_{19}$ 30 Å are rather square and asymmetric [5, 6]; the loops are shifted to an offset field $\mu_0 H_{\text{offset}}$ along the easy-axis x of typically 1.4 mT due to coupling of the pinned and the free layer. For simplicity, our convention is to perform systematically the correction of the easy-axis fields h_x to compensate this offset. The anisotropy field $\mu_0 H_{\text{k}}$ is typically 8 mT and arises from both the shape and the free-layer material. The reversal of the free-layer magnetization leads to a room-temperature GMR change of 5.6 %. In a manner that is very similar to *Russek* et al. [9, 10], a high-bandwidth coplanar waveguide is integrated onchip to generate the transverse magnetic field pulses h_y [26]. The h_y pulse durations T_{pulse} can be adjusted between 140 ps and 10 ns (at half-maximum) with rise times down to 55 ps and maximum fields $\mu_0 H_y$ around 24 mT. In addition, an external coil allows application of inplane static fields, and a resaturation setup can apply easy-axis fields $\mu_0 H_x$ up to 10 mT of duration 15 ns to 150 ns with a kilohertz repetition frequency.

3.2 Evidence of the Quasiperiodic Nature of the Magnetization Motion

In our experiment we followed the theoretical predictions [2, 27], suggesting that switching could be achieved by short-duration field pulses h_y along the inplane hard-axis of a memory cell. We thus measured the effect of the repetitive application of h_y pulses of various durations in the 140 ps to 1 ns range.

As demonstrated in Fig. 4 for $T_{\text{pulse}} = 140$ ps, $\mu_0 H_y = 15.5$ mT, each pulse toggles the magnetization direction, in a fully symmetrical and reversible manner [5, 6]. Indeed each pulse changes the GMR by an amount ±5.6 % that corresponds exactly to the full remanent reversal measured in quasistatic loops. The zero-error reliability of this bascule switching process has been tested for up to 750 consecutive pulses. Hence this pulse duration matches with an odd multiple of the π/ω.

For slightly longer pulses ($T_{\text{pulse}} = 190$ ps, 19.5 mT) no reversible behavior was observed (Fig. 4). The pulse duration is not sufficiently matched with a multiple of π/ω. The GMR changes stochastically between 0.01 % and 1.5 %, indicating that the pulse of 190 ps has a very tiny effect on the magnetization state.

Finally, for 270 ps, 21.5 mT pulses, the sample is always found in the same resistance state (Fig. 4), meaning that the initial and final states are

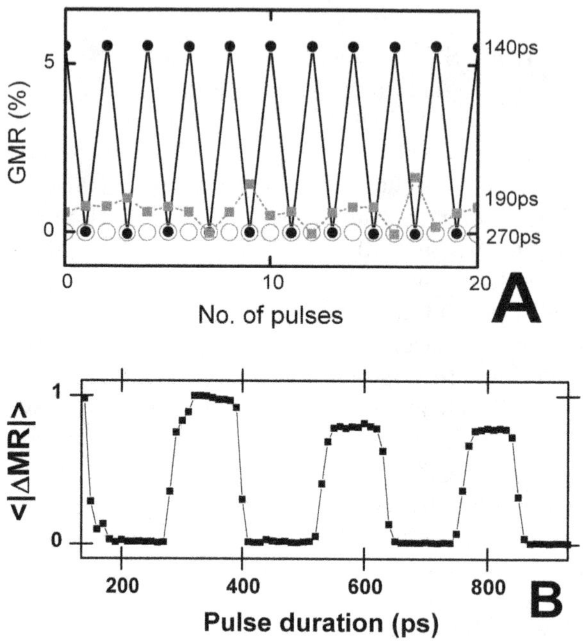

Fig. 4. (**a**) Precessional magnetization reversal by hard-axis field pulses. GMR measured after each pulse vs. pulse index at exact loop offset compensation, i.e., $H_x = 0$. Pulse parameters are $T_{\text{pulse}} = 140\,\text{ps}$, $\mu_0 H_{\text{pulse}} = 15.5\,\text{mT}$ (*black, full circles*); 190 ps, 19.5 mT (*gray squares*); and 270 ps, 21.5 mT (*gray, open circles*). The 140 ps hard-axis pulse reversibly toggles the free-layer magnetization with every pulse. $\Delta\text{GMR} = 5.6\,\%$ per pulse corresponds to full cell reversal. (**b**) Pulse duration dependence of the switching behavior of a $2 \times 4\,\mu\text{m}^2$ spin-valve cell of anisotropy field 8 mT. The field pulse is $\mu_0 H_y = 23\,\text{mT}$. The $\langle|\Delta\text{GMR}|\rangle$ is the normalized average of $|\Delta\text{GMR}|$ per pulse as a function of T_{pulse} at exact loop offset compensation, i.e., $H_x = 0$. Four regions of reversible hard-axis switching around $T_{\text{pulse}} = 140\,\text{ps}$, 350 ps, 590 ps and 800 ps are found. This figure is a horizontal cut of Fig. 5 at $\mu_0 H_y = 23\,\text{mT}$

essentially identical (no effective switching). Note that for a given applied field strength, the pulse duration leading to no effective switching ($T_{\text{pulse}} = 270\,\text{ps}$) is twice that leading to a full effective switching ($T_{\text{pulse}} = 140\,\text{ps}$, see Fig. 5). They correspond, respectively, to a half- ($T_{\text{pulse}} = \pi/\omega$) or a full precession period ($T_{\text{pulse}} = 2\pi/\omega$).

For zero damping, a further increase of T_{pulse} should lead to multiple precession cycles of $\boldsymbol{M}$ around H_y, hence to a periodic reoccurrence of the switching/no switching alternation. In the limit of strong pulses ($H_y \gg H_{\text{k}}$) and weak damping, switching is expected whenever the pulse terminates out of phase with the precession, i.e., when $T_{\text{pulse}} \approx (n + 1/2) \times 2\pi/\omega$, n being an integer defining the order of the switching process and T_{prec} the time for one precession period. On the contrary, pulses parameters with $T_{\text{pulse}} \approx n \times 2\pi/\omega$

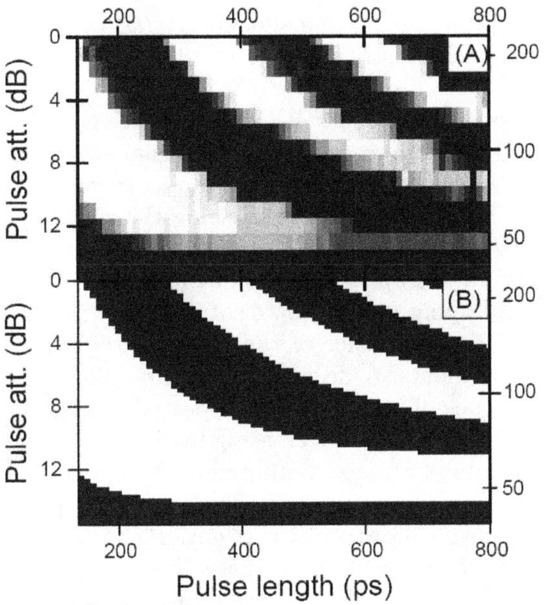

Fig. 5. Pulse field H_y and pulse duration T_{pulse} dependence of the precessional switching of a spin-valve cell of anisotropy field $\mu_0 H_{\mathrm{k}} = 8\,\mathrm{mT}$. (**a**) Experimental $\langle|\Delta\mathrm{GMR}|\rangle$ as a function of T_{pulse} and H_y at $\mu_0 H_x = 0\,\mathrm{mT}$. *Gray:* $\langle|\Delta\mathrm{GMR}|\rangle \leq 0.1$, *white:* $\langle|\Delta\mathrm{GMR}|\rangle \geq 0.8$. The field H_y is attenuated in steps of 1 dB. The *arrow* indicates that the fastest reversal can be triggered by pulse energies that are as low as 15 pJ (*arrow*). (**b**) Theoretical switching map of the same device obtained through a numerical integration of the LLG equations in the macrospin approximation with $\alpha = 0.03$ [5,6]. *Black:* no switching, *white:* switching. The order n of switching is indicated

will result in an effective nonreversal of the cell. Such phase-coherent, multiple-order reversal is indeed observed experimentally [5,6] as shown in Fig. 4b for a pulsed field strength $\mu_0 H_y \approx 23\,\mathrm{mT}$. To illustrate this point, we plot the degree of reversible switching as a function of the pulse duration T_{pulse}. As a measure of reversible switching reliability we use $\langle|\Delta\mathrm{GMR}|\rangle$, the average absolute resistance change per applied pulse normalized to full reversal, and averaged over a series of pulses. $\langle|\Delta\mathrm{GMR}|\rangle \approx 1$ indicates stable reversible switching. Figure 4b shows the measured values of $\langle|\Delta\mathrm{GMR}|\rangle$. Four regions of stable switching, well separated by stable nonswitching regions, are observed. Note again the zero-order switching ($n = 0$) for the shortest accessible pulses of $T_{\mathrm{pulse}} = 140\,\mathrm{ps}$. In the adjacent region up to $T_{\mathrm{pulse}} \approx 280\,\mathrm{ps}$ no switching takes place as $T_{\mathrm{pulse}} \approx T_{\mathrm{prec}}$. Additionally, higher-order switching processes ($n = 1, 2, 3$) are observed near 350 ps, 590 ps and 800 ps. For $n = 1$ full, stable switching is obtained. We note, however, a slight decrease of $\langle|\Delta\mathrm{GMR}|\rangle$ for higher reversal orders ($n = 2, 3$) due to less reliable or/and partial switching. Note that stable switching is observed over a relatively broad range of

T_{pulse} of about 100 ps indicating a large (slightly smaller than π/ω) tolerable phase mismatch of the coherent switching still allowing stable attraction to the reversed easy direction.

As shown previously, the precession frequency is expected to depend on the applied field strength [see (16)]. A further proof of precessional switching is to vary the pulse magnitude H_y for a fixed duration T_{pulse} to try and match T_{pulse} with π/ω. This was done [5, 6] by characterizing the switching properties over a wide range of H_y and T_{pulse}, as displayed in Fig. 5. The experimental $\langle|\Delta\text{GMR}|\rangle$ are found in the upper panel and the results of the macrospin simulation with finite damping and finite anisotropy are found in the lower panel. The simulated field dependence of the coherent regimes is in good agreement with the measurements. A decrease of H_y first shifts the switching regions towards larger values of T_{pulse}. This is expected from (16), since the precession period $2\pi/\omega$ increases with decreasing field. As we shall see below, the finite anisotropy results in a nonzero minimum switching field of about 5 mT.

In summary, when a field h_y is applied abruptly, the motion of the magnetization is quasiperiodic and the magnetization oscillates back and forth between $m_x = 1$ and $m_x = -1$. The frequency increases with the applied field strength. Switching off the field at the right moment is a way to toggle the magnetization direction.

3.3 Evidence of Quasiballistic Magnetization Switching Trajectories

In the former section, we looked solely at the final relaxed state, disregarding whether it was attained immediately at the fall of the pulse or after some time delay. However, the total time it takes to switch and to be again at rest is also of technological relevance. It determines the minimum time interval between successive programming of a memory cell, hence the maximum clocking of the random access memory. In this section, we show [5,6] that this minimum time interval can be below 400 ps; when $T_{\text{pulse}} \approx \frac{1}{2}T_{\text{prec}} = \pi/\omega$ the magnetization trajectory exhibits a quasiballistic behavior, i.e., the magnetization is almost at rest when the field h_y is switched off.

Figure 6 is an experimental illustration of such a ballistic behavior. In Fig. 6a and d two magnetization responses to short h_y pulses are calculated using a macrospin approximation for the material parameters of our spin valves. The magnetization trajectory is displayed in the m_x–m_z-plane. The pulse h_y in Fig. 6a has an amplitude of $\mu_0 H_{\text{pulse}} = 8.1$ mT (roughly h_k) and a duration of $T_{\text{pulse}} = 175$ ps. We have used the idealized pulse profile displayed in Fig. 6b (black line). During pulse application, $\boldsymbol{M}$ performs approximately a half-precession turn around H_y (see Fig. 6a). At pulse decay time, $\boldsymbol{M}$ is oriented near the reversed easy-axis direction $M_f = -M_i$ and relaxes towards it. Theoretically, the hard-axis pulse h_y is thus expected to induce a precessional switching of the magnetization. The experimental field pulse (gray) is displayed in Fig. 6b whereas the measured (gray) and calculated (black) magnetization responses are given in Fig. 6c where m_x is plotted as a function of time. The measured time evolution of m_x is well described by the simple macrospin simulation. The short pulse induces precessional switching of the magnetization with ultrashort measured reversal time ($-0.9m_x$ to $+0.9m_x$) $T_{\text{switch}} = 165\,\text{ps}$. Furthermore, no significant precession after pulse decay is present in m_x neither in the measured data nor in the simulation and long-wavelength magnetic excitations after pulse decay ("ringing") are suppressed. The pulse very nearly matches the half-precessional turn $T_{\text{pulse}} \approx \pi/\omega$ and, thus, the magnetization switches quasiballistically, i.e., with a close to optimum trajectory towards the reversed direction. This quasiballistic switching represents the fundamental ultrafast limit of field-induced magnetization reversal for the given field amplitude.

The trajectory in Fig. 6d is the calculated response to a higher and longer (20.5 mT, 240 ps) h_y field pulse. Due to the higher field and longer pulse duration time $\boldsymbol{M}$ now performs a full back-and-forth rotation about H_y before the pulse decays. Upon pulse termination, $\boldsymbol{M}$ is in this case oriented near the initial easy direction M_i and relaxes towards the latter as the result of the internal magnetic fields. Therefore, in spite of strong precession during pulse application no effective magnetization reversal takes place. The corresponding pulses and magnetization responses are found in Figs. 6e and f, respectively. Now, during pulse application, m_x oscillates from the initial direction to the reversed orientation and back. The pulse matches here a full precessional rotation $T_{\text{pulse}} \approx 2\pi/\omega$. Again, only little ringing is found following pulse termination.

The oscillatory nature of precessional switching by hard-axis field pulses h_y is well observed with the time-resolved experiments described in Fig. 7. In Fig. 7a, the measured response of m_x to a 305 ps, 27.2 mT pulse is plotted as a function of time. In Fig. 7b the response of m_x to a series of pulses with $\mu_0 H_y = 27.2\,\text{mT}$ and $T_{\text{pulse}} = 200, \ldots, 900\,\text{ps}$ (10 ps increments) is plotted as a gray-scale map, as a function of time and pulse duration. The curve in Fig. 7a is a section through the data in Fig. 7b along the horizontal dashed line. As seen in Fig. 7a, the 305 ps pulse induces a first-order ($n = 1$) precessional switch ($T_{\text{pulse}} \approx 3\pi/\omega$). $\boldsymbol{M}$ rotates one and a half times about H_y during pulse application corresponding to a triple change of sign of m_x. However, the pulse parameters are not well tailored to 1.5 precessional rotations. As a consequence, $\boldsymbol{M}$ is not fully aligned with the reversed easy direction upon pulse termination, resulting in a residual precession upon pulse termination [see arrows (1)]. For the given pulse, full alignment of $\boldsymbol{M}$ with the final easy-axis direction takes more than 1 ns. The precession-limited effective reversal time is thus considerably longer than the switching pulse duration of only 305 ps. This (purposely bad) example underlines the importance of a precise control of the pulse parameters to achieve ultrafast quasiballistic switching.

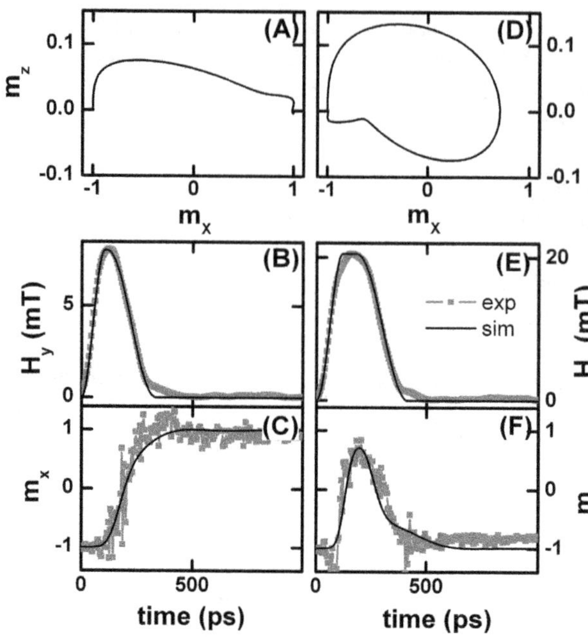

Fig. 6. Time-resolved precessional switching of a spin-valve cell. Precessional trajectories obtained by numerical integration of LLG in the macrospin approximation $\mu_0 H_\mathrm{k} = 8\,\mathrm{mT}$ and finite damping $\alpha = 0.03$ in the case of switching (**a**) and of non-switching (**d**) in the m_x–m_z-plane. For switching, the field pulse is stopped after a 180° precessional turn about H_y (**a**). ($\mu_0 H_y = 8.1\,\mathrm{mT}$, $T_\mathrm{pulse} = 175\,\mathrm{ps}$). Higher fields and longer pulse durations ($\mu_0 H_y = 20.5\,\mathrm{mT}$, $T_\mathrm{pulse} = 240\,\mathrm{ps}$) induce a full 360° precessional rotation (**d**) and $\boldsymbol{M}$ relaxes towards the initial easy-axis direction after pulse decay (no switch). Field (**b**) and magnetization component m_x (**c**) versus time for the $\mu_0 H_y = 8.1\,\mathrm{mT}$, $T_\mathrm{pulse} = 175\,\mathrm{ps}$ pulse. *Gray dots:* experiment, *black lines:* simulation. m_x switches within $T_\mathrm{switch} = 165\,\mathrm{ps}$. After pulse decay no residual precession is found, indicating optimum quasiballistic reversal. Field H_y (**e**) and m_x (**f**) vs. time for the $\mu_0 H_y = 20.5\,\mathrm{mT}$, $T_\mathrm{pulse} = 240\,\mathrm{ps}$ pulse; m_x toggles towards the reversed direction and back during pulse application: $\boldsymbol{M}$ is not reversed upon pulse termination

The multiple oscillations of $\boldsymbol{M}$ about H_y for longer pulses can also be seen in Fig. 7b. The oscillation maxima of m_x (light regions running vertically) are marked by the black dots and numbers 0 to 3 on the upper border of the gray-scale plot. The pulse duration is indicated in the data by the inclined dotted line. Pulse decay at a maximum of m_x (i.e., with $\boldsymbol{M}$ oriented near the reversed easy direction) inevitably leads to relaxation to the reversed easy-axis direction, i.e., to high-order precessional switching (white horizontal regions after pulse termination with switching order n). On the contrary, pulse decay at a minimum of m_x (i.e., with $\boldsymbol{M}$ oriented near the initial easy direction) always leads to relaxation towards the initial direction of $\boldsymbol{M}$ (gray

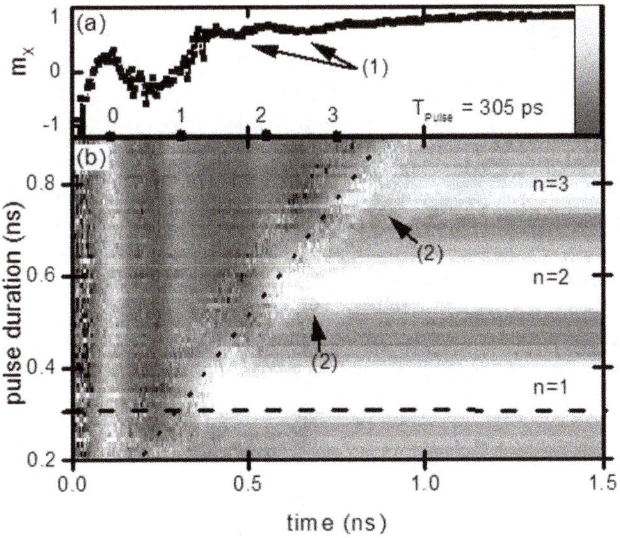

Fig. 7. Time-resolved, multiple-order precessional switching of a spin-valve cell. (**a**) First-order switch. Measured m_x vs. time for $\mu_0 H_y = 27.2\,\mathrm{mT}$, $T_{\text{pulse}} = 305\,\mathrm{ps}$. After pulse decay, residual precession occurs [*arrows* (1)]. (**b**) Gray-scale encoded map of m_x as a function of time and pulse duration T_{pulse}. *White:* $m_x = 1$, *dark gray,* $m_x = -1$. Nominal pulse amplitude $\mu_0 H_y = 27.2\,\mathrm{mT}$. Pulse field H_y decays to zero along the inclined *dotted line.* (**a**) Corresponds to a section through (**b**) Along the *dashed horizontal line.* Higher-order switching (*white horizontal regions*, switching order n is indicated) occurs in phase with the precession at pulse cutoff. Zero-order switching ($n = 0$) is not accessible for the given pulse amplitude

horizontal regions) and no effective switching is monitored. Again, near the transitions from switching to nonswitching the alignment of $\boldsymbol{M}$ and the final easy-axis direction is poor and a pronounced ringing of the magnetization upon pulse decay occurs [see, e.g., arrows (2)].

In summary, we have given experimental evidence [5,6] that for $H_y > H_\mathrm{k}$ and $T_{\text{pulse}} \approx (n + \frac{1}{2})T_{\text{prec}}$ with n being small, quasiballistic switching is observed, i.e., no significant ringing occurs after h_y pulse termination. The larger n, the less ballistic. The larger the phase mismatch $|T_{\text{pulse}} - (n + \frac{1}{2})2\pi/\omega|$, the less ballistic.

4 Precessional Switching at Moderate Hard-Axis Fields and Zero Damping

In the previous sections, we have shown that precessional switching could be obtained at high fields and that its major characteristics could be accounted for by simple precession-frequency arguments (16). However, at moderate fields, the simplified models do not satisfactorily account for the magnetic behavior. In practical applications, it is worth reversing the magnetization using moderate fields, i.e., of the order of the anisotropy field. This section is devoted to the description of this specific regime where the competition between the applied field h_y and the anisotropy field h_k leads to an increased complexity in the magnetic behavior [13]. In contrast to the isotropic case [12] (Sect. 2.3), we shall see that the finite anisotropy results in a finite minimal switching field that is $H_\mathrm{k}/2$. Also, for $H_y < H_\mathrm{k}$, there are two possible equilibrium states. The existence of these two degenerate attractors will substantially deform the magnetization trajectories.

The damping is still considered as negligible in this section. The LLG equations or equivalently the energy conservation yield the exact magnetization trajectory [13]:

$$m_x^2 = 1 - \frac{2h_y}{1+h_k} m_y - \frac{1}{1+h_k} m_y^2 , \tag{17}$$

$$m_z^2 = \frac{2h_y}{1+h_k} m_y - \frac{h_k}{1+h_k} m_y^2 . \tag{18}$$

Depending on the relative magnitudes of h_y and h_k, two types of trajectory can occur [13]. They are sketched in Fig. 8.

4.1 Classification of the Magnetization Trajectories

High fields ($h_y > h_k/2$) lead to switching trajectories (so-called field-dominated trajectories (Sect. 4.1.1)). Low fields ($h_y < h_k/2$) do not permit the switching. They lead to anisotropy-dominated trajectories (Sect. 4.1.2).

4.1.1 Field-Dominated (Switching) Trajectories

1. The simplest case happens when the energy landscape has an absolute minimum at $m_y = 1$, which is the unique stable equilibrium position. This is the case of rather high applied fields, i.e., when $h_y \geq h_k$. A typical trajectory is displayed in Fig. 8 for $h_y = h_k = 0.01$. In this range of applied fields higher than the anisotropy field, the m_z versus m_x magnetization trajectory is a stadium-shaped trajectory (Fig. 8), that looks very much like the trajectory corresponding to an isotropic thin film (Fig. 2). The maximum of m_z is $m_z \approx \pm\sqrt{2h_y - h_k}$, and it is reached when $m_x = 0$. The trajectory reaches systematically the full reversal, i.e., $m_x = -1$. The m_z versus m_y component (Fig. 8a) is an open path that looks like a half-ellipse. For such applied fields, the equilibrium magnetization is along y, and the trajectory rotates around that axis (Figs. 8b and c).
2. When the applied field is such that $h_k > h_y > h_k/2$, i.e., smaller than the anisotropy field but still at least half of it, the energy landscape has a saddle point at $m_y = 1$, and the initial energy is above it. The trajectory is significantly deformed compared to the case 1 of high applied fields.

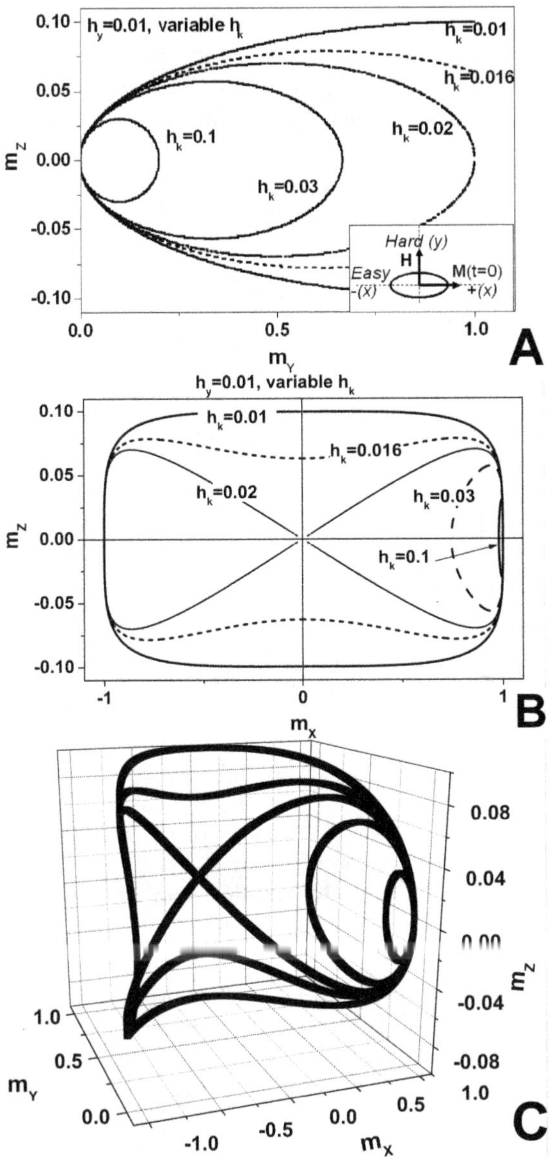

Fig. 8. Magnetization trajectory (17), (18) of a uniaxial nonlossy macrospin film of initial magnetization along x when subjected to a transverse field $h_y = 0.01$ applied along y, with $N_z = 1$. The uniaxial anisotropy along x. The anisotropy field is varied from $h_k = 0.01$, 0.016, 0.02 (twice the applied field: bifurcation trajectory), 0.03 and 0.1. The two latter h_k values correspond to anisotropy-dominated trajectories. (**a**) Trajectory in the zy-plane. (**b**) Trajectory in the xz-plane. (**c**) Vector $\boldsymbol{m}$ trajectory when initial magnetization is along x. *Inset:* Definition of the axes. The ellipse stands for the anisotropic macrospin

The m_z versus m_x trajectory takes a bone shape (Fig. 8b). It bends back to the easy plane when approaching the direction of the applied field y. The trajectory reaches systematically the full reversal ($m_x = -1$). For such applied fields, there are two degenerate equilibrium magnetization states, that have $m_y = h_y/h_k$ and $m_x > 0$ or $m_x < 0$. Qualitatively, the trajectory rotates alternatively around each of these two attractors. For instance, the maximum of m_z is obtained when $m_y = h_y/h_k$ and takes the value:

$$m_z^{\max} = \pm \frac{h_y}{\sqrt{h_k(1+h_k)}} \simeq \pm \frac{h_y}{\sqrt{h_k}} ; \tag{19}$$

m_z has a local minimum when $m_x = 0$. At this point:

$$m_y^{\max} = \sqrt{1 + h_k + h_y^2} - h_y \approx 1 - h_y - h_k/2 \quad \text{and} \quad m_z \approx \sqrt{2h_y - h_k} . \tag{20}$$

The m_z versus m_y is an open path that looks like a truncated ellipse (Fig. 8a). Note again, that in this latter situation we have $H_y < H_\mathrm{k}$, i.e., the switching fields can be lower than the classical quasistatic switching threshold.

The trajectories for applied fields $h_y > h_k/2$ will be referred to hereafter as field-dominated trajectories, in contrast to the anisotropy-dominated trajectories of the following section.

4.1.2 Anisotropy-Dominated (Nonswitching) Trajectories

The situation is topologically different when the field is too low to allow switching, i.e., when it is less than half of the anisotropy field ($h_y < h_k/2$). In that case the initial energy is below that of the saddle point at $m_y = 1$, such that the magnetization vector can never overcome the hard-axis. Magnetization stays in the $m_x > 0$ half-space. It circulates about the sole nearest-equilibrium magnetization position, the attractor obeying $m_y = +h_y/h_k$. The m_y component oscillates between 0 and $2h_y/h_k$. The m_x^2 oscillates between 1 and $1 - 4hy^2/h_k^2$ and the m_z versus m_x trajectory is ovoid-like with a maximum radius of curvature at $m_x = 1$. The m_z versus m_y has an ellipsoidal shape (Fig. 8a).

While the duration of the anisotropy-dominated trajectories is known from the ferromagnetic resonance theory [25], and the duration of high-field switching was estimated formerly in Sect. 2.3.2), the duration of a switching event at moderate field (case 2, Sect. 4.1.1) deserves to be studied.

4.2 Characteristic Times Involved in a Precessional Switching Event Near and Below H_k

In this section, we search for the scaling laws of the switching frequency [13]. Since the switching frequency was already derived in the regime $h_y \gg h_k = 0$

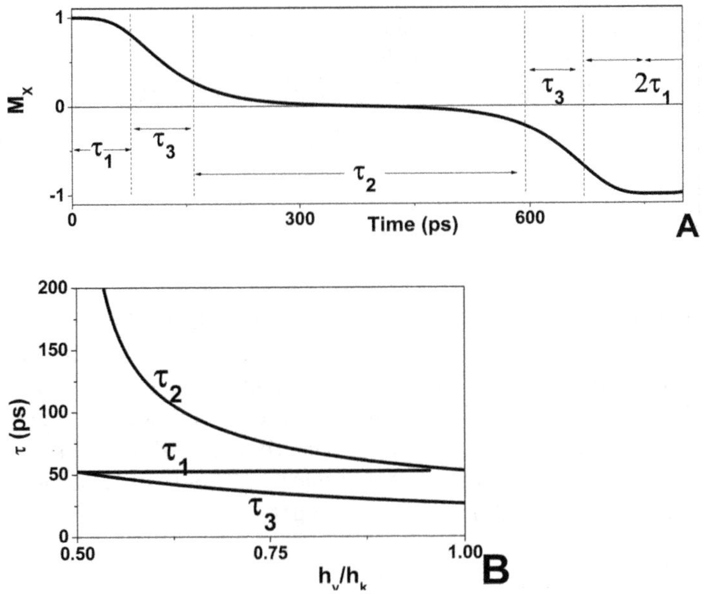

Fig. 9. Definition of the characteristic time intervals of a precessional switching event. (**a**) Numerical integration of the m_x component for $N_z = 1$, $h_k = 0.0148$, $\mu_0 M_S = 1.08\,\mathrm{T}$, $h_y = 1.0005 h_k/2$ and $\alpha = 0$. The initial delay τ_1 (21), the maximum speed time τ_3 (24) and the slow-down time τ_2 (23) are qualitatively superimposed on the time evolution. (**b**) Dependence of these three times for applied fields h_y between $h_k/2$ and h_k, for $N_z = 1$. The material parameters are those of Permalloy with $h_k = 0.01$

in Sect. 2.3.2, we now concentrate more specifically on the case $h_k/2 < h_y < h_k$. To this aim we introduce three qualitative times τ_1, τ_2 and τ_3 that govern the order of magnitude of the switching frequency. The reader should consider these times as conceptual guides. They will be reused in Sect. 5 to assess the effect of finite damping and to calculate the final magnetic state.

4.2.1 Initial Delay, Slow-Down Time and Maximum Speed Time

We first define the "initial delay" τ_1 during which m_x does not react much to the applied field, then the "slow-down time" τ_2 during which the system stays in the vicinity of the hard axis, and finally "maximum speed time" τ_3 the typical time during which there exists a large demagnetizing field. Their definition is illustrated in Fig. 9a, which displays $m_x(t)$ as computed from an exact numerical integration of the LLG equation.

First, it takes a time delay τ_1 for the system to set its maximum demagnetizing field. Indeed, the initial rate of change of m_x and m_y is zero (3a), (3b), which can be seen as an effective delay in the response of the inplane projection of the magnetization to the applied field. In experiments sensitive to only

m_x and m_y, this time delay during which the sole m_z changes significantly is difficult to measure. We define this initial delay τ_1 such that $\tau_1 \dot{m}_z = m_z^{\max}$, where the derivative is evaluated at $t = 0$, i.e., $m_x = 1$. Using (19), we get:

$$\tau_1 \approx \frac{1}{\gamma_0 M_S} \frac{1}{\sqrt{h_k(h_k+1)}} \quad \text{with} \quad h_k > h_y > h_k/2\,, \tag{21}$$

$$\tau_1 \simeq \frac{\sqrt{2h_y - h_k}}{\gamma_0 M_S h_y} \quad \text{if} \quad h_y > h_k\,. \tag{22}$$

For common soft alloys, the initial delay τ_1 during which the inplane projection of the magnetization does not react much is typically $\tau_1 = 50\,\text{ps}$.

On the other hand, the reversal of the m_x component occurs mainly due to the demagnetizing field and is thus dominated by the value of m_z when $m_x \approx 0$. We define the "slow-down time" τ_2 as the time it takes for m_x to pass from $+0.5$ to -0.5 as a result of the demagnetizing field, so that τ_2 is such that $\tau_2 \dot{m}_x|_{m_x=0} \sim 1$. Using (3c) and (20), we get

$$\tau_2 \approx \frac{1}{\gamma_0 M_S} \frac{1}{\sqrt{2h_y - h_k}} \quad \text{with} \quad h_y > h_k/2\,. \tag{23}$$

Using a slightly different approach, we can convert the demagnetizing field of (19) into a frequency, such that we can write τ_3 the typical time spent by the system around $m_{z\max}$. During this "maximum speed time" τ_3, there exists the largest demagnetizing field, driving the magnetization motion in a fast manner

$$\tau_3 = \frac{\sqrt{h_k}}{2\gamma_0 M_S h_y}\,. \tag{24}$$

Note τ_1, τ_2 and τ_3 are only qualitative conceptual guides: each of them may need to be multiplied by a numerical prefactor of the order of one. Their numerical prefactors, especially the one of τ_3, will be reconsidered in Sect. 5, by doing a feedback comparison with exact numerical calculations of the minimal switching field when damping is finite. Let us use these three times to guess the form of the precession frequency.

4.2.2 Rule of Thumb for the Precession Frequency

From Fig. 9b, it is straightforward to see that $\tau_2 > \tau_1 > \tau_3$ holds almost always (in fact for $0.51 h_k < h_y < h_k$). This justifies our naming convention for these times. In addition, the slow-down time τ_2 tends to infinity near the bifurcation criteria $h_y = h_k/2$, such that we expect the characteristic frequency to be limited by τ_2 and thus to scale with $\gamma_0\sqrt{M_S(H_y - H_k/2)}$. The latter precession frequency ω was calculated numerically. Plots of ω^2

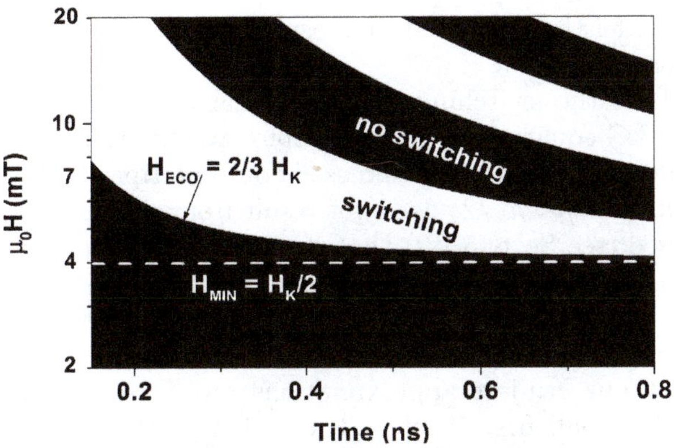

Fig. 10. Analytical evaluation (25) of the final magnetic state after having applied a magnetic field pulse h_y along the hard axis, of variable duration and variable amplitude. The sample has an effective anisotropy field of 8 mT and a magnetization $\mu_0 M_S = 1.3\,\mathrm{T}$. The color scale is the one of Fig. 5b: *white* (respectively, *black*) *area:* the magnetization of the final state is antiparallel (resp., parallel) to that of the initial state

versus $-h_k$ or h_y (not shown) are linear and can be very satisfactorily fitted by a generalization of (16) [13]:

$$\omega \approx 0.847\gamma_0\sqrt{M_S(H_y - H_k/2)}\,. \tag{25}$$

A comparison between the isotropic (16) and anisotropic (25) cases illustrates the "braking" effect of the finite anisotropy on the switching speed.

The above rule of thumb for the precession frequency (25) is useful for a better understanding of the experimental result of Fig. 5a. Indeed it can be used to simulate the expected final state that the magnetic system ends in if the field is switched off after a given duration T_{pulse}, as is done in Fig. 10. If the pulse is switched off between $\frac{\pi}{2} < \omega T_{\text{pulse}} < 3\frac{\pi}{2}$ (modulo 2π) like in Fig. 6a, the system has switched forth once more than it has switched back, such that the reversal is successful. Conversely, if the pulse is switched off between $3\frac{\pi}{2} < \omega T_{\text{pulse}} < 5\frac{\pi}{2}$ (modulo 2π) then the system has reversed back and forth the same number of times, and the final and initial states are identical.

A comparison between the switching diagram obtained experimentally (Fig. 5a) and obtained from this criteria and the rule of thumb for the frequency (Fig. 10) indicates that (25) describes very satisfactorily both the moderate- and high-field regimes, but poorly describes the very low field limit ($h_y \approx h_k/2$). Experimentally, there is a poor reproducibility at small fields: $\langle|\Delta\mathrm{GMR}|\rangle$ remains well below 1, indicating only partial and unreliable switching. In addition, the switching area is less broad than expected from the model (Fig. 5b and Fig. 10). Hence a relevant piece of physics must be missing in our above model that assumed macrospin, zero rise time, zero temperature and zero energy loss.

A thorough look back at the switching diagram obtained by an exact numerical solution of the LLG equation in the macrospin approximation *with finite damping* (Fig. 5b) indicates that the weakness of our description of the very low switching field limit ($h_y \approx h_k/2$) does not result from the neglected damping. Also, we shall see (see Sect. 5.1.2) that if our experimental finite rise time were too short, it would delay the reversal in all field regimes, which we do not observe. The missing piece of physics is not related to our finite rise time.

We are convinced that our crudest approximations are $T = 0\,\mathrm{K}$, and the macrospin behavior. We shall first (Sect. 4.3) see that in the very low switching field limit, it is the *topology* of the (*near bifurcation*) magnetization trajectories that makes them quite (thermal) noise sensitive, leading to an intrinsic probabilistic behavior near $h_y \approx h_k/2$. Then, we devote section Sect. 4.4 to the effect of a nonmacrospin behavior.

To summarize, the analytical expressions derived in this section to describe magnetization motion of a nondamped, anisotropic macrospin thin film can account exactly for its magnetization trajectories. Those are either field dominated and allow switching when $h_y > h_k/2$, or anisotropy dominated with no switching. At moderate switching fields (i.e., $h_y < h_k$), the magnetization rotates in three qualitative steps with durations $\tau_2 > \tau_1 > \tau_3$, which are, respectively, the slow-down time τ_2, the initial delay τ_1, and the maximum speed time τ_3. The characteristic switching frequency tends to zero when $H_y = H_\mathrm{k}/2$ and scales with the square root of the excess field above $H_\mathrm{k}/2$. This description is satisfactory except very near $h_y \approx h_k/2$.

4.3 Minimal Switching Field, Bifurcation Trajectory and its Intrinsic Noise Sensitivity

The case $h_y = h_\mathrm{bif} = h_k/2$ deserves a particular comment because it is a bifurcation trajectory [28]. If the applied field h_y is slightly higher, the system energy is above the hard-axis energy, the magnetization can overcome the hard-axis and the trajectory is field dominated, i.e., very extended and m_x reaches both $+1$ and -1. In contrast, if the applied field is slightly lower, the initial energy is below that of hard-axis magnetization and the trajectory is anisotropy dominated and stays in the $m_x > 0$ half-space.

The bifurcation trajectory $h_y = h_\mathrm{bif} = h_k/2$ sets the *minimal field required for a precessionnal switching event in a transverse applied field.*

This bifurcation trajectory $h_y = h_\mathrm{bif} = h_k/2$ corresponds to a total adiabatic transfer of the initial anisotropy energy in the Zeeman energy. Neither demagnetizing field, nor anisotropy field are present at this bifurcation point. Since the applied field h_y is then parallel to $\boldsymbol{M}$, the torque is zero. Note that even if the demagnetizing tensor incorporates nonvanishing N_x and N_y, this result still holds. As predicted formerly by *Acremann* et al. [28], this bifurcation point is thus also a *stationary* point where $\dot{m}_{x,y,z} = 0$. In the absence of damping, a system fulfilling (26) will go *ballistically* (without ringing) to that point and stay there forever. This correlates with the singularity in the precession frequency (25) at $h_y = h_{\mathrm{bif}} = h_k/2$. Numerical integration of the LLG equation (Fig. 9a) for anisotropy conditions very near the bifurcation criterion confirms a drastic slow down of the trajectory when approaching $m_x = 0$.

Note that the required field $H_{\mathrm{bif}} = H_{\mathrm{k}}/2$ is half of the anisotropy field, i.e., only half of the field needed to align the magnetization along the hard-axis in a quasistatic evolution. Hence, this point is *stationary but unstable:* in the case of a small perturbation, the resulting torque repels the magnetization away from this point. In real systems, the thermal agitation or any source of magnetic noise will kick magnetization away from this unstable point, and the final state will evolve randomly to either $m_x > 0$ or $m_x < 0$. This is why poor switching reproducibility is obtained (Fig. 5a) when one experimentally [5, 6] attempts to reach the promised minimal field cost $H_{\mathrm{k}}/2$.

This poor reproducibility is further exacerbated when attempting to approach the minimal reversal field (when $h_y \to h_k/2$), because the magnetization exhibits an increasingly drastic slow down ($\tau_2 \to \infty$) the nearer to the unstable position ($m_y = 1$, Fig. 9b). Since the system stays very long near that particularly noise-sensitive position, it is sensitive to any fluctuation with an enhanced susceptibility. As a result, the theoretical minimal reversal field is thus very noise sensitive and cannot be deterministic in real systems.

This is also the case when the initial magnetization is not strictly along x or when it departs from the macrospin case, as discussed below.

4.4 Arbitrary Initial Magnetization and Consequences for Nonmacrospin Samples

An important situation in real devices is when the initial magnetization is not at rest along x, for instance because it is still ringing as a result of a previous switching not performed under ballistic conditions. For an initial magnetization ($0 < m_{x_0} < 1$, m_{y_0} and m_{z_0}), the trajectories are simply obtained by substituting $\{m_x^2 - 1, m_y, m_y^2, m_z^2\}$ by $\{m_x^2 - m_{x_0}^2, m_y - m_{y_0}, m_y^2 - m_{y_0}^2, m_z^2 - m_{z_0}^2\}$ in (17) and (18). The generalized bifurcation criterion separating anisotropy-dominated and field-dominated trajectories is then

$$h_{\mathrm{bif}} = (1 + m_{y_0})\frac{h_k}{2} - \frac{m_{z_0}^2}{1 - m_{y_0}}\frac{1 + h_k}{2} \,. \tag{26}$$

Let us discuss more specifically the case $m_{z_0} = 0$ (Fig. 11).

If $m_{y_0} < 0$, the switching is feasible at a reduced applied field strength. Qualitatively, it is because the systems feels a Zeeman torque creating the demagnetizing field during a longer time interval than when $m_{y_0} = 0$.

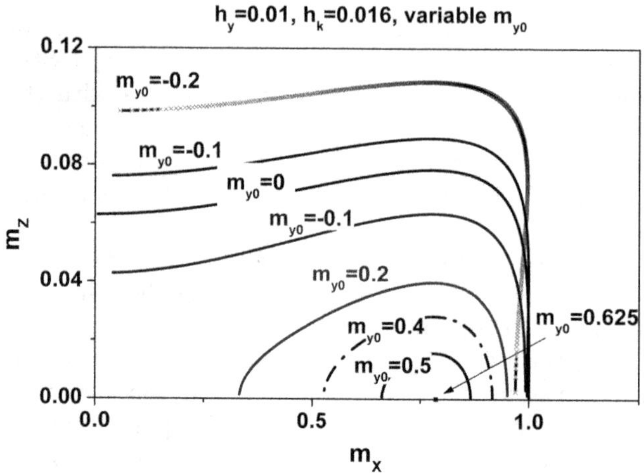

Fig. 11. Magnetization trajectories for various initial states in a thin film with uniaxial anisotropy $h_k = 0.016$ along x, when a transverse field is applied along y with magnitude $h_y = 0.01$. The initial magnetization m_{y0} is varied between -0.2 and 0.5. The equilibrium position at infinite time would correspond to $m_{y0} = 0.625$. Only one quadrant of the trajectory is displayed

If $m_{y0} > 0$, the switching requires more field. The initial Zeeman torque is reduced because the magnetization is nearer to its infield equilibrium position. The torque reduction is maximal when $m_{y0} = h_y/h_k$, i.e., when $\boldsymbol{m}_{(t=0)}$ is along the infield equilibrium position. In that case, setting the field has strictly no effect onto the magnetization (see Fig. 11, $m_{y0} = 0.625$)!

The former generalized bifurcation criterion (26) can also provide qualitative information on the expected behavior of *high remanence (though nonmacrospin) samples.* In a nonmacrospin sample, there are gradients of both the local m_{y0}s and of the local internal fields [29, 30]. In our formalism, the local internal fields are gathered in the terms h_y (which is homogeneous) and the inplane demagnetizing field (a part of the effective $\boldsymbol{h}_k$, most often heterogeneous). As a result of these gradients of m_{y0}s and $\boldsymbol{h}_k$, different parts of a nonmacrospin sample may feel a field above or below the generalized bifurcation criteria (26).

Generally speaking, the local inplane demagnetizing field (in our formalism, a part of h_k) is smaller in the central part of a sample than near its edges [29, 30] where magnetization divergence (magnetic pseudocharges) is present. The precessional susceptibility is maximal in the central part of these samples. If the pulse strength h_y is matched to a spatial average of the generalized bifurcation criteria, the pulse induces a complex micromagnetic state. The central part magnetization switches, but the edges magnetization does not (see for instance Fig. 27 of [9, 10]). A delicate balance determines whether the system relaxes to the initial or to the reversed magnetization state.

To illustrate this, let us consider a platelet magnetized in an idealized S-state [2]. The central part of the platelet is magnetized along x, while the two end domains are magnetized with $m_{y0} > 0$ due to the magnetostatics (i.e., the state is $\nearrow\rightarrow\nearrow$). For didactic purposes, we simplify and assume that these end domains obey locally $m_{y0} = h_y/h_k$ and do not have any precessional susceptibility. The field pulse switches the magnetization of the center of the platelet, while the magnetic configuration of the end domains remains unchanged. The final state (i.e., $\nearrow\leftarrow\nearrow$) is no longer an S-state and is unlikely to be stable, since it stores a lot of exchange energy in the form of two domain walls. The stability of this reversed state is questionable.

This complicated behavior is likely to happen in our micrometer-sized spin-valves when trying to approach the minimal switching field. The dispersion of local h_k and m_{y0} may then not be negligible against the applied field. Note that higher applied fields may be much greater that the dispersion of local h_k and m_{y0}. Hence, the departure from a macrospin state has a less critical effect at high applied fields. This correlates to the very fact that there is little field margin for a successful reversal near $h_k/2$ but large field margin when far above $h_k/2$, as was stated experimentally (Fig. 5a).

In conclusion, nonuniformities of the magnetization direction and of the internal fields should be avoided because they reduce the switching reliability margin, especially near the generalized bifurcation field (26). In that sense, having material-related anisotropy is preferable to shape anisotropy for a robust low-field precessional switching, because shape anisotropy often leads to magnetostatic patterns (end domains, ripple, etc.). Note that independently, material-related anisotropy (and circular MRAM cells) is also desirable for an optimal thermal stability at a given fixed switching field [22].

4.5 Minimal Reversal Field Versus Minimal Energy Cost

In Sect. 4.3, we have shown that the minimal reversal field was basically $h_{\mathrm{bif}} = h_k/2$. However, the associated characteristic switching frequency (25) indicates that the reversal lasts an infinite time ... which certainly costs a lot of energy! Hence, the minimal switching field does not correspond to the minimal energy cost of the reversal, which deserves to be studied separately.

The energy cost $C_{\mathrm{eff}} = \xi H_y^2 T_{\mathrm{pulse}}$ of a precessional switching event is related to the Joule power loss in the line creating the pulsed magnetic field. This energy cost scales with ξH_y^2 ($\xi > 0$) multiplied by the needed duration T_{pulse} of the applied field. Using the precession frequency, straightforward algebra leads to the "*most economic*" switching scenario:

$$H_{\mathrm{eco}} = \tfrac{2}{3} H_{\mathrm{k}} \quad \text{and} \quad T_{\mathrm{eco}} \simeq \frac{\pi\sqrt{2}}{\gamma_0 \sqrt{H_{\mathrm{k}} M_{\mathrm{S}}}} \,. \tag{27}$$

Note that the pulse duration could be corrected by using Serpico's more exact result (25). The previous equation describes clearly the energy optimization strategy. Taking, for instance, the material parameter of our experimental spin-valve systems, the minimal reversal cost (27) yields $T_{\mathrm{eco}} = 256\,\mathrm{ps}$, $\mu_0 H_{\mathrm{eco}} = 5.3\,\mathrm{mT}$ (see arrow in Fig. 10) and an energy cost in the range of 15 pJ. This is both fast and energy cost effective.

4.6 Perspectives for the Speed Scaling of Precessional MRAM

An interesting issue is how the reversal energy cost C_{eff} evolves with the clock frequency of a precessional MRAM. For that purpose, we express C_{eff} versus the explicit needed pulse duration T_{pulse} (we still use the simplified expression of the precession frequency with the 3/4 factor)

$$\frac{C_{\mathrm{eff}}}{\xi} = \frac{T_{\mathrm{pulse}} H_{\mathrm{k}}^2}{4} + \frac{4\pi^2 M_{\mathrm{S}} H_{\mathrm{k}}}{3(\gamma_0 M_{\mathrm{S}})^2} \frac{1}{T_{\mathrm{pulse}}} + \frac{16\pi^4 M_{\mathrm{S}}^2}{9(\gamma_0 M_{\mathrm{S}})^4} \frac{1}{T_{\mathrm{pulse}}^3} . \tag{28}$$

The $1/T_{\mathrm{pulse}}^3$ dependence of C_{eff} makes it clear that scaling the speed of a precessional MRAM deep into the sub-100 ps regime (more precisely into $T_{\mathrm{pulse}} \ll \frac{1}{\gamma_0 \sqrt{M_{\mathrm{S}} H_{\mathrm{k}}}}$) will be very challenging for energy dissipation. An additional drawback is that the anisotropy field may need to be increased because of the onset of superparamagnetism when scaling down the sizes far in the submicrometer regime. To conclude, this predicted catastrophic increase of the energy cost in the sub-100 ps and sub-100 nm regime should give a strong motivation to search for other switching trajectories costing less energy. Spin-transfer switching [31] has proven a solution to restore the MRAM scalability towards the sub-100 ps and the sub-100 nm regime [32, 33].

So far, we modeled the precessional switching with an optimistic faith that the damping did not play a major role. We reconsider this in the following section.

5 Damped Nanomagnets: Relaxation-Dominated Precessional Switching

This section is a step further towards realistic samples: we evaluate the effect of finite damping. We first describe its qualitative consequences: the finite energy-loss rate, the loss of periodicity of the magnetization motion, and the subsequent requirements on the field rise time. We then introduce the relaxation-dominated precessional switching, for which damping plays a key role. It leads to an improved estimate of the minimal reversal field.

Since our aim is to access the technological potential of precessional switching, we restrict the analysis to field-dominated trajectories ($h_y > h_k/2$), i.e., to those trajectories where energy considerations do not forbid the switching event. Let us first list the consequences of the finite damping.

5.1 Consequences of Finite Damping

5.1.1 Loss of Energy and of Periodicity

Introducing damping into the LLG equation makes it impossible to solve it analytically in most cases. Therefore, we introduce damping a posteriori as the rate at which the energy is dissipated. From [27] the damping is linked to the decay rate of the energy, i.e.,

$$\frac{1}{\mu_0 M_S^2}\frac{\mathrm{d}E}{\mathrm{d}t} = -\alpha(\dot{m}_x^2 + \dot{m}_y^2 + \dot{m}_z^2)\frac{1}{\gamma_0 M_S}\,. \tag{29}$$

With finite damping, the magnetization $\boldsymbol{m}(t)$ will converge at $t = +\infty$ to one of the infield equilibrium positions, which obeys $m_y = h_y/h_k$ and $m_z = 0$. Hence, the magnetization motion is no longer periodic. This loss of periodicity was already seen in Fig. 5 just above the switching field limit $H_k/2$, where macrospin simulations (Fig. 5b) with $\alpha = 0.03$ predict switching *independent* of T_{pulse}. Experiments (Fig. 5a) also provide evidence for nonperiodic trajectory switching for $\mu_0 H_y \approx 5\,\text{mT}$ and $T_{\text{pulse}} > 300\,\text{ps}$. Thus, the coherence criterion (Sect. 3.2) related to the matching of T_{pulse} with the π/ω does not apply just above $H_k/2$. In that case, $\boldsymbol{M}$ overcomes the hard-axis only once during the pulse and precesses then around the $m_x < 0$ equilibrium direction. Owing to the damping, oscillation back across the hard-axis is inhibited and m_x changes sign only once. Thus, independently of T_{pulse}, switching will be completed through relaxation towards $m_x = +1$ once the pulsed field decays. This case where the damping plays a key role is termed the "relaxation-dominated precessional switching". We study it in Sect. 5.3.

5.1.2 Rise-Time Constraints

The damping and the related finite energy loss add requirements on the field rise time $H_y/\dot{H}_y$. The rise time is too "slow" when the energy communicated to the magnetic system by the external field h_y grows at a rate smaller than the damping ability to dissipate it (see Sect. 2.1). Considering that the Zeeman energy to be dissipated grows typically with $\mu_0 M_S\, \mathrm{d}H_y/\mathrm{d}t$, the energy-loss rates that we shall estimate here from (31) and (32) indicate that the precessional effects dominate if the rise time obeys $\dot{H}_y/H_y \gg \alpha\gamma_0 M_S$. This sets a rather strong demand on the field rise time, since it means that the rise time should be typically *much* smaller than 160 ps, for instance, for $\alpha = 0.03$.

Let us now study qualitatively the effect of an intermediate rise time $H_y/\dot{H}_y$. In such a case, the trajectory is initiated at a low applied field. Hence, the z-incursion is reduced. When the field has reached its nominal value, the magnetization state is $m_{x0} < 1$, $m_{y0} > 0$, and $m_{z0} < 0$. Owing to the energy already lost, the switching can no longer be ballistic since the maximum negative incursion of $|m_x|$ is below 1.

As a consequence, the effect of finite rise time is not only a delay in the magnetization evolution but it is also a gradual reduction of the magnetization incursion, and an increased postpulse ringing. Also, the generalized bifurcation criteria (26) puts more demand on the applied field strength for the success of the switching.

5.2 Perturbation Treatment of Finite Damping

Let us now treat the damping in a perturbative manner, i.e., by integrating the work of the damping torque (29) along the trajectories that were calculated in the absence of damping (17), (18). In between initial ($m_x = +1$) and final states, the system bifurcates when the total energy gets smaller than that of the saddle point of the energy landscape, i.e., the hard-axis energy. This arises at the time when an energy ΔE_{bif} has already been dissipated:

$$\frac{\Delta_{\mathrm{bif}}}{\mu_0 M_{\mathrm{S}}^2} = h_y - h_k/2 \,. \tag{30}$$

The next section evaluates the energy loss per precession cycle $\Delta E_{\mathrm{cycle}}$ and then deduces the number of trajectories $\mathrm{NT} = \Delta E_{\mathrm{bif}}/\Delta E_{\mathrm{cycle}}$ being gone through before the system bifurcates to an anisotropy-dominated trajectory and gets trapped in one of the $m_x > 0$ or $m_x < 0$ half-spaces.

5.2.1 Energy Loss Per Unit Cycling of the Trajectory

When m_x is 0 or ± 1, the decay rates of the energy are found using (3c), (3a), (17) and (18), respectively

$$\begin{aligned} \left.\frac{1}{\mu_0 M_{\mathrm{S}}^2}\frac{\mathrm{d}E}{\mathrm{d}t}\right|_{m_x=1} &= -\alpha\gamma_0 M_{\mathrm{S}} h_y^2 \quad \text{and} \\ \left.\frac{1}{\mu_0 M_{\mathrm{S}}^2}\frac{\mathrm{d}E}{\mathrm{d}t}\right|_{m_x=0} &\approx -\alpha\gamma_0 M_{\mathrm{S}}(2h_y - h_k) \,. \end{aligned} \tag{31}$$

But most of the dissipation is likely to occur during the fastest parts of the trajectory, i.e., when the demagnetizing field is maximal during τ_3. At that point, both $\dot{m}_x$ and $\dot{m}_y$ are large, and $m_x^2 \approx 1 - h_y^2/h_k^2$. Using $1 \gg h_y$, (29) leads to

$$\left.\frac{1}{\mu_0 M_{\mathrm{S}}^2}\frac{\mathrm{d}E}{\mathrm{d}t}\right|_{m_z^{\max}} \simeq -\alpha\gamma_0 M_{\mathrm{S}} h_y^2/h_k \,. \tag{32}$$

We use $4\tau_1$ (21), $2\tau_2$ (23) and $4\tau_3$ (19) for the typical time [13] intervals where the dissipations of (31) and (32) occur. The factor 4 before τ_1 corresponds to the initial delay, the delay when m_x approaches -1, the following delay to set the opposite demagnetizing field, and the final slow down when m_x

approaches back to +1. The factor 2 before τ_2 corresponds to back-and-forth reversal. The factor 4 in $4\tau_3$ correspond to the number of occurrences of $m_z^{\max}$ in the case $h_y < h_k$. The total energy loss for the first trajectory is thus approximately:

$$-\Delta E_{\text{cycle}} \approx 4\tau_1 \left.\frac{\mathrm{d}E}{\mathrm{d}t}\right|_{m_x=1} + 2\tau_2 \left.\frac{\mathrm{d}E}{\mathrm{d}t}\right|_{m_x=0} + 4\tau_3 \left.\frac{\mathrm{d}E}{\mathrm{d}t}\right|_{m_z^{\max}}, \tag{33}$$

which reduces to

$$\frac{\Delta E^i_{\text{cycle}}}{\mu_0 M_S^2} \approx \alpha \left(\frac{4h_y^2}{\sqrt{k_k}} + 2\sqrt{2h_y - h_k} + 2\frac{h_y}{\sqrt{h_k}} \right). \tag{34}$$

In (34), the third term is by far the dominant energy-loss contribution for near-bifurcation trajectories. The second and the third terms are comparable only if h_y approaches h_k. The first term is negligible. Note that the above expression is valid only for $h_k/2 < h_y < h_k$. The third term would be discarded if the case $h_y \gg h_k$ was considered. However, we shall seek the smallest reversal fields and thus focus on moderate fields ($h_y \leq h_k$).

5.2.2 Number of Trajectories Gone Through Before Bifurcation

The (noninteger) number of trajectories NT gone through before bifurcation occurs is thus:

$$\mathrm{NT} \approx \frac{\Delta E_{\text{bif}}}{\Delta E_{\text{cycle}}} \approx \frac{1}{\alpha} \frac{(h_y - h_k/2)\sqrt{h_k}}{2h_y} \quad \text{for} \quad h_k/2 < h_y < h_k\,. \tag{35}$$

The physical meaning of the latter equation is transparent. The number of trajectories NT gone through before bifurcation is the "excess field" ($h_y - h_k/2$) above the bifurcation criteria divided by some mean energy-loss rate $2\alpha h_y/\sqrt{h_k}$.

It is worth noticing that the product NTα is almost [13] constant. As expected, the smaller α, the more trajectories the system goes through before being trapped. NT diverges when $\alpha = 0$, restoring the periodicity. The characteristic time τ_{bif} it takes for the system to bifurcate and circulate around a single attraction point is the number of trajectories divided by the slowest characteristic frequency (τ_2), plus the delays

$$\tau_{\text{bif}} \approx \frac{\sqrt{2}}{4\alpha\gamma_0 M_S} \frac{\sqrt{h_k(2h_y - h_k)}}{h_y} + \tau_1 + \tau_3\,. \tag{36}$$

For $T_{\text{pulse}} > \tau_{\text{bif}}$, there is no further change in the final magnetic state, which correlates with the experimentally constant final state for $T_{\text{pulse}} > 300\,\text{ps}$ near $h_y = h_k/2$ (Fig. 5a).

From (36), when very near to the bifurcation criterion (i.e., when $2h_y - h_k \ll h_y$), the trapping process is immediate after the delays. The slight dissipation suffices to make the trajectory bifurcate and prevent the switching. The holy grail of the minimal transverse switching field $h_y = h_k/2$ is thus out of reach as soon as $\alpha > 0$. Let us determine how much the switching field increases when α is finite.

5.3 Bifurcation Field in the Presence of Finite Damping

Our first comment concerns the bifurcation singularity $h_y = h_k/2$. Since it is the minimal cost in applied field for a switching event, it would seem interesting to try to switch the magnetization using a trajectory as close as possible to the bifurcation trajectory.

An important case is that where h_k, $h_y > h_k/2$ and the damping α are such that the hard-axis is passed by once, but may not be passed by a second time because of damping. The system falls then in the nearest and thus most attractive half-m_x space. The reversal is achieved, even if the pulse lasts longer than τ_{bif}.

Such a behavior was already obtained [5,6] in Fig. 5. Formerly reported as "relaxation-dominated reversal", the present authors prefer to quote it as the *relaxation-dominated precessional reversal,* to recall that its nature is very different from the Stoner–Wohlfarth (relaxation-dominated) reversal [15]. The requirements on h_y, h_k and α are obtained by considering [13] that the number of turns (35) should be $\mathrm{NT} = \frac{1}{2}$ for a full switch. With finite damping, the minimal bifurcation field h_{bif} is:

$$h_{\text{bif}} \approx \frac{h_k}{2} + \frac{1}{2}\alpha\sqrt{h_k}\,. \tag{37}$$

This approximate expression on the right holds in the limit of soft thin element, i.e., for $\alpha \ll \sqrt{h_k}$, which holds always in practical MRAM free layers.

The main output of (37) is the affine dependence of the bifurcation field with the damping parameter. The proportionality constant is the half *square root* of the anisotropy field H_k, as could have been guessed from physical arguments [22]. Let us, for instance, apply a field H_y of typical value H_k. Then, the rotation speed (25) scales with $\sqrt{H_\text{k} M_\text{S}}$. Since the energy loss is of viscous origin, we expect it to scale with both the "viscosity" α and with the rotation speed, so with $\alpha\sqrt{H_\text{k} M_\text{S}}$.

The bifurcation field required for such a relaxation-dominated precessional switching event had been already calculated in [2] for a few values of α and for $\mu_0 H_\text{k} = 16\,\text{mT}$. Their (exact) numerical integrations give an influence of α on the minimal cost h_{bif} that is compatible with (37).

In order to gain a better accuracy in this comparison, we calculated *numerically* the effect of α on the minimal transverse switching field h_{bif}, and ameliorated our estimate of the prefactor of α in (37). As shown in Fig. 12, putting an empirical prefactor of 0.59 instead of 0.5 in (37) gives a better

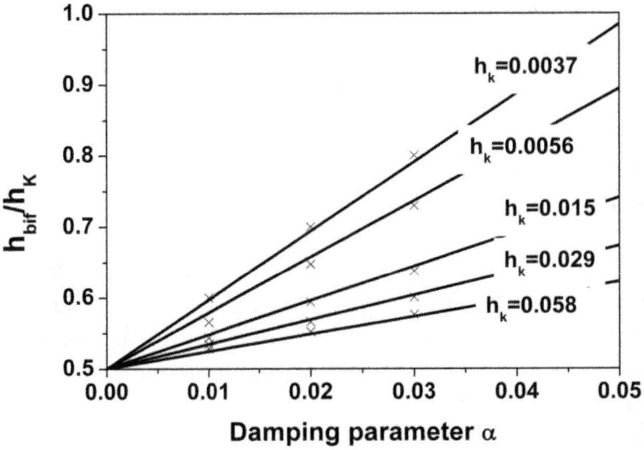

Fig. 12. Dependence of the bifurcation field upon the damping parameter α for relaxation-dominated precessional switching. The corrected analytical estimate [(38), (*full lines*)] is compared to the results of full numerical integration of the LLG equation (*cross symbols*) for several values of the anisotropy field, ranging from $h_k = 0.0037$ to 0.058, for $N_z = 1$

overall agreement with the (exact) results produced using direct numerical integration of the LLG equation. The minimal switching field can be satisfactorily accounted for by a modified (37), where we have added the demagnetizing factor N_z that was so far omitted for simplicity

$$H_{\mathrm{bif}} = \frac{H_{\mathrm{k}}}{2} + 0.59\alpha\sqrt{H_{\mathrm{k}} N_z M_{\mathrm{S}}}\,, \tag{38}$$

which summarizes the effect of the damping parameter on the minimal field required to switch the magnetization in a precessional way [13].

To summarize this section 5, we have analyzed the effect of damping in a perturbative manner for fields $h_y > h_k/2$. Damping both destroys the periodicity of the precession and puts additional constraints on the field rise time. The decay rate of the energy determines the effective number of extended precession cycles gone through. Criteria were derived for the analytical assessment of the switching ability of a given set of experimental parameters. A special focus was dedicated to the relaxation-dominated precessional switching, where the minimal cost in applied field (38) was shown to increase linearly with the damping constant, the slope of this dependence scaling with the square root of the anisotropy energy. For these switching conditions, the magnetization toggles only once whatever the pulse duration above a given threshold (36).

Toggling the magnetization of a memory cell is an interesting phenomenon. However, in most memory applications, a direct-write method is required. The next section is devoted to a generalization of precessional switching, which renders possible the direct-write of a free layer with uniaxial effective anisotropy.

6 Direct-Write: Biasing the Precession with an Easy-Axis Field

MRAM are promising candidates to provide nonvolatile and relatively dense memories [34, 35]. Information is stored in a magnetic "free layer", i.e., a soft thin magnetic cell with uniaxial inplane anisotropy. The cell magnetization can be written by the combination of two orthogonal magnetic field pulses generated by a word line (field H_y) and a bit line (field H_x) that cross near the cell. A sole pulse cannot switch a cell magnetization, which ensures the addressing capability in an array of cells. While industrial prototypes have satisfactory density and power consumption, their magnetization-reversal mechanism relies on quasistatic energy minimization and lasts typically a few nanoseconds [34, 35]. Achieving a sub-ns write time requires benefiting at best from the high frequencies associated with magnetization precession. Hence, in this last section, we study whether precessional switching could be used to advantage in MRAMs.

6.1 Precessional Switching in Magnetic Random Access Memories

So far in this Chapter, a sole transverse field h_y was used to toggle the magnetization direction; this scheme cannot directly write the intended final magnetic state. The precessional programming of an MRAM cell with a sole transverse applied field would thus require the prereading of the memory state. If this slow [34, 35] prereading were implemented, it would waste the interesting speed of precessional switching. In addition, no selectivity method has been experimentally demonstrated so far with a sole transverse field.

To render possible the direct-write in precessional switching, an easy-axis field H_x must be added [14] to the fast-rising hard-axis field pulse (H_y) that triggers a fast precession of the magnetization as described thoroughly above. The hard-axis field $h_y = H_y/M_S$ is always positive and is the excitation field. When high, it triggers the fast precession of the magnetization. When small, it still gives the system a finite susceptibility to easy-axis fields.

The superimposed bipolar easy-axis field ($\pm H_x$) is the programming field selecting which final magnetic state $m_x = \pm 1$ shall be written. It increases/lowers the precession amplitude to bifurcate/or not, and thus to select which final magnetization states are accessible. We shall see that a cell not excited by H_y cannot be programmed by H_x, which provides selectivity in a crosspoint architecture.

Together with the speed, we emphasize that precessional switching has an additional interest. The required writing fields $\{H_x, H_y\}$, constituting the so-called dynamical astroid are lower [16, 17] than in the quasistatic reversal (Stoner–Wohlfarth) scenario ([15], Fig. 1). This was predicted numerically by time integration of the LLG equation both in the macrospin approximation [2–4] and in a few more general cases [27]. Fast direct-write was confirmed experimentally in the limit of small programming fields H_x and high exciting field H_y and micrometer-sized cells [5, 6], but could not be obtained in larger cells [20, 21] because of a strong departure from the macrospin behavior.

The purpose of this last section is not to report more elaborate numerical results or additional experimental data [23], but rather to derive a complete set of *analytical* equations describing the dynamical astroid and its related time constraints in a simple, physically instructive formalism [14]. This is necessary for high-density MRAM where a lot of other phenomena (intercell dipolar coupling, cell-to-cell parameters fluctuations ...) render extremely cumbersome the purely numerical optimizations. We perform the analysis with the assumption of a given field-application timing (Fig. 1, inset). The easy-axis field h_x is switched on first at $t < 0$, and is maintained during the plateau of the hard-axis field pulse h_y. The latter is applied at $t = 0$ for a short duration T_{pulse}. Its rise and fall times are infinitely small, and occur during the plateau of h_x. We will see at the end of this section that this field timing is indeed optimal.

We first detail the magnetization trajectories (Sect. 6.2) that the system may go through as a result of the two magnetic fields. We discuss the nonprogramming and the adequate programming conditions. Studying the transition between these two regimes provides a description of the dynamical astroid (Sect. 6.3). In addition, by looking at some particular points of the trajectories (Sect. 6.4), the reversal process can be separated into three meaningful characteristic times [14] in a spirit very similar to the τ_1, τ_2 and τ_3 of Sect. 4.2.1. We finally (Sect. 6.5) comment briefly on the direct-write and the selectivity issues in a MRAM using precessional switching. References are finally suggested for more technical analysis regarding the size of the addressing window and the required memory architecture [22] for the density scaling of precessional MRAM.

Let us first classify the possible trajectories of the magnetization vector as a result of the two applied fields.

6.2 Magnetization Trajectories for Combined Easy-Axis and Hard-Axis Fields

A successful addressing in a memory array requires that half-selected cells, undergoing a sole h_x or h_y do not switch. Sufficient conditions are that $|h_x| < |h_k|$ and $|h_y| < h_k/2$. We restrict the discussion to these cases. Since $|h_x| < |h_k|$, we can assume that the initial state $m_x = +1$ is maintained until $t = 0$.

When the damping is negligible, conservation of the total energy $e(t=0) = -h_x - h_k/2$ yields:

$$m_z^2 = 2h_x(m_x - 1) + h_k(m_x^2 - 1) + 2h_y m_y \,. \tag{39}$$

Using $\boldsymbol{m}^2 = 1$, the resulting magnetization trajectory is:

$$m_y = -h_y \pm \sqrt{1 + h_k + h_y^2 + 2h_x(1 - m_x) - (1 + h_k)m_x^2} \,. \tag{40}$$

Only one of the two mathematical solutions of (40) is physical. Sorting out the right one can be done by checking that (39) gives a positive m_z^2, as is done systematically below. For a start, let us describe the situations with a total applied field that is roughly along the hard-axis.

6.2.1 Magnetization Trajectories for Nearly Transverse Applied Fields

Figure 13 displays some trajectories representative of the cases when the total field is almost transverse, with h_y slightly below $h_k/2$.

When h_x is positive and relatively large, the trajectory is a tiny lobe passing at the initial magnetization $m_x = 1$. Reducing h_x increases the lobe perimeter. When the Zeeman energy of the system is increased sufficiently, another (smaller) lobe starts growing in the $m_x < 0$ half-space. This indicates that at reduced $h_x > 0$, both the stable magnetization position (with $m_x > 0$ and $m_y > 0$) and the metastable one (with $m_x < 0$ and $m_y > 0$) have an energy smaller than that of the initial energy of $m_x = 1$. These two lobes constitute degenerate trajectories (both energetically allowed), but only the one lobe passing at the initial magnetization is gone through (Fig. 13). Also, since the magnetization stays in the $m_x > 0$ half-space, it relaxes systematically to $m_x = 1$ no matter when the fields are switched off. In summary, for h_y below $h_k/2$ and $h_x > 0$, the state $m_x = 1$ is always rewritten.

When h_x is driven negative, the circumference of the lobe $m_x < 0$ extends until a point where the two lobes become tangential and merge in a single, much-extended trajectory. Upon further increasing $-h_x$, the tangency point transforms into two saddle points in the $\{m_x, m_z\}$ projection (Fig. 13a). We refer to these points as the β points. With the parameters of Fig. 13, i.e., $h_k = 0.01$, $h_y = 0.45h_k$, the lobe merging occurs for $h_x = -0.053h_k$. To switch the magnetization, h_x must be above this point of the dynamical astroid and its duration should be such that magnetization has passed in the $m_x < 0$ half-space when h_y is turned off. Finally, when $|h_x|$ is increased to a largely negative value that we shall determine to be the Stoner–Wohlfarth astroid, the β points disappear and the presence of two merged lobes cannot be guessed any longer (Fig. 13a, $h_x = -0.003$) from the trajectory shape.

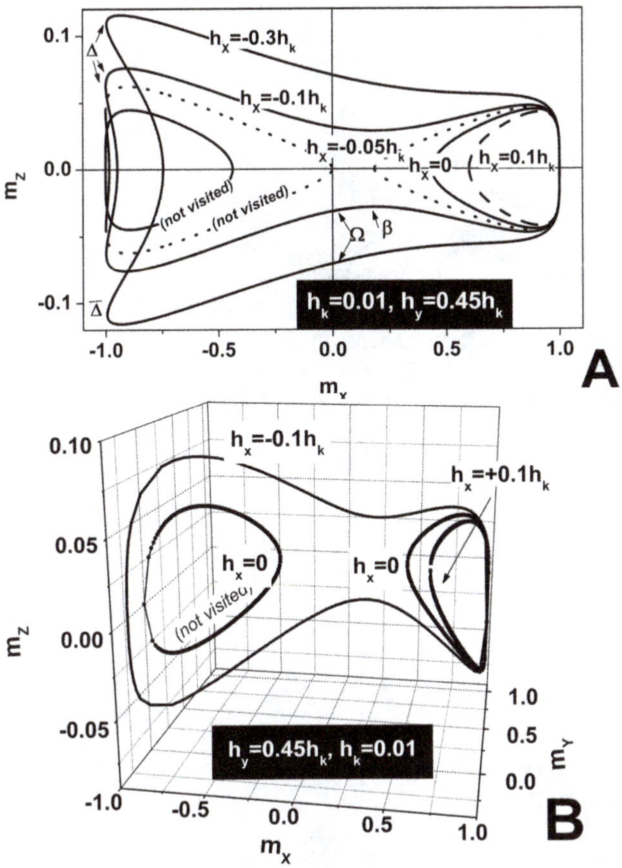

Fig. 13. Representative trajectories for nearly transverse total applied fields (i.e., $h_y \approx h_k/2$ and $h_x \ll h_y$) and anisotropy field $h_k = 0.01$. (**a**) Projection in the (m_z, m_x) plane of the magnetization trajectories for cells excited with $h_y = 0.0045$, when half-selected ($h_x = 0$), when rewritten ($h_x = +0.001$) and when switched ($h_x = -0.001$ or -0.003). In the two first cases, only the lobe in the $m_x > 0$ half-space is visited by the system. The β point is defined as the point where $\dot{m}_z = 0$; i.e., it is the reminiscence of the lobe-merging position. The Ω points are defined as the loci of maximum m_y incursions. The Δ and $\bar{\Delta}$ points are the loci of maximum negative m_x incursion. (**b**) Three-dimensional view of the trajectory. Note that the $h_x = 0.001$ trajectory has no lobe in the $m_x < 0$ half-space

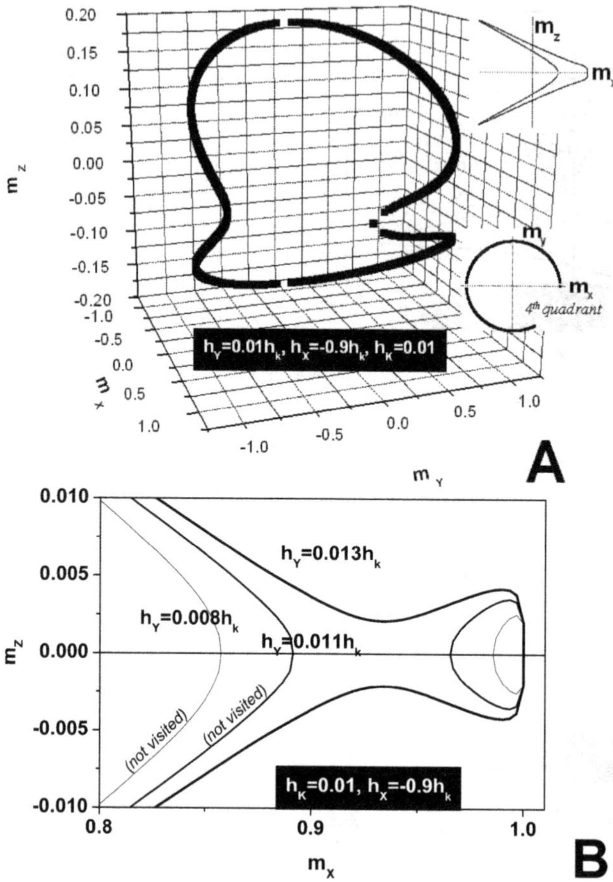

Fig. 14. Representative trajectories for total applied fields nearly antiparallel to the initial magnetization (i.e., $h_x \approx -h_k = -0.01$ and $h_y \ll h_x$). (**a**) Full magnetization trajectory and its projections in the case of $h_x = -0.009$ and $h_y = 0.0001$. (**b**) Projection in the (m_z, m_x) plane, focused near the merging of the lobes for $h_x = -0.009$. When the trajectory consists of two lobes, only the lobe in the $m_x > 0$ half-space is visited by the system

6.2.2 Magnetization Trajectories for Nearly Antiparallel Applied Fields

Figure 14 displays a more traditional way of reversing the magnetization, by using a total applied field mainly (anti) parallel to the initial magnetization. We consider h_x near $\pm h_k$, and much greater than the transverse field h_y. Although the field thresholds are very different from the previous case, the trajectories still consist of lobes, with their confluence ruling the possibility for the magnetization to switch.

Again, a half-selected cell stays on a given lobe keeping $m_x > 0$, and so that the magnetization always relaxes to its initial magnetization $m_x = 1$. Again also, a cell undergoing $h_x > 0$ always ends in the rewritten $m_x = 1$ state.

When switching is wanted ($h_x = -0.9h_k < 0$), the reversed magnetization is accessible only if the lobes have merged, which occurs if $h_y > 0.013h_k$ (Fig. 14b). Note that the lobe near $m_x = 1$ is tiny, and in fact it reduces to a single point when $h_x = -h_k$ and $h_y = 0$. This reduction to a single point coincides with the disappearing of the astroid curve at that specific point $h_x = -h_k$ and $h_y = 0$. Switching requires that the hard-axis field be switched off when the magnetization is in the $m_x < 0$ half-space. Let us emphasize that in this case of near easy-axis total applied field, the magnetization goes through a trajectory of very large inplane rotation amplitude (see inset Fig. 14a). Due to the inertia of the large precession, the magnetization switches too much: in *one* precession cycle the systems goes *twice* in the $m_x < 0$ half-space and twice in $m_x > 0$. In the $\{m_x, m_y\}$ plane, $\boldsymbol{m}$ rotates so much that it enters the 4th quadrant $\{m_x > 0, m_y < 0\}$, before boomeranging back to pass a second time near the reversed state and then finally come to the initial state. In the following, we refer to such trajectories as *trajectories with bounce.*

6.2.3 General Case: Lobe Centers and Bounce-Occurrence Criteria

Trajectories resulting from intermediate field orientations still consist of one or two lobes, with a merging phenomenon occurring at a bifurcation field that governs the possibility for the magnetization to reverse.

When the trajectory consists of two topologically separated lobes, the switching is not possible. Only the one lobe passing through the initial state ($m_x = 1$) is visited by the system. When h_x and the initial magnetization point in the same direction (rewriting of the cell), the lobe located opposite to h_x is smaller and disappears for sufficiently large h_x (see Fig. 13b, $h_x = 0.001$). This disappearance means that at sufficiently high h_x, the initial energy of the system is below that of the metastable state (having $m_x < 0$), if any.

When both fields are high enough (above the dynamical astroid) and when the easy-axis field is antiparallel to the initial magnetization, there is a single connex trajectory (i.e., two merged lobes, Fig. 13a and Fig. 14b). The system can visit any point along such a trajectory. The latter extends in both $m_x > 0$ and $m_x < 0$ half-spaces, so that depending on the h_y pulse duration T_{pulse}, the system can end either in the initial or in the reverse $m_x < 0$ half-space. In the latter case, magnetization relaxes to the reversed state as a result of the anisotropy and demagnetizing fields helped by h_x before its decay.

It is interesting to note the physical meaning of the trajectory points where $\dot{m}_z = 0$. Using (3c), this condition is

$$\frac{h_y}{m_y} - \frac{h_x}{m_x} = h_k\,. \tag{41}$$

From [15], this equation also describes the magnetization equilibrium (stable, metastable or unstable) positions in the total applied field $\boldsymbol{h}_y + \boldsymbol{h}_x$ when $m_z = 0$. Hence, the maximum out-of-plane incursions occur exactly above the equilibrium positions. Qualitatively, *each lobe is a portion of the trajectory where the magnetization rotates mainly around one attracting point* that is an infield equilibrium position.[1]

It is also worth evaluating the field conditions leading to a trajectory with a bounce, i.e., when the magnetization enters in the $\{m_x > 0,\, m_y < 0\}$ quadrant (Fig. 14a). The required fields can be found by writing (40) for $m_y = -1$, which leads to the very simple "bounce-occurrence" condition

$$h_x < -h_y - h_k/2\,. \tag{42}$$

This curve intersects the dynamical astroid at $h_x = -0.605 h_k$ and $h_y = 0.105 h_k$ (see Fig. 1).

This indicates that a total field at an angle $\theta < 135°$ with respect to the initial magnetization (i.e., x) never produces a bounce, whatever the anisotropy field. The *conventional optimal orientation* $\theta = 135°$ *is thus still safe when the reversal is of precessional nature.* Contrary to common thinking, *using a more negative* h_x (i.e., a more antiparallel applied field) makes the switching less safe.

Also, there is a spurious consequence of a bounce existence. At first glance, it could seem optimal to tune the h_y pulse duration to half of the precession period as was optimal in the case of a purely transverse applied field [2]. This leads to a robust reversal strategy when there is no bounce. However, if there

[1] This correlates with the fact that the lobes merge in a single connex (switching) trajectory when the field $h_x < 0$ is large enough so that the system energy in the initial state ($m_x = 1$) is increased above the energy saddle point corresponding to the unstable, infield equilibrium position (β point, small $m_x > 0$ see Fig. 13a). This bifurcation happens along the dynamical astroid. Also, when along the Stoner–Wohlfarth astroid, the unstable equilibrium position and the metastable one merge together and disappear, such that there is no longer a saddle point in the $\{m_x, m_z\}$ projection of the trajectory (no β point, see Fig. 13a, $h_x = -0.003$).

When above the Stoner–Wohlfarth astroid, there exists only one infield equilibrium position. The consequence is that for both signs of h_x there is only one lobe. But the reciprocal argument is not true: there may be two infield equilibrium positions though only a single lobe. This may happen either because the initial energy is below the energy of the metastable equilibrium position (the case of $h_x > 0$, just below the Stoner astroid), or because the lobes have merged (the case of $h_x < 0$ above the dynamical astroid).

is a bounce, the magnetization is in the 4th quadrant $\{m_x > 0, m_y < 0\}$ when h_y is switched off. The magnetization then relaxes to $m_x = 1$ and the switching fails. Therefore much care should be taken when using fields in the bounce-existence region.

6.3 The Dynamical Astroid

6.3.1 Derivation of the Dynamical Astroid

The previous descriptions provide a method to derive the dynamical astroid. For a given h_y, we need to evaluate the bifurcation criteria, i.e., the minimum $|h_x|$ giving a connex-extended trajectory. We thus use the equations describing the lobe tangency, i.e., when the point β (41) has its m_z as zero (39).

Let us first estimate a simple equivalent function describing the relation between h_x and h_y when $h_y \approx h_k/2 \gg h_x$ (the case of Fig. 13). In that case, the lobes merge near the point $m_y = 1$. Using $\boldsymbol{m}^2 = 1$, we can write $m_x = \varepsilon$ and $m_y = 1 - \varepsilon^2/2$. Using (41), $m_z = 0$ and (39), we can eliminate ε and get the tangent of the dynamical astroid near $\{h_x \approx 0, h_y \approx h_k/2\}$

$$\frac{h_y}{h_k} = \frac{1}{2} + \frac{h_x}{h_k} . \tag{43}$$

Using a similar method [14] near the other end of the astroid $\{h_x \approx -h_k \gg h_y\}$, we get a simple equivalent of the dynamical astroid:

$$\frac{h_y}{h_k} = \frac{1}{2}\left(\frac{h_x}{h_k} + 1\right)^{\frac{3}{2}} . \tag{44}$$

Equations (43), (44) are compared in Fig. 1 to a numerical evaluation [2] of the dynamical astroid that was obtained for $\mu_0 H_\mathrm{k} = 16.5\,\mathrm{mT}$ and $\alpha = 0.001$. The tangency is excellent at both ends of the dynamical astroid. In between, (43) and (44) can be interpolated with one simple analytical expression having the right values and the right tangents at $h_x = 0$ and $h_y = 0$ and an overall precision of at worst $0.007 h_k$. We write this interpolation of the dynamical astroid as $h_y = \mathrm{DynAst}(h_x)$, with the DynAst function being

$$\frac{h_y}{h_k} = \frac{\left(1 + \frac{h_x}{2h_k}\right)\left(1 + \frac{h_x}{h_k}\right)}{2 - \frac{h_x}{h_k}} . \tag{45}$$

6.3.2 Smallest Switching Field and Extrema of the Dynamical Astroid Curve

A useful particular point along the dynamical astroid is the smallest switching field $\boldsymbol{h}_\mathrm{min}$ (see arrow in Fig. 1). Its coordinates are $h_x = -0.224 h_k$ and $h_y = 0.315 h_k$. Its magnitude and orientation are:

$$\|\boldsymbol{h}_\mathrm{min}\| = 0.387 h_k \, , \{\boldsymbol{h}_\mathrm{min}, x\} = \pm 125^\circ . \tag{46}$$

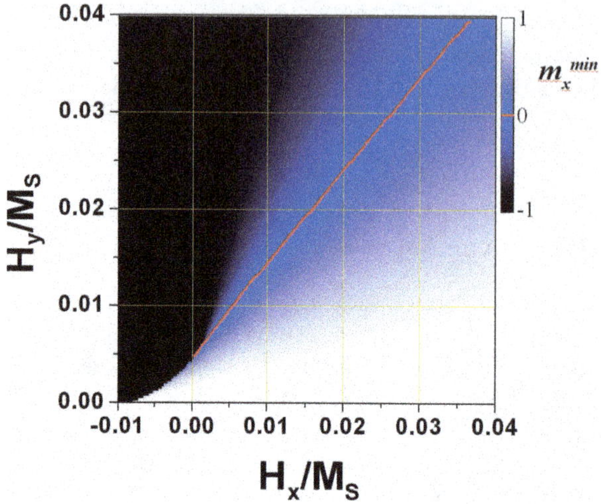

Fig. 15. Switching/no switching curve: map of the maximum approach towards $m_x = -1$ along a magnetization trajectory, calculated numerically for a lossless ($\alpha = 0$) thin ($N_z = 1$) platelet of Permalloy ($\mu_0 M_S = 1.08\,\mathrm{T}$), with effective anisotropy field $h_k = 0.01$ (i.e., 10.8 mT). The color scales with the maximum negative incursion of m_x. The red line corresponds to the zero of the m_x incursion, i.e., the switching/no switching curve separation

Hence the smallest switching field in precessional switching is less antiparallel to the initial magnetization than it is in a Stoner–Wohlfarth scenario [15], for which it is 135°.

To summarize this section, we have derived analytically the tangents of the astroid near-hard-axis and near-easy-axis field orientations. In between, a simple function $h_y = \mathrm{DynAst}(h_x)$ (45) interpolates the dynamical astroid with better than 1 % accuracy. Together with the "bounce-occurrence condition" (42), the smallest switching field (46) and the Stoner–Wohlfarth astroid, the corresponding curves are gathered in Fig. 1.

A last important comment is that the dynamical astroid curve does end at the point $\{0, h_k/2\}$, but extends infinitely in the $h_x > 0$ region; when the applied field makes a sufficient angle with the initial magnetization, the magnetization can still pass the hard-axis. This angle tends asymptotically to 45° when the applied field strength tends to infinity. To illustrate how the dynamical astroid prolongates in the $h_x > 0$ region, we have mapped in Fig. 15 the maximum of $-m_x$ along a trajectory defined with $\{h_x, h_y\}$, i.e., the maximum approach of the magnetization towards the reversed state $m_x = -1$. The red curve in Fig. 15 is the switching curve: it is constructed with the pairs $\{h_x, h_y\}$ leading to the trajectory passing exactly at $m_y = 1$.

6.4 Characteristic Duration of a Precessional Direct-Write Event

So far in this section we have ignored the rate of change of the magnetization.

Our first comment relates to the topology of the trajectories for field combinations along the dynamical astroid. In these trajectories, the system bifurcates at the unstable equilibrium position, where the total torque is zero, and the motion is stationary. The associated switching time is theoretically infinite, and in practice a random final state is determined by the actual magnetic noise. Since this situation should be avoided in practice, we assume that the applied fields are sufficiently above the dynamical astroid curve to disregard this singularity.

To estimate the needed duration of h_y, we introduce four characteristic times. The first is a generalized version of the "initial" delay τ_1', which takes into account the near-vanishing initial torque occurring when the total applied field is near antiparallel to the initial magnetization. The second is the corrected "main rotation time" τ_Ω that the system spends in the neighborhood of the hard-axis. The τ_Ω is the generalization of τ_2 (see Sect. 4.2.1) for $h_x \neq 0$. The third is the "alignment time" $\tau_{\Omega\Delta}$ that the system needs to pass from near hard-axis magnetization to near reversed magnetization. The last characteristic time is the "pause interval" $\tau_{\Delta\bar{\Delta}}$ that the system spends around the reversed position before being launched back to the initial magnetization, in the case when no bounce occurs. This pause correlates with the fact that whenever the magnetization is inplane, it is stationary [$\dot{m}_x = \dot{m}_y = 0$ see (3)]. This pause interval $\tau_{\Delta\bar{\Delta}}$ is a counterpart of the final delay τ_1 (see Sect. 4.2.1) when $h_x \neq 0$. This pause interval expresses how precisely the duration of the applied field h_y should be controlled.

6.4.1 Generalized Initial Delay

It is convenient to call Ω the loci of maximum m_y incursions, when the system is near the hard-axis. The conditions $\dot{m}_y^\Omega = 0$ and $m_z^\Omega =$ are (3b)

$$m_x^\Omega = \frac{-h_x}{1+h_k} \approx -h_x \,. \tag{47}$$

Using $m_y^\Omega \approx 1$, $h_x, h_y, h_k \ll 1$ and (40) at Ω, we can obtain self-consistently m_y^Ω and m_z^Ω

$$m_z^\Omega \approx \sqrt{2(h_y - h_x - h_k/2)} \quad \text{and} \quad m_y^\Omega \approx 1 - (h_y - h_x - h_k/2)\,. \tag{48}$$

As was formerly done in Sect. 4.2.1, we can define a generalized initial delay τ_1' by setting: $\dot{m}_z \tau_1 = m_z^\Omega$. This leads to

$$\tau_1' \approx \frac{\sqrt{2(h_y - h_x - h_k/2)}}{\gamma_0 M_S h_y} \quad \text{i.e.,} \quad \tau_1' \approx \frac{\sqrt{2M_S(H_y - H_x - H_k/2)}}{\gamma_0 M_S H_y}\,. \tag{49}$$

The divergence of τ_1' when $h_y = 0$ recalls the vanishing of the torque in such field conditions: the initial susceptibility only depends on h_y.

6.4.2 Main Rotation Time

It is worth noting that the coordinates of the Ω point generalize (20) when $h_x \neq 0$. They express the amount of energy stored in the demagnetizing field when magnetization passes around the hard axis: $H_\mathrm{d}^\Omega = \sqrt{2M_\mathrm{S}(H_y - H_x - H_\mathrm{k}/2)}$. The $1/\gamma_0 H_\mathrm{d}$ defines a first characteristic time. As pointed out by the numerical study performed by *Albuquerque* [27], $\pi/\langle\gamma_0 H_\mathrm{d}\rangle$, where $\langle\rangle$ denotes the time average, is the reversal duration when in ballistic conditions. This argument can be used also for our present study. Indeed, when the magnetization is near Ω (near the hard axis), the rotation of the inplane component of the magnetization occurs mostly due to the demagnetizing field. We thus define τ_Ω as the time it takes typically for m_x to pass from +1 to circa −1 so that $\pi\tau_\Omega \dot{m}_x^\Omega \approx 2$. Using (3) and (48), we can get τ_Ω. However, to maintain the consistency with the precession frequency in a transverse applied field (25), it is preferable to change the prefactor and write:

$$\frac{1}{\tau_\Omega} = \frac{0.847}{\pi}\gamma_0\sqrt{M_\mathrm{S}(H_y - H_x - H_\mathrm{k}/2)}\,, \tag{50}$$

(meaningful only for $h_y \gg |h_x|$).

Note that this main rotation time τ_Ω is the generalization of τ_2 when an easy-axis field is present. Equation (50) gives the relevant physical insight only when the points β and Ω are not far from each other, i.e., in the case of $h_y \gg h_x$, i.e., mostly transverse total applied field (Fig. 13). When the lobe merging occurs far from the hard-axis (case of Fig. 14), τ_Ω does not take properly into account the drastic slow down occurring near the β point, and poorly estimates the real rotation time.

However, the comparison of (50) and (43) is instructive; when physically justified (i.e., when $h_y \gg |h_x|$), *the main rotation time diminishes as the square root of the excess field above the dynamical astroid curve. This constitutes a generalization of the precession frequency that was obtained in a purely transverse field (25).*

6.4.3 Alignment Time

Once the system has passed the hard-axis, it is also worth establishing how long it then takes for the system to reach a *near reversed state,* i.e., the "alignment time". This information is partly redundant with the former main rotation time τ_Ω. We need to define the near-reversed states as the Δ and $\bar{\Delta}$ points (see Fig. 13a), which are the loci of the maximum negative m_x incursion. Using (3a), these points satisfy $m_y^{\Delta} = -h_y$. Using (40), their other coordinates are:

$$m_x^{\Delta} = \frac{-h_x - \sqrt{1 + 2h_k + h_y^2 + h_x^2 + h_k^2 + h_k h_y^3}}{1 + h_k} \approx -1 - h_x \,, \tag{51}$$

$$(m_z^{\Delta})^2 = -\frac{2h_x}{1 + h_x} - h_y^2 - \frac{h_x^2}{(1 + h_k)^2} \approx -2h_x \,. \tag{52}$$

The latter approximation is valid for *not* transverse fields, i.e., typically when $-h_x \gg h_y^2$.

The rotation from the Ω (roughly above the hard-axis) to the Δ (near the easy-axis) point is due to the Zeeman torque between $\boldsymbol{m}$ and h_x and to the vertical demagnetizing field. Since the demagnetizing contribution was already included in the definition of τ_Ω, we define the alignment time $\tau_{\Omega\Delta}$ as only due to the Zeeman torque by using $\tau_{\Omega\Delta}\dot{m}_y^{\Delta} \approx m_y^{\Omega}$. This leads to the alignment time

$$\frac{1}{\tau_{\Omega\Delta}} = \gamma_0\sqrt{-H_x M_\mathrm{S}} \quad \text{for} \quad |h_x| \gg h_y^2 \,. \tag{53}$$

Due to the former approximations, the above equation describes correctly the alignment time only when $|h_x| \gg h_y^2$.

6.4.4 Pause Interval

Finally, let us determine how long the system stays near the reversed state (making somehow a "pause") before precessing back to the initial (not reversed) state. This will be the analogy of the final delay τ_1 (Sect. 4.2.1). For this, we estimate the time it takes to go from one near-reversed state to the other, i.e., from Δ to $\bar{\Delta}$. When h_y is not too small, this evolution results mainly from the Zeeman torque between h_y and $\boldsymbol{m}$, we can write. $\tau_{\Delta\bar{\Delta}}\dot{m}_z^{\Delta} \approx m_z^{\bar{\Delta}} - m_z^{\Delta}$. Note that this expression is not valid when a bounce (42) is present or when the evolution from the Δ point to $\bar{\Delta}$ arises from a significant contribution of the demagnetizing field.[2]

The "pause interval" $\tau_{\Delta\bar{\Delta}}$ is thus:

$$\tau_{\Delta\bar{\Delta}} = \frac{\sqrt{-8H_x M_\mathrm{S}}}{\gamma_0 M_\mathrm{S} H_y} \quad \text{for} \quad h_y > |h_x| \,. \tag{54}$$

6.4.5 Precession Frequency for Arbitrary Field Orientation

As underlined above, the previous expressions of τ_1', τ_Ω, $\tau_{\Omega\Delta}$ and $\tau_{\Delta\bar{\Delta}}$ are not valid (or meaningful) for all field orientations. The expression of $\tau_{\Delta\bar{\Delta}}$ makes

[2] In the case when (42) is satisfied (occurrence of a bounce in the trajectory) it is easy to find by a similar analysis that the main contribution in $\tau_{\Delta\bar{\Delta}}$ is approximately $2\tau_{\Omega\Delta}$.

sense only when there is no bounce. The expression of τ_Ω scales with the real switching duration only when the total applied field is close to the hard-axis (i.e., $|h_x| \ll h_y$). The expression of $\tau_{\Omega\Delta}$ applies only when the total applied field is far from being transverse (i.e., $|h_x| \gg h_y^2$). Note that the four times are reasonably valid along the direction of the smallest reversal field $\boldsymbol{h}_{\min}$ (46).

Although of questionable generality when departing from an applied field orientation in the vicinity of $\boldsymbol{h}_{\min}$, these characteristic times contain consistent information. Indeed, a higher $-h_x$ stabilizes the reversed state that results in an increase in the pause interval $\tau_{\Delta\bar{\Delta}}$ and an acceleration of the alignment (lowering of $\tau_{\Omega\Delta}$). Conversely, when $-h_x$ is small, the system does not pause significantly in the reversed state (low $\tau_{\Delta\bar{\Delta}}$) because Δ and $\bar{\Delta}$ converge to each other. When $-h_x$ is small, the reversed and initial states have almost the same energy such that the alignment process is slow (high $\tau_{\Omega\Delta}$).

Concerning the effect of the hard-axis field, a high h_y generates a high-frequency precession, which fixes the main rotation time (54). A high h_y gives also some transverse susceptibility to an easy-axis magnetization, which consecutively decreases the pause interval and the generalized initial delay.

The above analysis is too simplified to account quantitatively for the optimal pulse duration for any arbitrary field orientation. The qualitative trend is that the optimal pulse duration is related to $\tau_1' + \tau_\Omega + \frac{1}{2}\tau_{\Delta\bar{\Delta}}$. For a successful reversal T_{pulse} should be controlled with a tolerance scaling typically with $\pm\frac{1}{2}(\tau_{\Omega\Delta} + \tau_{\Delta\bar{\Delta}})$. The ringing is minimized and the reversal is quasiballistic if T_{pulse} is controlled with a tighter tolerance related to $\pm\frac{1}{2}\tau_{\Delta\bar{\Delta}}$.

Before ending with the subject of precession frequency, we report in Fig. 16 the half-precession period in the $\{h_x, h_y\}$ plane, as calculated numerically for a Permalloy thin film of anisotropy $h_k = 0.01$ ($\mu_0 H_\text{k} = 10.8\,\text{mT}$). As expected, the precession period diverges when the fields are along the dynamical astroid and when $H_y = 0$. We emphasize that the astroid curve prolongates slightly in the $H_x > 0$ region along a curve where the precession period diverges. This "*period-divergence curve*" ends at the Γ point $\Gamma = \{0.14h_k, 0.66h_k\}$, where one of the merging lobes formerly described reduces to a single point (not shown). Note that this period-divergence curve is distinct from (and slightly above) the switching/no-switching curve that prolonged the dynamical astroid (Sect. 6.3) in the $H_x > 0$ region.

Another important point is that since the Γ point corresponds to a field orientation near the hard-axis, (43) holds and the "*period-divergence curve*" is almost linear from $\{0, 0.5h_k\}$ to the Γ point (see Fig. 16). As a result τ_Ω describes reliably the precession period in that region.

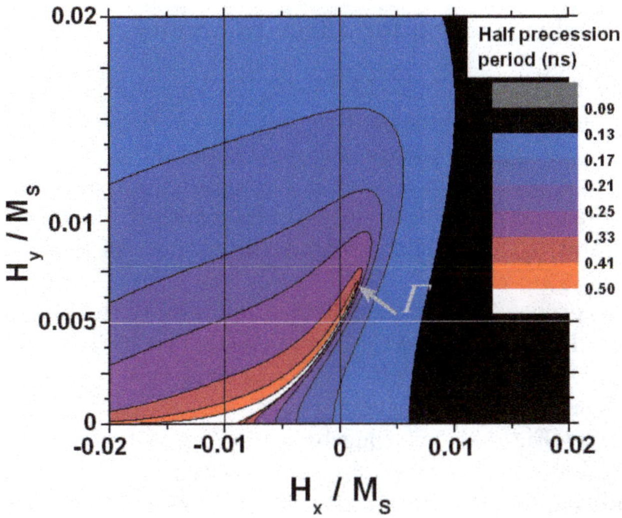

Fig. 16. Map of the half-precession period calculated numerically for a lossless ($\alpha = 0$) thin ($N_z = 1$) platelet of Permalloy ($\mu_0 M_S = 1.08\,\mathrm{T}$), with effective anisotropy field $h_k = 0.01$ (i.e., 10.8 mT). In the $H_x < 0$ region, the precession-period divergence occurs along the dynamical astroid and along the $H_y = 0$ curve. In the $H_x > 0$ region, the precession-period divergence continues until the so-called $\Gamma = \{h_x = 0.14h_k, h_y = 0.66h_k\}$ point, where the period becomes finite

6.5 Strategies for the Cell Writing and Selection in a Practical Memory Array

6.5.1 The Selectivity and Direct-Write Issues in Precessional MRAM

A successful direct-write requires that the fields be above the dynamical astroid (45) and that the pulse duration be matched to the adequate interval around half of the precession frequency (Fig. 16).

A successful addressing in a memory array requires that h_x and h_y be such that half-selected cells, undergoing only one field, do not switch. From the above study, the conditions $|h_x| < h_k$ and $|h_y| < h_k/2$ are *sufficient conditions* to ensure selectivity.

Strictly speaking, these conditions are not necessary conditions. Playing on the difference between the (slow) switching time in a purely transverse field (25) and a (faster) switching time in superimposed orthogonal fields (50) could be used to ensure the selectivity. However, the tiny difference in switching time would certainly make it technically difficult and would bring the writing stage far away from ballistic conditions, both facts being undesirable.

6.5.2 Optimal Field Synchronization for Jitter Immunity

Our second comment concerns the choice of the $\{h_x, h_y\}$ fields timing, for an optimal immunity to the jitter. We confess that we initially chose to set first the easy-axis field (inset in Fig. 1) just to ease the mathematical treatments of this section. Anyway, first setting on h_x has a strong advantage in the macrospin case: there is a large (theoretically infinite) tolerance in the delay between the onset of h_x and h_y, since the susceptibility is zero as long as $h_y = 0$.

In contrast, the delay tolerance (or equivalently the immunity to the jitter) would be drastically reduced if the reversed timing was used, i.e., if the hard-axis field h_y was set first at $t = t_{\text{hard}}$. To be able to program the final magnetic state, the easy-axis field h_x should be then switched on at t_{easy} early enough to connect the two lobes (see Fig. 13), i.e., long before $\omega(t_{\text{easy}} - t_{\text{hard}}) = \frac{\pi}{2}$, with ω defined by (25). The least worse would be to switch on h_x before $t_{\text{easy}} - t_{\text{hard}} = \tau_1$. In addition to this jitter immunity, this reversed timing would generate another complexity: each given delay corresponds to a slightly different dynamical astroid, hence reducing the overall tolerances.

We are thus led to the conclusion that *for macrospin samples,* the optimal field timing is to first set on the easy-axis field (Fig. 1 inset). This ensures good jitter immunity. However, in large real samples that are far from being in a macrospin state, setting on first the easy-axis field h_x may be dangerous; a large $t_{\text{easy}} - t_{\text{hard}} > 0$ may greatly amplify the already present magnetostatic patterns (end domains and ripples) and prevent the proper (coherent) excitation by the hard-axis field. This was observed experimentally on $20 \times 7\,\mu\text{m}^2$ free layers [20,21] but is not likely to happen in the far submicrometer regime.

6.5.3 Size of the Addressing Window

In a dense MRAM array, the net magnetic moment of a given cell interacts with the dipolar field generated by the other cells. The sum of these dipolar fields can be modeled as a perturbation magnetic field $\Delta\boldsymbol{H}(t)$ that is superimposed on the controlled sources of magnetic fields [word-line excitation field $H_y(t)$ and bit-line programming field $H_x(t)$].

This intercell dipolar coupling has several implications.

1. If $\Delta\boldsymbol{H}(t)$ is high, it may contribute to a significant reduction of the data-retention time.
2. The fast-rising part of $\Delta\boldsymbol{H}(t)$, resulting from the reversal of a cell can contribute to write neighboring semiaddressed cells, leading to a kind of avalanche of reversal.
3. The quasistatic part of $\Delta\boldsymbol{H}(t)$ can add to or oppose the addressing fields, and may prevent addressed cells being properly written to. When they oppose, they can prevent the switching, whereas when they add, they can trigger an unwanted bounce.

A technical analysis of effect 3 [14] has given an upper estimate of the maximum tolerable intercell dipolar coupling. In a square array of cells, an intercell random dipolar coupling field higher than 18 % of the anisotropy field defeats the reliability of the memory-writing process. This dipolar contribution adds to the other limitations arising in high-density arrays, such as line-to-cell crosstalk [36] and cell thermal stability [37].

A further discussion of the density scaling of precessional MRAM and of the required memory architecture can be found in reference [22]. To cope with the relatively low selectivity margin resulting from the intercell coupling, several encoding algorithms (a high-pass encoding and a balanced encoding [22]) can be used to effectively decrease the intercell dipolar field magnitude without changing the cell geometry and the organization of the cell array.

7 Conclusion

We have reported our advances on the physics of the precessional switching of uniaxial magnets made of soft thin films. We have combined magneto-transport investigations [5, 6] having ps-scale time resolution with a comprehensive, mostly analytical framework [11–14]. Our major outputs concern:

1. The precession frequency ω (25) when the field is purely transverse, and the consequences of the precession frequency: the intrinsic reversal speed (precessional switching proceeds within $\pi/\omega \approx 100\,\text{ps}$ to $400\,\text{ps}$), the coherence criteria [successful reversal requires $\frac{\pi}{2} < \omega T_{\text{pulse}} < 3\frac{\pi}{2}$ (modulo 2π)], the ballistic conditions (the optimal pulse duration T_{pulse} is π/ω), and the most energy-cost-effective toggling scenario [$H_{\text{eco}} = 2H_{\text{k}}/3$, (27)].
2. The bifurcation phenomenon and its consequences: the minimal transverse reversal field in the presence of finite damping (38) is above $H_{\text{k}}/2$ by an extra field cost linked to the damping constant times the square root of $M_{\text{S}}H_{\text{k}}$. When trying to approach that bifurcation field, the corresponding relaxation-dominated precessional switching slows down drastically and becomes intrinsically noise sensitive.
3. The optimal combination of easy-axis and hard-axis fields for magnetic random access memory applications. Jitter immunity requires a field timing (Fig. 1, inset) in which the easy-axis field is switched on first and switched off last. Robust direct-write requires that the easy-axis and hard-axis fields be above the dynamical astroid (45) and be at an orientation not too antiparallel to the initial magnetization, so as to avoid the occurrence of a bounce (42). Selectivity is more reliable when the easy-axis field is below H_{k} and the hard-axis field is below $H_{\text{k}}/2$. We have studied the duration of a precessional direct-write event, and shown that it needs field pulses in the 200 ps range (50), (53), (54). We have also identified the prolongation of the dynamical astroid in the $H_x > 0$ region:

the "switching/no switching curve" (Fig. 15) and the distinct "precession period–divergence curve" (Fig. 16).

In addition, we have emphasized technical criteria to assess the switching ability of realistic experimental parameters. The hard-axis field rise time should obey $\frac{\dot{H}_y}{H_y} \gg \alpha\gamma_0 M_S$. Nonuniformities of the magnetization direction and of the internal fields resulting, for instance, from the magnetostatics should be avoided because they reduce the switching reliability, especially near the generalized bifurcation field (26). In that sense, having material-related anisotropy is preferable to shape anisotropy for a robust switching, in addition to allowing a higher thermal stability [22].

In future precessional MRAMs, caution should be paid to several factors. First, the intercell dipolar interactions may become a major source of error at the bit programming and considerably affect the data-retention time. Strategies are currently explored to minimize the cell-to-cell coupling [38]. Second, Joule-energy dissipation in the word and bit lines may soon become a concern since this energy cost of the reversal increases (28) as the cubic clocking frequency when above 10 GHz. Finally, we stress that a large effort must be engaged to be able to tune the local damping parameter so as to match optimally the ballistic conditions, which will be very challenging in a complete MRAM architecture.

Acknowledgements

This work is supported by the European Communities Human Potential programme under contract number HRPN-CT-2002-00318 "ULTRASWITCH", by the NEDO contract "NANOPATTERNED MAGNETS" and by the French Ministry of Research through the RMNT programme "MAGMEM". The authors are grateful to the "Equipe de Recherche Technologique" of Paul Crozat for their microwave measurement expertise, to Ricardo Sousa and Paulo Freitas for their advanced sample preparation and to Jacques Miltat, André Thiaville and Burkard Hillebrands for their having been a fruitful proposition force.

References

[1] C. H. Back, R. Allenspach, W. Weber, S. S. P. Parkin, D. Weller, E. L. Garwin, H. C. Siegmann: Minimal field strength in precessional magnetization reversal, Science **285**, 864 (1999)

[2] J. Miltat, G. Aburquerque, A. Thiaville: An introduction to micromagnetics in the dynamical regime, in B. Hillebrands, K. Ounadjela (Eds.): *Spin Dynamics in Confined Magnetic Structures*, vol. I (Springer, Berlin, Heidelberg 2001)

[3] J. Fassbender: Magnetization dynamics investigated by time-resolved Kerr effect magnetometry, in B. Hillebrands, K. Ounadjela (Eds.): *Spin Dynamics in Confined Magnetic Structure*, vol. II (Springer, Berlin, Heidelberg 2003)

[4] M. Bauer, J. Fassbender, B. Hillebrands, R. L. Stamps: Switching behaviour of a Stoner particle beyond the relaxation time limit, Phys. Rev. B **61**, 3410 (200)
[5] H. W. Schumacher, C. Chappert, R. C. Sousa, P. P. Freitas, J. Miltat: Quasi-ballistic magnetization reversal, Phys. Rev. Lett. **90**, 017204 (2003)
[6] H. W. Schumacher, C. Chappert, R. C. Sousa, P. P. Freitas, J. Miltat, J. Fassbender, B. Hillebrands: Phase coherent precessional magnetization reversal in microscopic spin valve elements, Phys. Rev. Lett. **90**, 17201 (2003)
[7] T. Rasing, H. van den Berg, T. Gerrits, J. Hohlfeld: Ultrafast magnetization and switching dynamics, in B. Hillebrands, K. Ounadjela (Eds.): *Spin Dynamics in Confined Magnetic Structures*, vol. II (Springer, Berlin, Heidelberg 2003)
[8] T. Gerrits, H. van Den Berg, J. Hohlfeld, L. Bär, T. Rasing: Ultrafast precessional magnetization reversal by picosecond magnetic field pulse shaping, Nature **418**, 6897 (2002)
[9] S. E. Russek, R. D. McMichael, M. J. Donahue: High speed switching and rotational dynamics in small magnetic thin film devices, in B. Hillebrands, K. Ounadjela (Eds.): *Spin Dynamics in Confined Magnetic Structures*, vol. II (Springer, Berlin, Heidelberg 2003)
[10] S. Kaka, S. E. Russek: Precessional switching of submicrometer spin-valves, Appl. Phys. Lett. **80**, 2958 (2002)
[11] T. Devolder, M. Belmeguenai, H. W. Schumacher, C. Chappert, Y. Suzuki: Precessional strategies for the ultrafast switching of soft and hard magnetic nanostructures, MRS Proc. **746**, 189 (2003)
[12] T. Devolder, C. Chappert: Spectral analysis of the precessional switching of the magnetization in an isotropic thin film, Solid State Commun. **192**, 97 (2004)
[13] T. Devolder, C. Chappert: Precessional switching of thin nanomagnets: analytical study, Eur. Phys. J. B **36**, 57 (2003)
[14] T. Devolder, C. Chappert: Cell writing selection when using precessional switching in a magnetic random access memory, Appl. Phys. **95**, 1933 (2004)
[15] E. C. Stoner, E. P. Wohlfarth: A mechanism of magnetic hysteresis in heterogeneous alloys, Philos. Trans. R. Soc. London Ser. A **240**, 599 (1948) reprinted in IEEE Trans. Magn. **27**, 3475 (1991)
[16] L. He, W. D. Doyle, H. Fujiwara: High speed coherent switching below the Stoner–Wohlfarth limit, IEEE Trans. Mag. **30**, 4086 (1994)
[17] L. He, W. D. Doyle: Theoretical description of magnetic switching experiments in picosecond pulses, Appl. Phys. **79**, 6489 (1996)
[18] E. L. L. Landau: On the theory of the dispersion of magnetic permeability in ferromagnetic bodies, Phys. Z Sowjetunion **8**, 153 (1935)
[19] T. L. Gilbert: A lagrangian formulation of the gyromagnetic equation of the magnetization field, Phys. Rev. **100**, 1243 (1955)
[20] W. K. Hiebert, L. Lagae, J. De Boeck: Spatially inhomogeneous ultrafast precessional magnetization reversal, Phys. Rev. B **68**, 020402R (2003)
[21] L. Lagae: IMEC Belgium, private communication
[22] T. Devolder, C. Maufront, J.-V. Kim, H. W. Schumacher, C. Chappert, R. Fournel: Bit selection scheme and dipolar interactions in high density precessionnal MRAMs, IEEE Sci. Meas. Tech. **152**, 196–200 (2005)
[23] C. Maunoury, T. Devolder, C. K. Lim, P. Crozat, C. Chappert, J. Wecker, L. Bär: Appl. Phys. **97**, 74503 (2005)

[24] G. B. C. Serpico, I. D. Mayergoyz: Analytical solutions of Landau–Lifshitz equation for precessional switching, Appl. Phys. **93**, 6909 (2003)
[25] C. Kittel: *Introduction to Solid State Physics*, 5th ed. (Wiley, New York 1976)
[26] H. W. Schumacher, C. Chappert, P. Crozat, R. C. Sousa, P. P. Freitas, M. Bauer: Coherent suppression of magnetic ringing in microscopic spin valve elements, Appl. Phys. Lett. **80**, 3781 (2002)
[27] G. M. B. Albuquerque: *Magnetization precession in confined geometry: physical and numerical aspects*, Ph.D. thesis, Université Paris Sud, Orsay (2002) URL `http://www.lps.u-psud.fr/Collectif/gr_25/Publications.htm`
[28] Y. Acremann, C. H. Back, M. Buess, D. Pescia, V. Pobrovsky: Bifurcation in precessional switching, Appl. Phys. Lett. **79**, 2228 (2001)
[29] P. Bryant, H. Suhl: Thin-film magnetic patterns in an external field, Appl. Phys. Lett. **54**, 2224 (1989)
[30] C. Bayer, S. O. Demokritov, B. Hillebrands, A. N. Slavin: Spin-wave wells with multiple states created in small magnetic elements, Appl. Phys. Lett. **82**, 607 (2003)
[31] J. Miltat, M. Stiles: in B. Hillebrands, K. Ounadjela (Eds.): *Spin Dynamics in Confined Magnetic Structures*, vol. III (Springer, Berlin, Heidelberg 2004)
[32] A. A. Tulapurkar, T. Devolder, K. Yagami, P. Crozat, C. Chappert, A. Fukushima, Y. Suzuki: Appl. Phys. Lett. **85**, 5358 (2004)
[33] T. Devolder, C. Chappert, P. Crozat, A. Tulapurkar, Y. Suzuki, J. Miltat, K. Yagami: Appl. Phys. Lett. **86**, 62505 (2005)
[34] M. Durlam, P.-J. Naji, A. Omair, M. DeHerrera, J. Calder, J. M. Slaughter, B. N. Engel, N. D. Rizzo, G. Grynkewich, B. Butcher, C. Tracy, K. Smith, K. W. Kyler, J. J. Ren, J. A. Molla, W. A. Feil, R. G. Williams, S. Tehrani: IEEE J. Solid. State. Circ. **38**, 769 (2003) announcement of 1Mbit prototype MRAM by Motorola
[35] IBM-Infineon alliance: *Announcement of 128-kbit prototype MRAM by the IBM-Infineon alliance at the Symposium on VLSI Technology, Kyoto* (2003)
[36] Y. Nozaki, K. Matsuyama: Size dependence of switching current and energy barrier in the magnetization reversal of rectangular magnetic random access memory cell, Appl. Phys. **93**, 7295 (2003)
[37] R. P. Cowburn: Superparamagnetism and the future of magnetic random access memory, Appl. Phys. **93**, 9310 (2003)
[38] J.-V. Kim, T. Devolder, C. Chappert, C. Maufront, R. Fournel: Appl. Phys. Lett. **85**, 4094 (2004)

Index

Spin-Wave Excitations in Finite Rectangular Elements

Christian Bayer[1], Jörg Jorzick[1], Sergej O. Demokritov[2], Andrei N. Slavin[3], Konstantin Y. Guslienko[4], Dmitry V. Berkov[5], Natalia L. Gorn[5], Mikhail P. Kostylev[1,6], and Burkard Hillebrands[1]

[1] Fachbereich Physik and Forschungsschwerpunkt MINAS, Technische Universität Kaiserslautern, Erwin-Schrödinger-Straße 56, 67663 Kaiserslautern, Germany
{cbayer,jorzick,hilleb}@physik.uni-kl.de

[2] Institut für Angewandte Physik, Westfälische Wilhelms-Universität Münster, Corrensstr. 2–4, 48149 Münster, Germany
demokrit@uni-muenster.de

[3] Department of Physics, Oakland University, Rochester, Michigan 48309, USA
slavin@oakland.edu

[4] Materials Science Division, Argonne National Laboratory 9700 S. Cass Ave., Argonne, IL 60439, USA
gusliyenko@anl.gov

[5] INNOVENT e.V. Jena, Pruessingstraße 27B, 07745, Jena, Germany
DB@innovent-jena.de

[6] St. Petersburg Electrotechnical University, 197376 St. Petersburg, Russia
kostylev@mail.ru

Abstract. A review on recent Brillouin light scattering work on spin-wave modes in arrays of micrometer-size magnetic $Ni_{80}Fe_{20}$ stripes and rectangular elements is given. Several effects caused by the lateral confinement in the stripes are reviewed: 1. lateral quantization of dipole-dominated Damon–Eshbach spin-wave modes in a longitudinally magnetized stripe due to its finite width, 2. localization of exchange-dominated spin-wave modes near the edges and dipole-dominated spin-wave modes near the center of a transversely magnetized long magnetic stripe due to the inhomogeneity of its internal magnetic field, 3. combination of quantization and localization effects for the spin-wave modes in rectangular elements. The observed effects are analyzed using an analytical approach and numerical simulations.

1 Introduction

Recent developments in the field of magnetic storage and magnetic sensors have created much interest into the dynamic properties of small magnetic elements [1–3]. Growing demand for higher lateral density in magnetic recording and for higher speed of memory operation has focused the attention of many researchers on the spin dynamics of thin-film magnetic elements with lateral dimensions of 0.1 μm to 1 μm operating in the microwave frequency range (3 GHz to 30 GHz). It turns out that the frequencies and wavelengths of dipole-exchange spin waves (that happen to be dynamic eigenmodes of nano-sized magnetic elements) lie inside these mentioned spatial and frequency

B. Hillebrands, A. Thiaville (Eds.): Spin Dynamics in Confined Magnetic Structures III, Topics Appl. Physics **101**, 57–104 (2006)

intervals. Thus, the understanding of spin-wave eigenmodes of small magnetic elements is of critical importance both for fundamental reasons and for modern applications in magnetic recording and sensor technology. The problem is not trivial. For the magnetic elements studied here, both dipole–dipole and exchange interactions must be taken into account simultaneously. An additional difficulty is created by the fact that most of these small magnetic elements (e.g., thin rectangular prisms) have a nonellipsoidal shape and, therefore, the magnetic field inside such elements is strongly inhomogeneous when a homogeneous external magnetic field is applied.

These particular circumstances result in several qualitatively new physical effects that have been discovered recently: spin-wave quantization in longitudinally magnetized magnetic stripes and tangentially magnetized magnetic discs [4–9], and localization of spin-wave modes in the regions of inhomogeneity of internal magnetic field (spin-wave wells) in transversely magnetized magnetic stripes and rectangular elements [10–12]. Many of these effects have been discovered experimentally and described theoretically since the publication of the last review on this topic [13]. Therefore, we devote this review to the description of these new results.

The study of quantized and localized spin-wave modes of small magnetic elements is very important also from another point of view. Each confined magnetic element can be considered as a magnetic resonator. The thermal noise can be observed at the resonance frequencies of such a resonator that are in the microwave frequency range. If this element, e.g., is used as a part of a magnetic reading head, or as a memory element at microwave frequencies thermal magnetic noise limits the figures of merit of those devices [14–16]. Therefore, it is very important to know the resonance frequencies. Recently, it has been found [14–16] that the frequency spectrum of the magnetic noise is quantized, i.e., shows well-defined maxima at some frequencies. To understand this effect and to be able to have control for improving the properties one needs to study the properties of thermally excited spin-wave modes in such laterally confined magnetic elements.

The problem of the spin-wave spectrum of ellipsoidal elements can be solved analytically [17,18]. Therefore, for many decades, experimentally studied magnetic elements relevant for applications were approximated by ellipsoidal elements with corresponding demagnetizing factors [19]. Recently, it has been found that a nonellipsoidal shape drastically affects the dynamic properties [10]. Because of the nonellipsoidal shape both the static and the dynamic internal fields are inhomogeneous due to demagnetizing effects. We will discuss the influence of the field inhomogeneity on the eigenmodes in the elements, gradually increasing the complexity of the problem.

First, the case of long stripes magnetized along their length will be considered. In such a stripe the static internal field is homogeneous (if one neglects the areas close to the stripe ends). Nevertheless, the dynamic demagnetizing field associated with the precession of magnetization is still inhomogeneous. Although the mean direction of magnetization lies along the stripe axis and, therefore, does not cause a demagnetizing field, in the course of precession the magnetization tips out of the stripe plane and generates dynamic dipolar fields that oscillate with the frequency of the precession. These dipolar fields are essentially inhomogeneous and cause a significant change in the mode profiles of the spin-wave modes that leads to the so-called dipolar pinning [20]. We will provide experimental evidence for this effect and will compare the calculated magnetization distributions with experimental results.

Second, for the case of a transversely magnetized stripe (or a stripe magnetized along its width) the main effects are produced by a strongly inhomogeneous static internal magnetic field. This inhomogeneity causes the creation of so-called "spin-wave wells" [10–12]. We will discuss this phenomenon in detail and will present a semianalytical approach to calculate the spin-wave spectrum and the distribution of the dynamic magnetization. We will also formulate an approximate analytical method for the calculation of the eigenfrequencies.

Third, the more complicated case of a rectangular magnetic element will be considered. Here, confinement takes place for *both inplane* coordinates. We will show that the combination of the effects of lateral quantization and spatial localization of spin-wave modes takes place in this case and we will formulate a simple analytical approach to the calculation of spin-wave eigenfrequencies. This approach is based on the factorization of the two-dimensional magnetization distribution in the element into one-dimensional spin-wave eigenfunctions of a longitudinally and transversely magnetized stripe.

To check the results of the above-mentioned approximate analytical methods for long stripes and finite-size rectangular prisms we will report on numerical calculations of three types: direct solution of the integrodifferential equation describing spin-wave eigenfrequencies of the element, thermodynamic simulations based on Langevin dynamics, and field-pulse simulations at zero temperature based on the object-oriented micromagnetic framework (OOMMF) [21].

All these analytical and numerical results will be compared to the results of laboratory experiments performed using Brillouin light scattering spectroscopy.

This review is organized as follows: In Sect. 2 the experimental technique will be briefly reviewed. Section 3 is devoted to the experimental results, while Sect. 4 describes the analytic approach and Sect. 5 the micromagnetic simulations. A strict separation between the experimental and theoretical part is made so as to distinguish accurately between experiment and interpretation. Nevertheless, at some points results from the theoretical section have to be mentioned in the experimental parts.

2 Samples and Experimental Setup

The investigated samples were prepared from Permalloy ($Ni_{80}Fe_{20}$) films, thermally evaporated on Si(111) substrates in UHV. Patterning was performed by means of e-beam lithography and ion-beam etching. Stripes with a thickness of $L = 30$ nm to 40 nm, a width of $w = 1$ µm and 1.8 µm, a length of 90 µm, and a distance between the stripes of 0.3 µm to 0.8 µm were prepared. Rectangular elements were prepared with lateral dimensions of 1 µm × 1.75 µm, a thickness of 35 nm and an element spacing of 0.1 µm. The elements were arranged in arrays of dimensions of 500 µm × 500 µm. The high quality of the patterning process has been confirmed by scanning electron and atomic force microscopy, as illustrated in the preceding review in SDI of this series [13]. The used technique guarantees a high-quality patterning process, which provides a superb flatness of the stripe boundaries and reproducibility of the stripe widths.

The spin-wave spectrum of a confined system can be investigated by various techniques: ferromagnetic resonance [22], time-resolved Kerr magnetometry [12, 23–25], and Brillouin light scattering spectroscopy (BLS) [26]. In the field of patterned structures BLS has a number of advantages. It combines the possibility to study the dynamics of patterned systems in the frequency range up to several 100 GHz with a high lateral resolution of 30 µm to 40 µm defined by the size of the laser-beam focus. Moreover, BLS can be used in the so-called "Fourier microscope" mode-resolving lateral structures of feature sizes down to 120 nm to 130 nm by Fourier analysis of intensity versus wavevector [7, 13]. Since no additional excitation technique is required, BLS is especially successful in studying complicated, strongly confined spin-wave modes, which are excited thermally.

The BLS process in an unconfined medium is illustrated in Fig. 1a: monoenergetic photons (visible light, usually the green line of an Ar^+ laser, 514.5 nm, or a solid-state laser) with the wavevector $\boldsymbol{q}_\mathrm{I}$ and frequency $\omega_\mathrm{I} = cq_\mathrm{I}$ interact with the elementary quanta of spin waves (magnons), described by $\boldsymbol{q}$ and ω. Due to conservation laws resulting from the time- and space-translational invariance of the system the scattered photon gains an increase or decrease in energy and momentum,

$$\hbar\omega_\mathrm{S} = \hbar(\omega_\mathrm{I} \pm \omega) \tag{1}$$

$$\hbar\boldsymbol{q}_\mathrm{S} = \hbar(\boldsymbol{q}_\mathrm{I} \pm \boldsymbol{q})\,, \tag{2}$$

if, using the language of second quantization, a magnon as the quantum of a spin wave is annihilated or created. Measuring the frequency shift of the scattered light one obtains the frequency of the spin wave participating in the BLS process. From (2) it is evident that the wavevector $\boldsymbol{q}_\mathrm{S} - \boldsymbol{q}_\mathrm{I}$, transferred in the scattering process, is equal to the wavevector $\boldsymbol{q}$ of the spin-wave. Changing the scattering geometry one can sweep the value of q and measure the corresponding ω. Thus, the spin-wave dispersion $\omega(q)$ can be

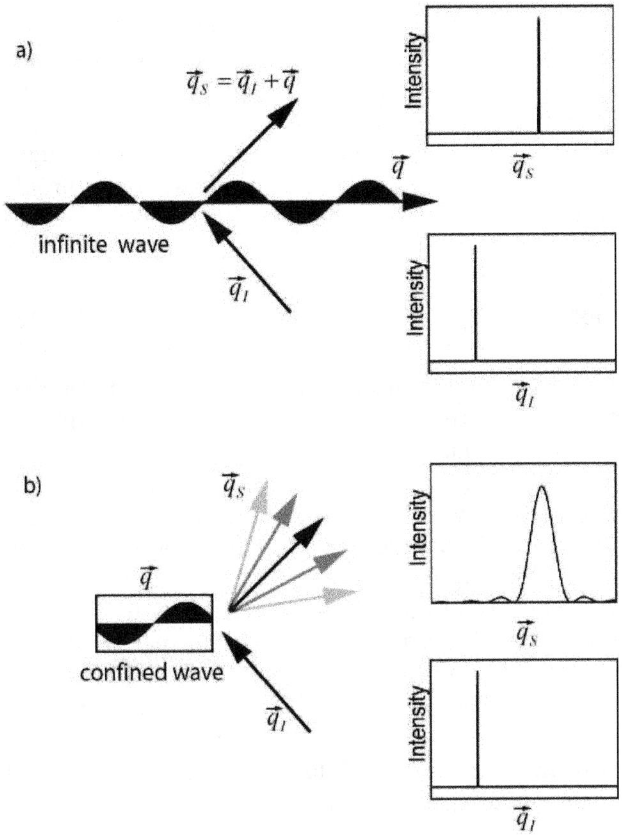

Fig. 1. Scheme of the light-scattering process for (**a**) infinite scattering volume (plane wave) and (**b**) confined scattering volume. q_I stands for the wavevector of the incoming light, while q_S for that of the scattered light

studied. For room temperature ($T_\mathrm{R} \gg \hbar\omega/k_\mathrm{B} \approx 1\,\mathrm{K}$) both processes have about the same probability.

For a film, the translational invariance is broken along the direction perpendicular to the film plane. Thus, the conservation conditions are fulfilled only for the components of the wavevector that are along the unconfined coordinates: the inplane wavevector component $\boldsymbol{q}_{\|}$ is conserved. In BLS experiments with the backscattering geometry the value of the inplane wavevector $\boldsymbol{q}_{\|}$, transferred in the light-scattering process, can be varied by changing the angle of light incidence θ measured from the film surface normal: $q_{\|} = (4\pi/\lambda_\mathrm{Laser}) \times \sin\theta$. The absolute value of $q_{\|}$ varies in the typical range of $(0\text{–}2.5)\times10^5\,\mathrm{cm}^{-1}$. The frequency resolution of the technique is about 100 MHz to 200 MHz.

For the laterally confined structures investigated here additionally the *inplane* translation invariance is broken. In the case of a long stripe only one component of $\boldsymbol{q}$ parallel to the long stripe axis is conserved. For the rectangular elements no *inplane* component is conserved at all. Therefore there is a given uncertainty in the scattered wavevector as is shown in Fig. 1b, the value of the uncertainty being determined by the lateral size of the element. It was shown in the preceding review [13], that the intensity of the scattered light as a function of the transferred wavevector q is connected with the profile of the dynamic magnetization, $m(y)$ of the spin-wave mode confined in the element

$$I(q) \propto \left| \int_{-w/2}^{w/2} m(y) \times \exp(-\mathrm{i}qy)\mathrm{d}y \right|^2 , \tag{3}$$

where w is the stripe width and the integration is performed along the stripe width. Thus, the light-scattering intensity is proportional to the squared Fourier transform of the dynamic magnetization, $m(y)$. As the integration is performed over the interval $[-w/2, w/2]$ and the dynamic magnetization is essentially a sine function with a wavevector k, the maximum in scattering intensity will be at $q = k$ and the broadening will be determined by the stripe width w. In the more complicated case of a rectangular element, a two-dimensional Fourier transformation is needed. If BLS is used in the "Fourier microscope" mode, the light-scattering intensity is measured as a function of q. On the basis of this information the mode profile $m(y)$ in the elements is reconstructed. In the "Fourier microscope" mode the spatial resolution, δ is determined by the accessible transferred wavevector interval $\pi/\Delta q$. For backscattering experiments it follows that $\Delta q = 2q_{\mathrm{I}}$ and $\delta = 120\,\mathrm{nm}$ to $130\,\mathrm{nm}$ for the green Ar^+ laser line.

3 Experimental Results

3.1 Longitudinally Magnetized Long Stripes

For the first type of sample studied here, long stripes, two different geometries can be considered, which allows us to investigate two different effects separately. In the first geometry the applied magnetic field is along the stripe length and therefore the internal magnetic field is homogeneous and equal to the applied external field. For this case the quantization of the Damon–Eshbach-like spin waves caused by the physical boundaries of the stripe, i.e., spatial confinement due to a finite width of the stripe, can be observed [6–8]. Therefore, in the following, such an orientation of the field will be called the Damon–Eshbach geometry [27]. Although this effect has been already reported in the preceding review in [13], a new thorough theoretical model has been formulated since then and thus we revisit this effect.

Figure 2 shows a typical BLS spectrum for a transferred wavevector of $q = 0.59 \times 10^5\,\mathrm{cm}^{-1}$ oriented along the width of the stripe and a magnetic field of 100 Oe applied along the stripe length. Several peaks corresponding

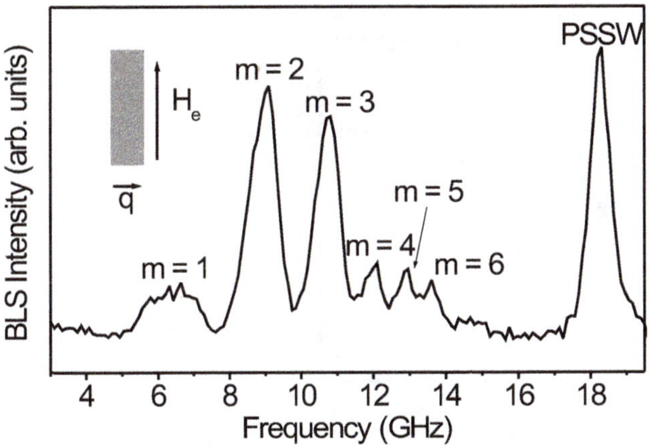

Fig. 2. BLS spectrum of 1 μm wide and 33 nm thick stripes in an external field of 100 Oe for a transferred wavevector of $q = 0.59 \times 10^5\,\mathrm{cm}^{-1}$. The *inset* shows the used experimental geometry. PSSW indicates "perpendicular standing spin wave". Numbers m are the mode numbers

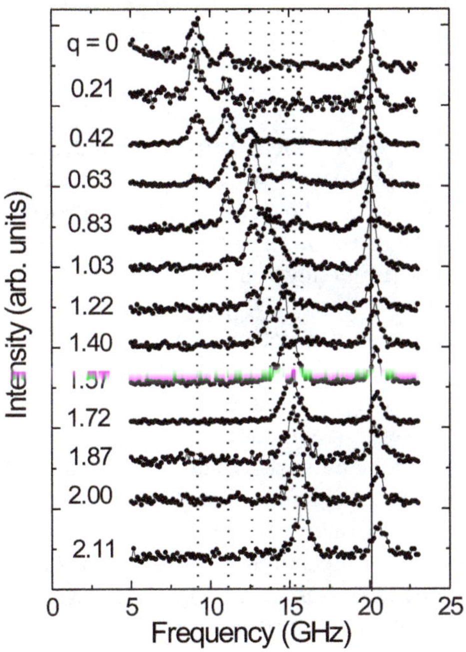

Fig. 3. BLS spectra for different transferred wavevectors of 1 μm wide and 35 nm thick stripes in an external field of 500 Oe. The wavevector $\boldsymbol{q}$ is given in units of $10^5\,\mathrm{cm}^{-1}$. The *vertical solid line* indicates the nearly fixed frequency of the PSSW mode, the *vertical dotted lines* indicate the fixed frequencies of the quantized Damon–Eshbach modes

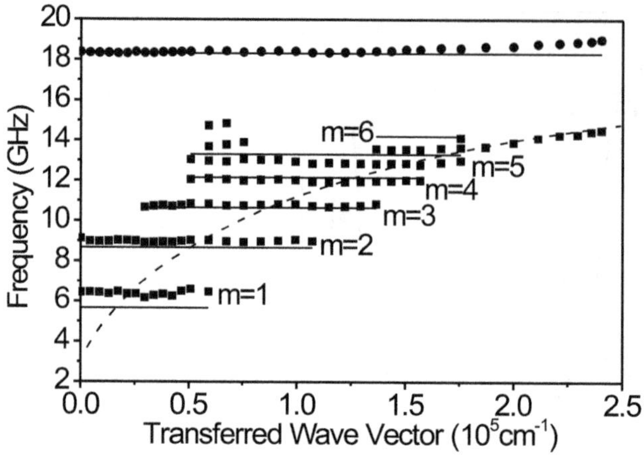

Fig. 4. Measured dispersion relation of a 1 μm wide and 33 nm thick stripe in an external field of 100 Oe applied along the length of the stripe. The *squares* correspond to Damon–Eshbach modes, the *circles* to perpendicular standing spin waves. The *dashed line* shows the dipole-exchange dispersion for spin waves in an unpatterned Permalloy film calculated with [28]. The *horizontal full lines* are theoretical values of discrete spin-wave mode frequencies calculated using (23) as explained in Sect. 4

to various quantized spin-wave modes can be identified. With $m = 1, 2, 3, \ldots$ we will count the quantized modes in this geometry. An example of spectra obtained for different wavevectors transferred in the light-scattering process is shown in Fig. 3. Each quantized mode can be observed over a continuous interval of the transferred wavevector. Its frequency remains constant while the intensity changes with the wavevector.

The measured frequencies of the modes as a function of the wavevector are shown in Fig. 4. The mode with the frequency close to 18.5 GHz (shown by circles in Fig. 4) can be identified as the perpendicular standing spin wave (PSSW) [7, 13]. The modes with lower frequencies are the Damon–Eshbach modes quantized due to confinement caused by the finite width of the stripe (squares). Values of these frequencies calculated as described in the theoretical section using (30) are shown by horizontal lines for comparison.

A comparison between the experimentally measured BLS intensities and the results of the calculation based on (3) is presented in Fig. 5. Shown are the light-scattering intensities from different quantized modes normalized to the intensity of the PSSW mode. It is assumed that the intensity of the PSSW modes is independent of the transferred wavevector as the total wavevector of the PSSW mode is dominated by its *out-of-plane* component (proportional to one over the thickness of the stripe) that is at least ten times larger than the maximum *inplane* wavevector for the described measurements. The only fitting parameter used in Fig. 5 is the scaling factor along the intensity axis, which is the same for all modes. Squares in Fig. 5 correspond to the experi-

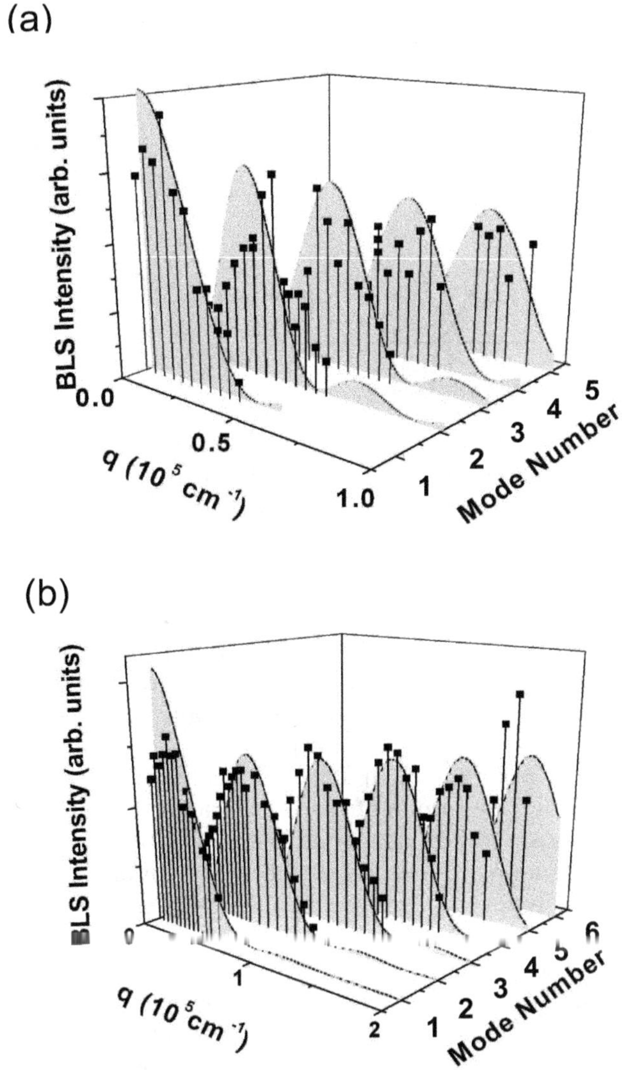

Fig. 5. Measured relative BLS intensities for different quantized modes (*black squares*) as a function of the transferred wavevector q and the mode-quantization number m in comparison to the calculated results based on (3) (*gray-colored curves*); (**a**) for the 1.8 μm wide stripes with rough edges in a field of 500 Oe where unpinned boundary conditions were assumed and (**b**) for 1 μm wide stripes in 100 Oe with dipolar boundary conditions. (From [7] and [30])

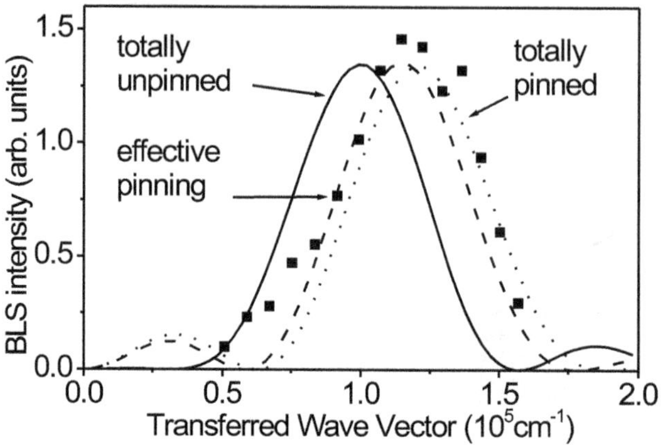

Fig. 6. Intensity of the scattered light as a function of the transferred wavevector of 1 μm wide stripes in a field of 100 Oe for the mode $m = 4$. The *squares* indicate the experimental values, while the *lines* correspond to the Fourier transforms of three different mode profiles: totally unpinned (*full line*), totally pinned (*dotted line*), effectively pinned (*dashed line*). (From [30])

mentally measured intensities for stripes with Fig. 5a 1.8 μm and Fig. 5b 1 μm width. The gray curves represent the calculated intensities using (3). For the mode profiles of the 1.8 μm stripes unpinned boundary conditions were assumed, while the intensities of the 1 μm stripes were calculated with the so-called "dipolar pinning" (see Sect. 4.1). The very good agreement between the experimental data and the results of the calculation confirms that the observed spin-wave modes are in fact quantized Damon–Eshbach modes and justifies the proposed boundary conditions. The comparison of Figs. 5a and 5b demonstrates that the wavevector with maximum light-scattering intensity for a particular mode is inversely proportional to the width of the stripe. This is a direct consequence of the Fourier theorem. In previous publications [7, 13] unpinned boundary conditions at the side edges of the stripes were assumed for the calculations of the mode frequencies and profiles. The assumption is based on a negligible surface anisotropy of Permalloy. However, the latest data obtained for samples with smoother side edges prepared using e-beam lithography cannot be described using unpinned boundary conditions. Recently, a new thorough theoretical analysis of the boundary condition has been performed [20]. It was found that in addition to the Rado–Wertman exchange pinning caused by surface anisotropy [29], "dipolar pinning" must be taken into account in the considered geometry. The origin lies in the fact that the oscillating magnetization leads to dynamic magnetic surface charges at the stripe edges that produce a dipolar field. So energy minimization requires the minimization of the surface charges at the side edges, i.e., vanishing magnetization (pinned boundary condition). On the other hand, energy minimization also favors larger wavelengths in this case that would

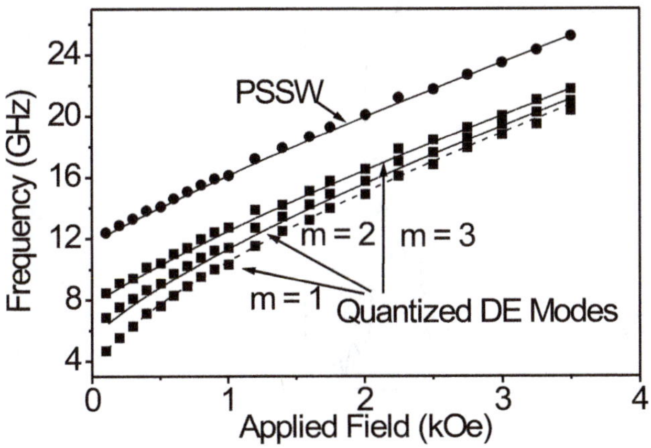

Fig. 7. Spin-wave frequency as a function of the applied field parallel to the stripes. The stripes are 40 nm thick and 1.8 μm wide. The transferred wavevector is $q = 0.3 \times 10^5\,\mathrm{cm}^{-1}$. (From [7])

require unpinned boundary conditions. The competition of both effects leads to the so-called effective "dipolar pinning". This new pinning mechanism will be discussed in detail below.

Figure 6 shows the experimentally measured BLS intensities for the mode with $m = 4$ in the stripes with 1 μm width as well as the calculated functions for three different profiles of the dynamic magnetization: a totally unpinned profile, a profile with the effective pinning from (24), which will be introduced later, and a completely pinned profile. It is evident from Fig. 6 that the measured curve does not fit to the unpinned profile. However, it is difficult to distinguish experimentally between the cases of the effective dipolar pinning and ideal complete pinning due to a strong value of the former for the considered stripe geometry.

With increasing applied magnetic field the frequencies of the quantized modes increase as is shown in Fig. 7. The solid lines shown in the figure are calculated based on the approach mentioned above using unpinned boundary conditions for the sample with rough edges. The detailed consideration in [20] shows that the pinning length ξ, and correspondingly the mode profile of a given mode do not change dramatically within the used field interval. This fact is experimentally confirmed by the observation that the relative Fourier intensities of the peaks corresponding to the different modes remain constant with increasing field.

3.2 Transversely Magnetized Long Stripes

If the applied field is directed along the width of a thin stripe, both the static and the dynamic internal field are strongly inhomogeneous. Spin waves propagating along the field are affected by confinement effects in this case. Since

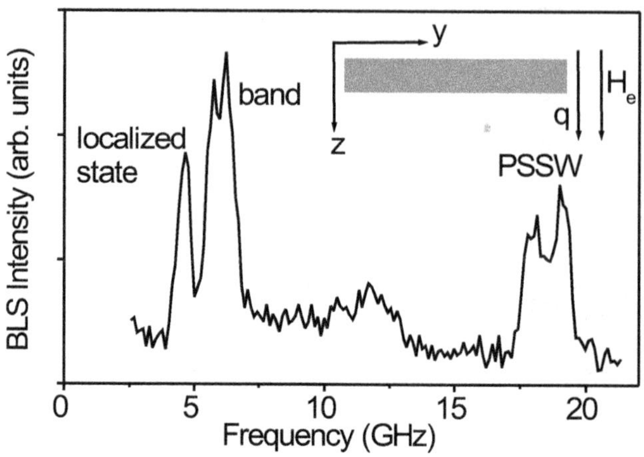

Fig. 8. BLS spectrum of a 1 μm wide stripe for a transferred wavevector $q = 1 \times 10^5\,\text{cm}^{-1}$ and an external field of $H_e = 500$ Oe. PSSW indicates the perpendicular standing spin-wave modes. The *inset* shows the used experimental geometry. (From [10])

in unconfined media a wave with $\boldsymbol{q} \| \boldsymbol{M}_S$ is called a magnetostatic backward volume wave (MSBVW), we refer to this experimental geometry as the MSBVW geometry. "Backward" comes here from the negative group velocity of this wave [27].

Figure 8 shows a typical BLS spectrum for the described geometry containing several peaks corresponding to spin-wave modes. The peak structure at high frequencies can be identified as being dominated by two exchange-dominated PSSW modes [7]. Since the frequencies of the PSSW modes are determined by the internal static field H_i, the intensity versus frequency distribution reflects the distribution of the internal field across the stripe.

As already mentioned, in the MSBVW geometry, both the static and the dynamic internal magnetic fields are strongly inhomogeneous in the stripe. Nevertheless, for a large applied field the magnetization is parallel to the applied field within almost the entire stripe. Therefore poles are created at the edges of the stripe that decrease the internal magnetic field in these regions. A detailed analysis reveals that the internal static field H_i has a broad maximum in the center of the stripe, while it vanishes completely near the edges of the stripe [31]. Figure 9 shows the effective field, H_{eff} together with the distribution of the static magnetization for a 2 μm wide stripe. The effective field consists of the internal field and the exchange field $H_{\text{eff}} = H_i + H_{\text{ex}}$. The contribution of the exchange field is only important at the edges of the stripe where the magnetization is inhomogeneous. Due to the definition of the exchange field it is antiparallel to the static magnetization [32]. We will discuss in Sect. 4.2 why we can restrict ourselves to the internal field and neglect the exchange contribution. The calculated profile of the internal field

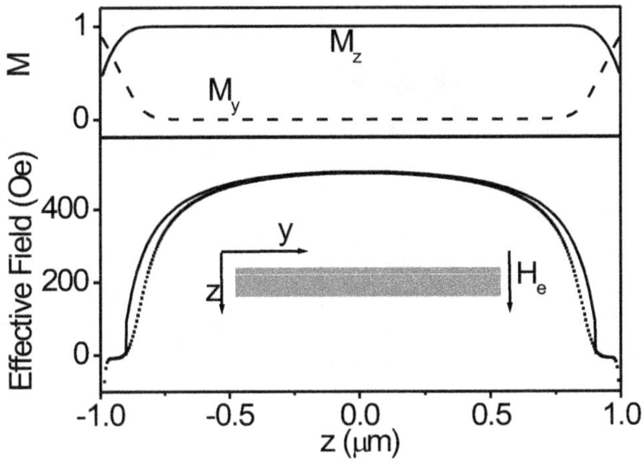

Fig. 9. *Lower panel*: Effective magnetic field for a Permalloy stripe with a width of 2 μm and a thickness of 30 nm in an applied field of 600 Oe. The *full line* is the approximation of the internal field by (32) and (33) from 4.2. The *dotted line* is the output for the effective field of OOMMF [21]. Note that the effective field calculated with OOMMF is negative at the edges, which is due to the definition of the exchange field [32]. *Upper panel*: The corresponding distribution of the static magnetization obtained with OOMMF. The *inset* shows the geometry for a transversely magnetized stripe

explains the observation of the two-peak structure of the PSSW mode in the above BLS spectrum (see Fig. 8): there is one mode existing in the saturated central region of the stripe, where H_i is close to its maximum value 300 Oe, and another mode existing in the unsaturated regions near the stripe edges ($\pm$ 0.5 μm), where H_i is close to zero.

In those regions where H_i is nonzero the magnetization is aligned along the internal field. By contrast, near the edges where H_i is close to zero the magnetization is not saturated: a complicated domain-wall-like structure is created. The inhomogeneity of the static internal field changes the spin-wave eigenspectrum qualitatively and causes the creation of a so-called "spin-wave well" near the edges of the stripe [10–12].

In fact, one can see two distinct peaks in the low-frequency region of Fig. 8. The stronger broad peak close to 6 GHz corresponds to a band of nonlocalized spin-wave modes as will be discussed in detail in Sect. 4. The peak with the lower frequency is a localized spin-wave mode corresponding to a spin-wave eigenstate existing in the spin-wave well. The spatial localization of the mode is illustrated by its dispersion shown in Fig. 10. First, the frequency of the mode does not depend on the transferred wavevector. But what is more important, the mode is observed over the entire interval of the accessible transferred wavevector and its BLS intensity shows no significant change as a function of the wavevector. Using the relation between the profile of

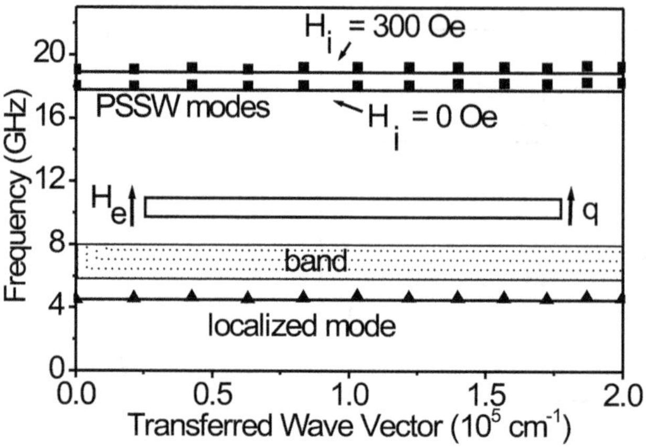

Fig. 10. Measured dispersion relation of a 1 μm wide stripe in an external field of 500 Oe applied along the width of the stripe. The *squares* correspond to perpendicular standing spin waves (PSSW), the *triangles* to localized modes. The *inset* shows the geometry. The *full lines* were calculated with [28] for the PSSW modes and with (39). (From [10])

the dynamic magnetization in real space and the BLS intensity profiles in q-space via the Fourier transformation discussed in the previous subsection one can easily estimate the upper limit of the localization width of the mode: $\Delta r = \pi/q_{\max} \simeq 120\,\text{nm}$. Note, this argument cannot be applied to the band of nonlocalized excitations: the structure of the band changes significantly with the transferred wavevector as some modes gain intensity while other modes lose it.

The formation of the spin-wave well and the localization of the lowest mode due to the inhomogeneity of the internal field can be basically understood as follows [10]: let us consider a spin wave of frequency ν_0, propagating towards the center of the stripe (wavevector $\boldsymbol{q} \| \boldsymbol{M}_\text{S} \| \boldsymbol{e}_z$) and entering the region of changing values of the internal static magnetic field $H_\text{i} = H_\text{i}(z)$. First, the wave keeps propagating through the inhomogeneous field, albeit with changing wavevector, $q = q(z)$ to fulfill the dispersion $\nu_0 = \nu[q(z), H_\text{i}(z)]$ [33]. However, it will be reflected at the point $z = z_l$, where H_i becomes so large that no solution $\nu_0 = \nu(q, H_\text{i})$ with a real q exists any longer. The described effect will be discussed in detail later. On the other hand, for higher frequencies, the solution of the dispersion equation can be found in any point of the stripe, i.e., the wave propagates through the stripe without reflection. The exact values of the frequencies of the localized modes are determined by the phase conditions as is described below. The frequency of the nonlocalized waves in this geometry are determined in a similar way to those of the quantized modes in the Damon–Eshbach geometry. But since the group velocity of the waves with $\boldsymbol{q} \| \boldsymbol{M}_\text{S}$ is much smaller than that of the Damon–Eshbach mode, the gaps due to quantization are much narrower, and not observable

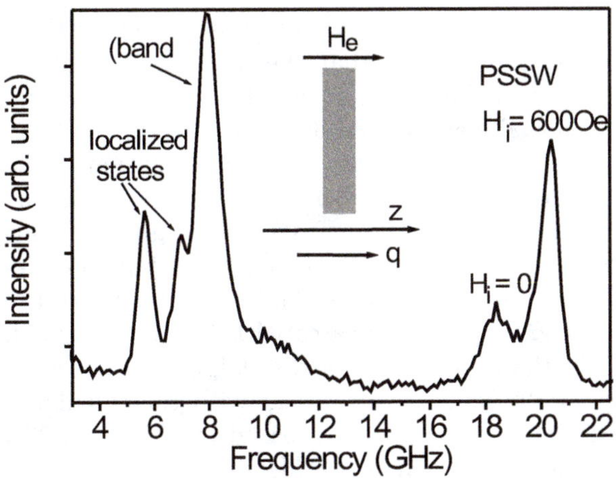

Fig. 11. BLS spectrum of a $1\,\mu$m wide stripe for a transferred wavevector $q = 0.47 \times 10^5\,\mathrm{cm}^{-1}$ and an external field of $H_\mathrm{e} = 800\,\mathrm{Oe}$. PSSW indicates the perpendicular standing spin wave. The *inset* shows the used experimental geometry. (From [11])

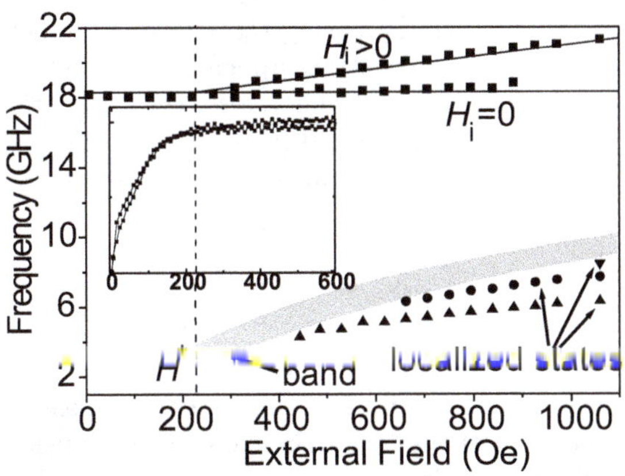

Fig. 12. Frequencies of the modes observed in the $1\,\mu$m wide stripe at $q = 0.47 \times 10^5\,\mathrm{cm}^{-1}$ as a function of H_e. The *vertical dashed line* marks the critical field H^* both in the main figure and in the *inset*. Note the constant frequency of 18 GHz of one of the PSSW modes and the increase in frequency for the other mode for $H_\mathrm{e} > H^*$. The *gray region* illustrates the band of nonresolved excitations. The *inset* shows the magnetization curve of the stripe for reference. (From [11])

experimentally. The modes form a band of nonlocalized excitations filling a frequency interval.

It is interesting to track the development of the observed modes with changing external applied field H_e, which is illustrated in Figs. 11 and 12.

Comparing the spectra shown in Figs. 8 and 11, which are recorded at $H_e =$ 500 Oe and 800 Oe, respectively, one can see that the number of localized modes increases with increasing field. Figure 12 shows the mode frequencies as a function of the applied external field H_e. While the magnitude of H_e is small and not sufficient to create a region of saturated magnetization in the stripe, only a single PSSW mode corresponding to $H_i = 0$ is present in the spectrum. At some critical field $H_e = H^*$ the PSSW peak begins to split. The first peak corresponding to $H_i = 0$ retains its frequency, but loses its intensity with increasing H_e (compare Figs. 8 and 11). The frequency of the second peak (which appears at $H_e = H^*$) increases with the applied field, indicating the increase of H_i in the central part of the stripe. The observed value of the critical field $H^* = 220$ Oe is in agreement with the calculated demagnetizing field in the stripe center, $H_d(z = 0)$, based on (32) and (33), and with static magnetization curves, shown as the inset in Fig. 12.

In order to understand the appearance of several well-resolved spin-wave modes corresponding to the multiple states in a spin-wave well, one should take into account that the depth of the well strongly depends on H_e. In fact, for applied fields smaller than the saturation field at the edges of the stripe the bottom of the well corresponds to $H_i = 0$ independent of H_e. The field, which determines the position of the top of the well, however, can be roughly estimated as $H_e - H_d(z = 0) \approx H_e - H^*$. With increasing H_e the well becomes deeper, and, as a result, more localized spin-wave states appear in the well, in analogy to the quantum-mechanical potential well. Therefore, the character of the modes changes with the increase of the applied field. A mode that is nonlocalized at a low field can become a localized one with increasing field.

3.3 Rectangular Elements

Contrary to long stripes, for the case of rectangular elements there is no external field orientation that provides a homogeneous static internal magnetic field. However, as will be shown below, such a complicated case can be treated in good approximation as a combination of the two stripe geometries considered above. Experimental results for rectangles with the dimensions of width $w = 1\,\mu$m, length $l = 1.75\,\mu$m, and thickness $L = 35$ nm will be discussed in detail. The applied field is directed along one of the edges of the rectangle. If the external field is large enough the magnetization is essentially homogeneous and parallel to the applied field. Nevertheless, the effective internal field is strongly inhomogeneous as is shown in Fig. 13, which is obtained using micromagnetic simulations as discussed in Sect. 5. It is seen that the inhomogeneity occurs mainly along the direction of the applied field (z-direction). The inhomogeneity along the y-direction is not so pronounced.

Figure 14 shows the dispersion relation of such an element. Here both the applied field and the transferred wavevector are oriented along the long edge of the element. But due to the finite aperture of the objective lens the

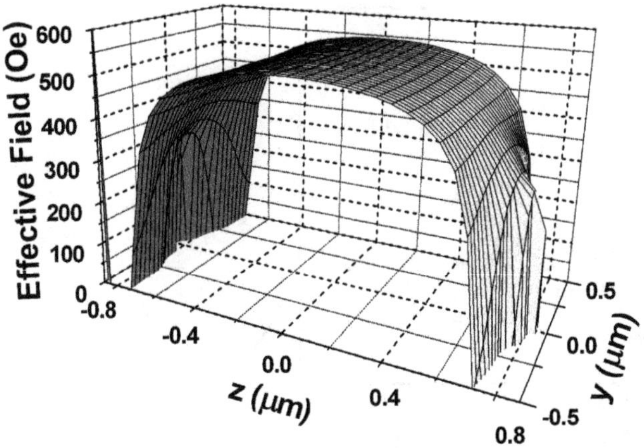

Fig. 13. Plot of the z-component of the effective field of a rectangular element with the dimensions of width $w = 1\,\mu\text{m}$, length $l = 1.75\,\mu\text{m}$, and thickness $L = 35\,\text{nm}$ for an applied field of 600 Oe calculated with OOMMF [21]. Note that, again, the effective field calculated with OOMMF is negative at the edges $z = \pm l/2$, which is due to the definition of the exchange field [32]. Here we show only positive values for clarity. The corresponding distribution of magnetization is shown in Fig. 27

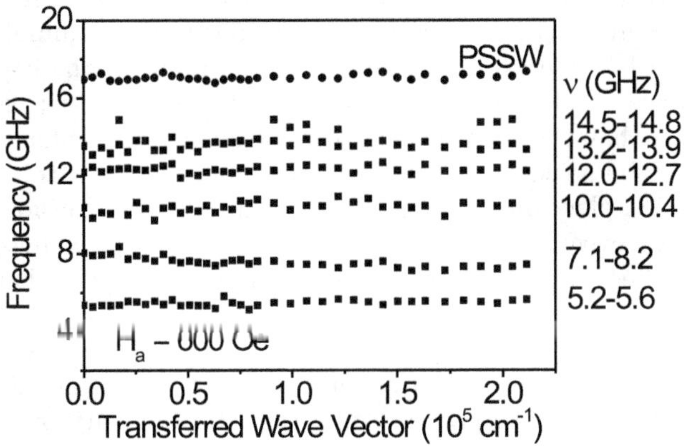

Fig. 14. Dispersion relation of the $1\,\mu\text{m} \times 1.75\,\mu\text{m}$ elements in an applied field of 600 Oe along the long side of the elements. The transferred wavevector is also oriented along the long axis of the elements. (From [34])

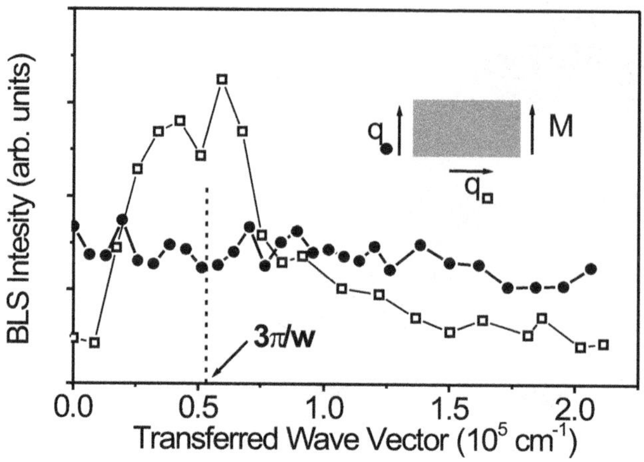

Fig. 15. BLS intensity for rectangular elements magnetized along their width and a transferred wavevector along the magnetic field for the lowest mode (*circles*) and perpendicular for the third mode (*squares*). Note, that the applied field in this special case is along the short element axis. (From [34])

wavevector resolution is also finite and thus also spin-wave modes with inhomogeneous-mode profiles along the direction perpendicular to the nominal transferred wavevector can be observed. This effect is also present if one investigates the stripes. However, due to the specific geometry of the stripes (the length is much greater than the width) the spin-wave quantization along the length of the stripe can be neglected. Contrarily, the quantization within a rectangular element taking place along both directions needs to be considered, therefore the finite wavevector resolution should be taken into account. For example, due to this effect modes quantized along the short edge of the element are seen in Fig. 14. In general, it is necessary to note that, for both directions of the transferred wavevector parallel to the long and to the short edge of the element, modes with the same frequencies are observed, albeit with different relative BLS intensities. By recording the BLS intensity of a given mode as a function of the transferred wavevector along these two directions, a two-dimensional "Fourier mapping" can be performed, which helps in the determination of the character of the modes.

As an example, Fig. 15 shows the BLS intensity of a mode as a function of the absolute value of the transferred wavevector parallel and perpendicular to the applied field. As can be seen, both curves indicate completely different dependences. For a transferred wavevector parallel to the applied field an almost constant BLS intensity over a wide interval of wavevectors indicates a strong localization of the mode in real space. Contrarily, for a transferred wavevector perpendicular to the applied field the change in intensity shows that this mode must be quantized over a wider spatial length as discussed before. Therefore one can determine the mode structure. The lowest mode is

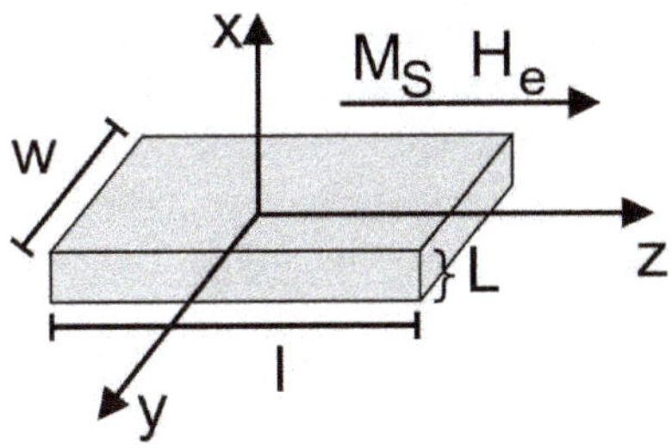

Fig. 16. Coordinate system for a rectangular element. The directions of the static magnetization $\boldsymbol{M}_S$ and the external field $\boldsymbol{H}_e$ are indicated as well as the dimensions $L \ll l, w$ of the element

a localized mode near the edges perpendicular to the applied field ($z = \pm l/2$). All the mode profiles can be expressed as a product of modes from the two cases of a transversely and a longitudinally magnetized stripes as will be discussed below.

4 Analytical Approach

In this section we will develop a general theory for the eigenmodes of homogeneously magnetized rectangles with an inhomogeneous internal field. First we will derive a general equation describing the excitations in such elements. This equation is two-dimensional and cannot be solved easily. Therefore we will proceed to the two special and easier cases already presented in the experimental part: a long stripe that is either magnetized along its length or its width. Having understood the properties of these particular cases we will come back to the rectangles. We will show that the eigenmodes of a rectangle have nearly the spatial distributions of the dynamic magnetization in the two *inplane* directions as the two stripe geometries. So the distribution of magnetization can be factorized. Based on this separation approach we will formulate an approximate theory to calculate the mode frequencies.

Let us consider a rectangular magnetic element schematically shown in Fig. 16. Note that in all considered geometries the x-direction is *perpendicular* to the film plane, while the z-axis is *parallel* to the external magnetic field.

We assume that the element is magnetized to saturation by the external field $\boldsymbol{H}_e = H_e \boldsymbol{e}_z$ that is parallel to one of the inplane sides of the rectangular element as shown in Fig. 16. In the following the static magnetization in the elements is constant $\boldsymbol{M}_{\mathrm{stat}} = M_S \boldsymbol{e}_z = \text{const.}$ everywhere, except in the narrow regions situated near the edges perpendicular to the applied field (edge domains). Both crystalline and surface anisotropy of the material are neglected, which is a good approximation for Permalloy.

Since we are interested in the linear oscillations of the magnetization, we assume that the dynamic component of the magnetization m_z directed along the field $\boldsymbol{H}_\mathrm{e}$ (and the magnetization $\boldsymbol{M}_\mathrm{stat}$) is much smaller than the transverse dynamic components, $m_z \ll m_x, m_y$. Further, we assume that the element is thin and we are interested only in the lowest spin-wave modes having a uniform distribution of the variable magnetization along the $\boldsymbol{e}_x$-axis, i.e., $\boldsymbol{M}(\boldsymbol{r},t) \rightarrow \boldsymbol{M}(\boldsymbol{\rho},t)$, where $\boldsymbol{\rho} = y\boldsymbol{e}_y + z\boldsymbol{e}_z$. So the magnetization vector can be expressed as

$$\boldsymbol{M}(\boldsymbol{\rho},t) = M_\mathrm{S}\boldsymbol{e}_z + m_x(\boldsymbol{\rho},t)\boldsymbol{e}_x + m_y(\boldsymbol{\rho},t)\boldsymbol{e}_y\,. \tag{4}$$

The effective field in this quasi-two-dimensional case can be expressed as

$$\boldsymbol{H}_\mathrm{eff}(\boldsymbol{\rho},t) = \boldsymbol{H}_\mathrm{e} + \frac{2\,\mathrm{A}}{M_\mathrm{S}^2}\nabla_\rho^2\boldsymbol{M}(\boldsymbol{\rho},t) + 4\pi\int\hat{\boldsymbol{G}}(\boldsymbol{\rho},\boldsymbol{\rho}')\cdot\boldsymbol{M}(\boldsymbol{\rho}',t)\mathrm{d}\boldsymbol{\rho}'\,, \tag{5}$$

where A is the exchange constant and $\nabla_{\boldsymbol{\rho}}^2 = \frac{\partial^2}{\partial y^2} + \frac{\partial^2}{\partial z^2}$ is the two-dimensional Laplace operator. $\hat{\boldsymbol{G}}(\boldsymbol{\rho},\boldsymbol{\rho}')$ is the Green's function of the dipolar interaction as discussed below and so the integral in (5) provides the demagnetizing field. Note that the effective field $\boldsymbol{H}_\mathrm{eff}(\boldsymbol{\rho},t)$ consists of a static and a small dynamic part similar to the magnetization (4),

$$\boldsymbol{H}_\mathrm{eff}(\boldsymbol{\rho},t) = H_\mathrm{i}(\boldsymbol{\rho})\boldsymbol{e}_z + h_{\mathrm{eff},x}(\boldsymbol{\rho},t)\boldsymbol{e}_x + h_{\mathrm{eff},y}(\boldsymbol{\rho},t)\boldsymbol{e}_y\,. \tag{6}$$

To calculate the spectrum of spin waves in the element we use the formalism of tensorial Green's functions developed for infinite inplane magnetized films in [28]. In the framework of this formalism the linearized Landau–Lifshitz equation of motion for the magnetization and the Maxwell equations in the magnetostatic limit are reduced to an integrodifferential equation for the dynamic magnetization $m_x(\boldsymbol{\rho},t) = m_x(\boldsymbol{\rho}) \times \exp(\mathrm{i}\omega t)$ and $m_y(\boldsymbol{\rho},t) = m_y(\boldsymbol{\rho}) \times \exp(\mathrm{i}\omega t)$. Inserting (4) and (5) into the Landau–Lifshitz equation and taking only the linear terms in m_x and m_y into account, after some calculus, one obtains the commonly used form

$$\begin{aligned}&\left[-\alpha\omega_M\nabla_{\boldsymbol{\rho}}^2 + \omega_{H_\mathrm{i}}(\boldsymbol{\rho})\right]\hat{\boldsymbol{I}}\boldsymbol{m}(\boldsymbol{\rho}) + \mathrm{i}\omega\hat{\boldsymbol{T}}\boldsymbol{m}(\boldsymbol{\rho})\\ &\quad -\omega_M\textstyle\int \mathrm{d}\boldsymbol{\rho}'\hat{\boldsymbol{G}}_{x,y}(\boldsymbol{\rho},\boldsymbol{\rho}')\boldsymbol{m}(\boldsymbol{\rho}') = 0\,.\end{aligned} \tag{7}$$

Here

$$\hat{\boldsymbol{I}} = \begin{bmatrix}1\,0\\0\,1\end{bmatrix},\ \hat{\boldsymbol{T}} = \begin{bmatrix}0 & -1\\1 & 0\end{bmatrix},\ \hat{\boldsymbol{G}}_{x,y} = \begin{bmatrix}G_{xx}\;G_{xy}\\G_{yx}\;G_{yy}\end{bmatrix},\ \boldsymbol{m} = \begin{pmatrix}m_x\\m_y\end{pmatrix}, \tag{8}$$

and $\omega_M = \gamma 4\pi M_\mathrm{S}$, $\alpha = \frac{A}{2\pi M_\mathrm{S}^2}$ is the exchange constant measured in cm^2,

$$\omega_{H_\mathrm{i}}(\boldsymbol{\rho}) = \gamma H_\mathrm{i}(\boldsymbol{\rho})\,, \tag{9}$$

where the internal field

$$H_\mathrm{i}(\boldsymbol{\rho}) = H_\mathrm{e} - N_{zz}(\boldsymbol{\rho}) \times 4\pi M_\mathrm{S}\,, \tag{10}$$

and the coordinate-dependent demagnetizing factor for a rectangular prism

$$N_{zz}(\boldsymbol{\rho}) = -\int \mathrm{d}\boldsymbol{\rho}' G_{zz}(\boldsymbol{\rho}, \boldsymbol{\rho}') \tag{11}$$

was first calculated by *Joseph* and *Schlömann* in [19].

In (7) the first term describes the exchange contribution. The second term $\omega_{H_\mathrm{i}}(\boldsymbol{\rho})$ describes the position-dependent static demagnetization field inside the nonellipsoidal rectangular magnetic element. The term containing ω results from the time-derivative term in the Landau–Lifshitz equation, while the last integral term describes the dynamic demagnetizing field in the dot.

Equation (7) is a two-dimensional integrodifferential equation for the magnetization amplitude $\boldsymbol{m}(\boldsymbol{\rho})$. It contains a differential as well as an integral operator. Furthermore, it has the form of an eigenequation: a solution consists of a pair of eigenvalue ω and corresponding eigenfunction $\boldsymbol{m}(\boldsymbol{\rho})$. It could be solved numerically in principle, but this solution is rather complicated as mentioned above. In fact, this solution is not substantially simpler than a direct numerical calculation of spin-wave modes using a micromagnetic approach as will be described below in Sect. 5. However, we can obtain a lot of insight from (7) if we consider it in certain particular cases, when it can be easily solved both numerically and analytically.

For completeness, first the components of the tensorial dipolar Green's function for a rectangular element will be derived from the most general form $\hat{\boldsymbol{G}}(\boldsymbol{r}, \boldsymbol{r}') = [G_{\alpha\beta}]$, $\alpha, \beta \in \{x, y, z\}$ with

$$G_{\alpha\beta}(\boldsymbol{r}, \boldsymbol{r}') = -\frac{1}{4\pi} \frac{\partial}{\partial \alpha} \frac{\partial}{\partial \beta'} \frac{1}{|\boldsymbol{r} - \boldsymbol{r}'|} \,. \tag{12}$$

As the magnetization has a uniform distribution across the thickness L, (12) can be simplified by integration across the thickness. Furthermore, the mean value of the field across the thickness is taken,

$$G_{\alpha\beta}(\boldsymbol{\rho}, \boldsymbol{\rho}') = \frac{1}{L} \int\limits_{-L/2}^{L/2} \int\limits_{-L/2}^{L/2} G_{\alpha\beta}(\boldsymbol{r}, \boldsymbol{r}') \mathrm{d}x \mathrm{d}x' \,. \tag{13}$$

Evaluation of (13) leads to

$$\hat{\boldsymbol{G}}(\boldsymbol{\rho}, \boldsymbol{\rho}') = \begin{pmatrix} G_{xx} & 0 & 0 \\ 0 & G_{yy} & G_{yz} \\ 0 & G_{zy} & \tilde{G}_{zz} \end{pmatrix} , \tag{14}$$

with

$$G_{xx}(\boldsymbol{\rho}, \boldsymbol{\rho}') = \frac{1}{2\pi L} \left(\frac{1}{\sqrt{L^2 + (\boldsymbol{\rho} - \boldsymbol{\rho}')^2}} - \frac{1}{|\boldsymbol{\rho} - \boldsymbol{\rho}'|} \right) , \tag{15}$$

and

$$G_{\eta\zeta}(\boldsymbol{\rho},\boldsymbol{\rho}') = -\frac{1}{2\pi}\frac{\partial^2}{\partial\eta\partial\zeta'}R\left(\frac{|\boldsymbol{\rho}-\boldsymbol{\rho}'|}{L}\right), \tag{16}$$

where $R(\xi) = \sinh^{-1}(1/\xi) + \xi - \sqrt{1+\xi^2}$ and $\eta, \zeta = y, z$.

4.1 Longitudinally Magnetized Long Stripes

Let us consider a particular case of a longitudinally magnetized long stripe, i.e., the case when the length l (along the applied field) is much larger then the width w, or $l \to \infty$. In that case the problem becomes one-dimensional, i.e., translational invariant along the z-direction, and is substantially simplified.

In such a case (14) can be reduced by evaluating the integral along the stripe axis coordinate z',

$$G_{\alpha\beta}(y,y') = \int\limits_{-\infty}^{\infty} G_{\alpha\beta}(\boldsymbol{\rho},\boldsymbol{\rho}')\mathrm{d}z', \tag{17}$$

which leads, for simple symmetry reasons, to a vanishing of the dependence on z. Evaluation of (17) yields

$$\hat{\boldsymbol{G}}(y,y') = \begin{pmatrix} G_{xx} & 0 & 0 \\ 0 & G_{yy} & 0 \\ 0 & 0 & 0 \end{pmatrix}, \tag{18}$$

with

$$G_{xx}(y,y') = \frac{1}{2\pi L}\ln\left(\frac{(y-y')^2}{(y-y')^2+L^2}\right), \tag{19}$$

$$G_{yy}(y,y') = -\delta(y-y') - G_{xx}(y,y'). \tag{20}$$

$G_{zz}(y,y') = 0$ and, therefore, $N_{zz}(\boldsymbol{\rho}) = N_{zz}(y) = 0$ and the internal static field $\boldsymbol{H}_\mathrm{i} = H_\mathrm{i}\boldsymbol{e}_z$ is homogeneous and exactly equal to the external field $\boldsymbol{H}_\mathrm{e} = H_\mathrm{e}\boldsymbol{e}_z$. Thus, (7) is reduced to

$$\left(-\alpha\omega_M\frac{\mathrm{d}^2}{\mathrm{d}y^2} + \omega_H\right)\hat{\boldsymbol{I}}\boldsymbol{m}(y) + \mathrm{i}\omega\hat{\boldsymbol{T}}\boldsymbol{m}(y) - \omega_M\int\limits_{-w/2}^{w/2}\mathrm{d}y'\hat{\boldsymbol{G}}_{x,y}(y,y')\boldsymbol{m}(y') = 0. \tag{21}$$

Although the static demagnetizing field is vanishing and the internal field is homogeneous, the dynamic demagnetizing field described by the last integral term in (21) is substantially inhomogeneous. The inhomogeneous distribution of this field turns out to be such that, at the stripe edges ($y = \pm w/2$), the dynamic magnetization $\boldsymbol{m}$ is close to zero, i.e., the modes are nearly pinned at the side walls. Equation (21) can be solved numerically and the two lowest eigenmodes are shown in Fig. 17 as full lines. They are very similar to cosine-like profiles, which leads to the idea that they can be approximated by trigonometric functions.

Since a typical width of a Permalloy stripe ($w \cong 1\,\mu\text{m}$) is much larger than the exchange length in Permalloy ($l_{\text{ex}} = \sqrt{\alpha} \cong 5\,\text{nm}$), we can neglect in zeroth order the exchange contribution to the frequency and use as the eigenfunctions of (21) the eigenfunctions of the dipolar integral operator [last term in (21)]

$$\int_{-w/2}^{w/2} \mathrm{d}y' 4\pi \boldsymbol{G}_{xx}(y, y')\varphi_m(y') = \lambda_m \varphi_m(y) \,. \tag{22}$$

With this, the approximate dispersion equation for the spin-wave modes of a stripe can be calculated from (21) and (22) as

$$\left(\frac{\omega_m}{\omega_M}\right)^2 = \left(\frac{\omega_H}{\omega_M} + 1 + \frac{\lambda_m}{4\pi}\right)\left(\frac{\omega_H}{\omega_M} - \frac{\lambda_m}{4\pi}\right) . \tag{23}$$

Nevertheless, the eigenvalues λ_m and eigenfunctions $\varphi_m(y)$ are still unknown.

It was shown in [20] that solutions $\varphi_m(y)$ of the the integral eigenequation (22) obey for $p \ll 1$ approximate boundary conditions for the variable magnetization at the lateral edges of the stripe:

$$\left. \pm \frac{\partial \varphi_m(y)}{\partial y} + d(p)\varphi_m(y) \right|_{y=\pm w/2} = 0 \,, \tag{24}$$

where the effective “pinning” parameter

$$d(p) = \frac{2\pi}{p[1 + 2\ln(1/p)]} \tag{25}$$

depends only on the stripe aspect ratio $p = L/w$. For $p = 0$, i.e., an infinitely thin stripe, $d = \infty$ and $\varphi_m(\pm w/2) = 0$. In other words, this boundary condition is a way of taking the inhomogeneity of the dynamic demagnetizing field into account.

Due to the symmetry of the problem the eigenfunctions are either symmetric or antisymmetric. It was shown in [20] that for $p \ll 1$ the symmetric eigenfunctions $\varphi_m^{\text{s}}(y)$ and the antisymmetric eigenfunctions $\varphi_m^{\text{a}}(y)$ can be approximated by the above-mentioned cosine and sine functions

$$\varphi_m^{\text{s}}(y) = A_m^{\text{s}} \cos(\kappa_m y), \qquad \varphi_m^{\text{a}}(y) = A_m^{\text{a}} \sin(\kappa_m y) \,, \tag{26}$$

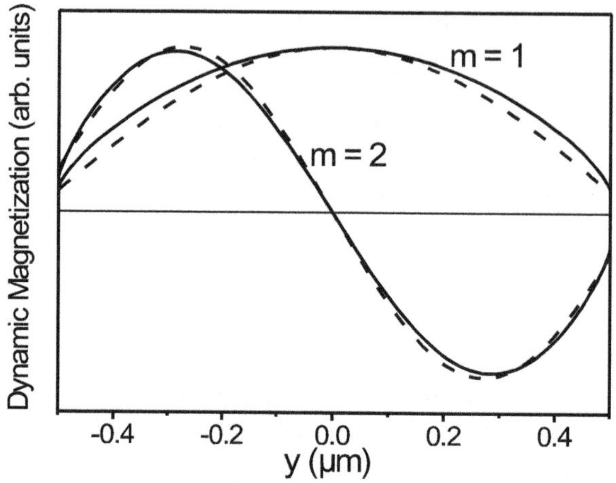

Fig. 17. Calculated mode profiles of the dynamic magnetization in a longitudinally magnetized $w = 1\,\mu\text{m}$ wide and $L = 33\,\text{nm}$ thick stripe. The *full lines* correspond to the direct solution of (21), while the *dashed lines* are the approximations from (26)

where

$$\kappa_m = \frac{m\pi}{w_{\text{eff}}}, \quad w_{\text{eff}} = w\frac{d(p)}{d(p)-2}, \quad m = 1, 2, 3, \ldots \tag{27}$$

and the eigenvalues λ_m of (22) can be approximately evaluated as

$$\lambda_m(p) = -4\pi + \pi^2(2m-1)p, \quad m = 1, 2, 3, \ldots \tag{28}$$

The profiles of the dipolar eigenmodes (26) are illustrated in Fig. 17 and compared to the exact solutions of (22). Substituting (28) for λ_m in (23) and using a linear approximation in p (for $p = \frac{L}{w} \ll 1$) we obtain the following simple expression for the discrete frequencies of the spin-wave modes in a longitudinally magnetized stripe

$$\omega_m^2 = \omega_H(\omega_H + \omega_M) + \omega_M^2\frac{\pi(2m-1)}{4}p, \quad m = 1, 2, 3, \ldots \tag{29}$$

Taking into account the small contribution of the exchange interaction to the frequency, the approximate dispersion equation for spin-wave modes of a stripe can be more closely approximated by

$$\left(\frac{\omega_m}{\omega_M}\right)^2 = \left(\frac{\omega_H}{\omega_M} + \alpha\kappa_m^2 + 1 + \frac{\lambda_m}{4\pi}\right)\left(\frac{\omega_H}{\omega_M} + \alpha\kappa_m^2 - \frac{\lambda_m}{4\pi}\right). \tag{30}$$

The results of comparison to experimental data are shown in Fig. 4. It is evident that Fig. 4 demonstrates a good quantitative agreement between theory and experiment.

4.2 Transversely Magnetized Long Stripes

Now we come to the second case and consider a long, transversely magnetized stripe where the size w becomes infinitely large ($w \gg l$ or $w \to \infty$), while the size l remains finite and the magnetization is along this direction. Note that in contrast to the previous subsection l denotes now the width of the stripe. This is done deliberately as the formulae of the two particular cases in this subsection will be combined in the following subsection for the case of rectangles.

In this particular case, (7) again becomes one-dimensional

$$\left[-\alpha\omega_M \frac{\mathrm{d}^2}{\mathrm{d}z^2} + \omega_{H_\mathrm{i}}\right] \hat{\boldsymbol{I}}\boldsymbol{m}(z) + \mathrm{i}\omega\hat{\boldsymbol{T}}\boldsymbol{m}(z) - \omega_M \int\limits_{-\frac{l}{2}}^{\frac{l}{2}} \mathrm{d}z' \hat{\boldsymbol{G}}_{x,y}(z,z')\boldsymbol{m}(z') = 0\,, \tag{31}$$

but in contrast to (21) the quantity $\omega_{H_\mathrm{i}}(z) = \gamma H_\mathrm{i}(z)$, proportional to the internal magnetic field $H_\mathrm{i}(z)$, is now substantially dependent on the coordinate z along the stripe width l due to the static demagnetization. In fact, the static demagnetization problem in this case has been solved analytically by *Joseph* and *Schlömann* [19] and the solution has the following form [analogous to (10)]

$$H_\mathrm{i}(z) = H_\mathrm{e} - N_{zz}(z) \times 4\pi M_\mathrm{S}\,, \tag{32}$$

where

$$N_{zz}(z) = \frac{1}{\pi} \times \left[\arctan\left(\frac{L}{2z+l}\right) - \arctan\left(\frac{L}{2z-l}\right)\right]\,. \tag{33}$$

Here, the demagnetizing field was calculated as if the whole stripe was homogeneously magnetized along the applied field. We chose the field at $x = 0$ as (33) becomes easier in this case compared to averaging the field across the thickness. This approximation is good for $L \ll l$. The tensorial Green function $\hat{\boldsymbol{G}}(\boldsymbol{\rho}, \boldsymbol{\rho}')$ after integration of (14) along y' is given by

$$\hat{\boldsymbol{G}}(z,z') = \begin{pmatrix} G_{xx} & 0 & 0 \\ 0 & 0 & 0 \\ 0 & 0 & G_{zz} \end{pmatrix}\,, \tag{34}$$

with

$$G_{xx}(z,z') = \frac{1}{2\pi L} \ln\left(\frac{(z-z')^2}{(z-z')^2 + L^2}\right)\,, \tag{35}$$

$$G_{zz}(z,z') = -\delta(z-z') - G_{xx}(z,z')\,. \tag{36}$$

The integrodifferential equation (31) with a coordinate-dependent coefficient (internal field) (32) can be solved numerically to find the spin-wave eigenmodes $\mu_n(z)$. However, before solving this equation it is necessary to make several comments on the additional new feature of this geometry: the inhomogeneity of the internal field H_i ((32) and (33)).

The field profile $H_\mathrm{i}(z)$ ((32) and (33)) is illustrated in Fig. 9. Near the stripe edges $z = \pm\frac{l}{2}$ (32) yields negative values for H_i that have no physical meaning. In reality, near the edges the static magnetization of the stipe becomes nonparallel to the field, the magnetization direction rotates in the stripe plane, and forms a so-called edge domain. In the region where the magnetization is parallel to the applied field the internal field is identical with the effective field and the simplest approximation for the internal field (32) gives a good agreement with the output of OOMMF, which is also shown in Fig. 9. In the edge domains the effective field cannot be approximated by (32) any longer. It has been shown by *Bryant* and *Suhl* [31] that it is reasonable to assume that within the edge domain the internal field is equal to zero. One can also conclude this from symmetry considerations: The magnetization will rotate as long as its crossproduct with the internal field $\boldsymbol{M} \times \boldsymbol{H}_\mathrm{i}$ does not vanish. The component of the internal field along y is always zero as no magnetic charges are created in this direction. Therefore only the field component along z remains, which consists of the applied field and the demagnetizing field. This internal field component is not parallel to the magnetization. Therefore it must be zero as well, as it would produce a torque otherwise.

The results of the calculation of $H_\mathrm{i}(z)$ with this assumption are shown in Fig. 9. In our further calculations we shall ignore the existence of the edge domains and will assume that the physical boundary for the spin-wave modes exists at the point where the internal field vanishes and the static magnetization starts to rotate. Therefore, we can use the approximation (32) for the internal field instead of the effective field, which does not have an easy analytic expression.

Using $H_\mathrm{i}(z)$ shown in Fig. 9 as a coordinate-dependent coefficient in (31) this equation was solved numerically, assuming a homogeneous distribution of magnetization across the stripe thickness as already mentioned above. We also assumed that due to negligible exchange surface anisotropy at the lateral edges of the stripe, the Rado–Weertman boundary conditions [29] for the variable magnetization are completely determined by the dipolar interaction. It turns out that these boundary conditions are very close to the conditions of the dipolar pinning given by (24). Also, since the stripe thickness L is much smaller than the stripe width l ($L \ll l$), for the lowest spin-wave modes these conditions are very close to the boundary conditions of total pinning of the variable magnetization at the edges $m|_{z=\pm\frac{l}{2}} = 0$.

The eigenfunctions $\mu_n(z)$ of the variable magnetization m_y obtained from the numerical solution of (31) are presented in Fig. 18 for $H_\mathrm{e} = 2000\,\mathrm{Oe}$. The

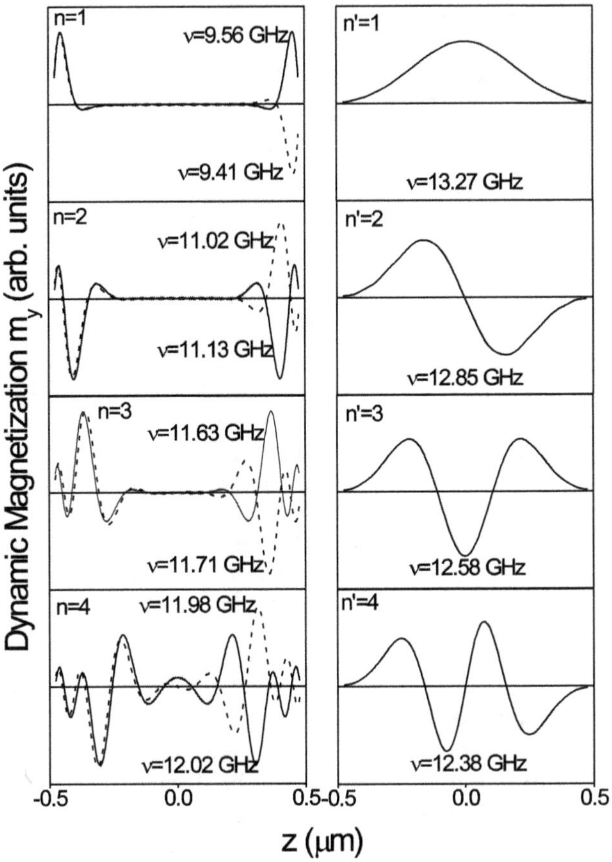

Fig. 18. Mode profiles and corresponding frequencies of a $l = 1\,\mu\text{m}$ wide and $L = 33\,\text{nm}$ thick stripe in a transversely applied magnetic field of 2000 Oe. The mode profiles $m_y(z)$ were obtained by solving (31) numerically. The index n denotes the exchange-dominated modes, while n' denotes the purely dipolar modes as discussed in the text. Note that two exchange-dominated modes corresponding to the same mode number n are symmetric and antisymmetric in z and nearly degenerate in frequency

spatial distribution of the component m_x is not shown as it is similar in shape, but one order of magnitude smaller due to the ellipticity of the precession.

In the case of a strong external bias magnetic field there are two different types of spin-wave modes in the spectrum: the "edge" modes numbered with the index n and localized in the narrow regions near the stripe edges, and quasicosinusoidal modes numbered with the index n' and localized near the center of the stripe. An approximate analytic treatment with analytic approximations of the dynamic magnetization of the localized "edge" modes

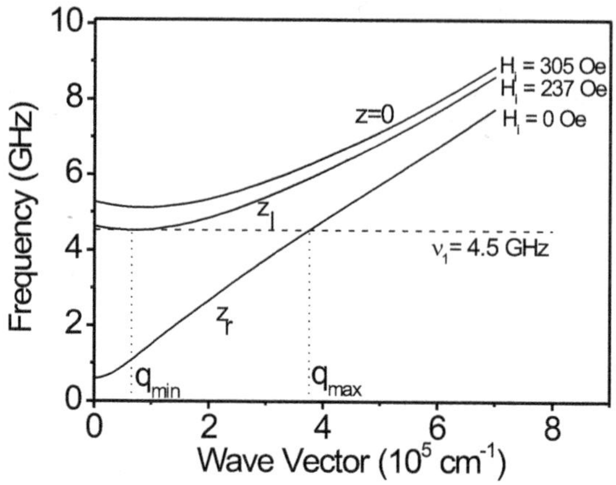

Fig. 19. Dispersion of plane spin waves in the MSBVW geometry at constant internal fields as indicated, calculated with (37) for an external field $H_e = 500$ Oe. $H_i = 0$ and $H_i = 237$ Oe correspond to the internal field at the right ($z = z_r$) and left ($z = z_l$) turning point, respectively. Furthermore, the minimum and maximum wavevectors of the first localized mode are indicated. The fixed frequency $\nu_1 = 4.5$ GHz corresponds to the experimental value (see Fig. 8)

is given in [35]. There the internal field is approximated by a cosine, which allows us to directly solve the spin-wave eigenequations.

The nature of the mode localization near the stripe edges in the regions of strong inhomogeneity of the internal magnetic field will be discussed now [10, 11]. The dispersion law for dipole-exchange spin waves, having a wavevector parallel to the magnetic field (quasimagnetostatic backward volume waves) in an *infinite* film [28], is given by

$$\left(\frac{\omega}{\omega_M}\right)^2 = \left(\frac{\omega_H}{\omega_M} + \alpha q^2\right)\left[\frac{\omega_H}{\omega_M} + \alpha q^2 + 1 - P(qL)\right] . \tag{37}$$

It is shown in Fig. 19 for several values of the internal field H_i, where $\omega_H = \gamma H_i$ and

$$P(qL) = 1 - \frac{1 - \exp(-qL)}{qL} . \tag{38}$$

These values of H_i were calculated using (33) for different positions within the stripe. The internal field $H_i(z)$ in a transversely magnetized stripe lies in the interval $0 < H_i < H_i^{max} = H_e - N_{zz}(z = 0)4\pi M_S$. It is evident from Fig. 19 that for a fixed frequency of the lowest spin-wave mode ($\nu_1 = 4.5$ GHz in this example, taken from the experiment, see Fig. 8) there is only a limited range of sufficiently small values of the internal field H_i in which a mode with such a frequency can exist ($0 < H_i < 237$ Oe). At $z = 0$ the internal field is $H_i = 305$ Oe and the fixed frequency of the spin-wave mode has no intersection

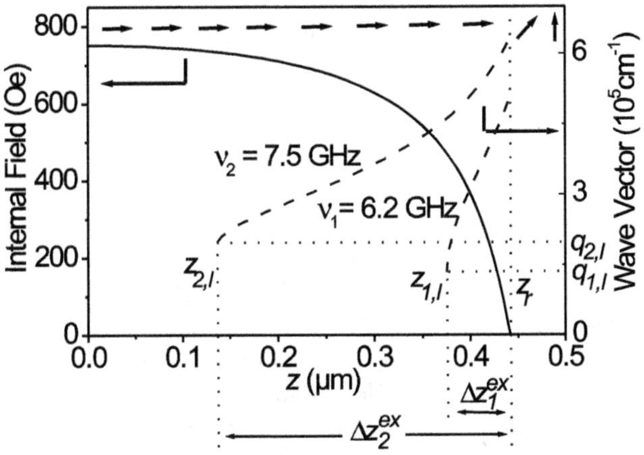

Fig. 20. Internal field (*solid line*) and wavevector of the observed states (*dashed lines*) calculated as functions of z for an applied field of 970 Oe. $z_{1,1}$, $z_{2,1}$, and z_{r} are the left and right turning points of the localized modes with numbers $n = 1$ and $n = 2$ with the frequencies as indicated. $\Delta z_1^{\mathrm{ex}} = z_{\mathrm{r}} - z_{1,1}$ and $\Delta z_2^{\mathrm{ex}} = z_{\mathrm{r}} - z_{2,1}$ are the localization lengths of the two modes. $q_{1,1}, q_{2,1}$ are the nonzero wavevectors of the waves near the turning points. The *arrows* at the top of the figure illustrate the orientation of the static magnetization. (From [10])

with the dispersion relation and therefore no real wavevector exists at $z = 0$. Thus, in accordance with Fig. 9, this mode can only be localized in the spin-wave well (region of "small" and rapidly changing internal field) near the stripe edge.

Knowing the fixed frequency of a localized mode and the spatial distribution of the internal field $H_{\mathrm{i}} = H_{\mathrm{i}}(z)$ (see, e.g., Fig. 9) one can solve (37) for the wavevector $q = q(z)$. The variation of this wavevector in space is shown in Fig. 20 for an applied field of 970 Oe. In accordance with the experiment [11] two localized modes can exist at this field (compare to experiment in Fig. 12). The solid line represents the internal field, while the dashed lines represent the wavevectors. The points indicated by $z_{1,l}$ and $z_{2,l}$ are the left turning points of the two lowest localized modes, respectively. At these points the solution of (37) no longer gives a real value of the wavevector, and the wave changes its character from a sinusoidal-like wave to an exponential-like decay towards the center of the stripe. The localization lengths for the two localized modes are shown as Δz_1^{ex} and Δz_2^{ex} in Fig. 20. This figure clearly demonstrates that the local wavevector of a localized mode varies with coordinate $q = q(z)$.

The quantization condition for spin-wave modes existing in the inhomogeneous internal field is similar to the quasiclassical quantization condition in quantum mechanics (see [10, 33]). We formulate it as a phase quantization condition in the form of

$$\Delta\phi_\mathrm{l} + \Delta\phi_\mathrm{r} + 2\int_{z_\mathrm{l}}^{z_\mathrm{r}} q[H_\mathrm{i}(z), \omega]\mathrm{d}z = 2n\pi\,, \tag{39}$$

where $n = 1, 2, 3, \ldots$ is the localized mode index. $\Delta\phi_\mathrm{l}, \Delta\phi_\mathrm{r} \in [0, \pi]$ are the (unknown) phase jumps at the left and right turning points, between which (37) has a real solution $q(z)$ for a fixed frequency ω_1. Henceforth, we will call the integral in (39) a quantization integral.

Equation (39) is formulated using the WKB approximation. In this method, we assume that the spin-wave mode profile $m(z)$ has the form $\exp(\mathrm{i}k(z)z)$ with a position-dependent wavevector $k(z)$, which is determined by the requirement that the frequency given by (37) is independent of position. However, in the case considered here, with a large change in wavevector on the length scale of the spin-wave wavelength, it is not "a priori" clear that this quantization condition can be used, as the WKB condition of the spin-wave wavevector varying "smoothly" in space is definitely not fulfilled. However, it has been shown in [36] that the quantization condition (39) leads nearly to the same results as the numerical solution of (31). Thus, (39) can be safely used for the qualitative description of the problem.

Figures 19 and 20 demonstrate that the localized modes discussed above are, indeed, exchange dominated. It follows from Fig. 19 that these modes are formed from dipole-exchange plane waves with positive group velocity crossing the lines $\omega = \omega_1^\mathrm{ex}$ and $\omega = \omega_2^\mathrm{ex}$ at the exchange-dominated region of the dispersion law (37).

It should also be noted that in the numerical solution of (31) the exchange-dominated localized modes appear as a nearly frequency-degenerate doublet consisting of a symmetric and an antisymmetric mode as shown in Fig. 18. The doublet is not completely degenerate due to the dynamic dipolar interaction between the left and right branches of the mode depending on the symmetry of the mode. Therefore, the symmetric mode has a slightly higher frequency than the antisymmetric one. This effect is most pronounced for the mode $n = 1$, while for higher modes the splitting is relatively small, of the order of the numerical accuracy.

The lowest exchange-dominated modes localized near the stripe edges (modes $n = 1, \ldots, 4$ in Fig. 18) increase their frequencies with n. In contrast, the lowest dipole-dominated modes are localized near the stripe center, and their frequencies decrease with the increase of the mode index n' (see modes $n' = 1, \ldots, 4$ in Fig. 18). This is due to the fact that the dipole-dominated localized modes are formed from plane spin waves (37) with negative group velocity that cross the lines $\omega = \omega_{n'}^\mathrm{dip}$ in the region of low values of the plane-wave wavevectors. As an example, in Fig. 21 we show by the dashed line the frequency of a dipolar mode ($n' = 1$ from Fig. 18) and dispersion curves (37) calculated for two different internal fields at the positions $z = 0$ and

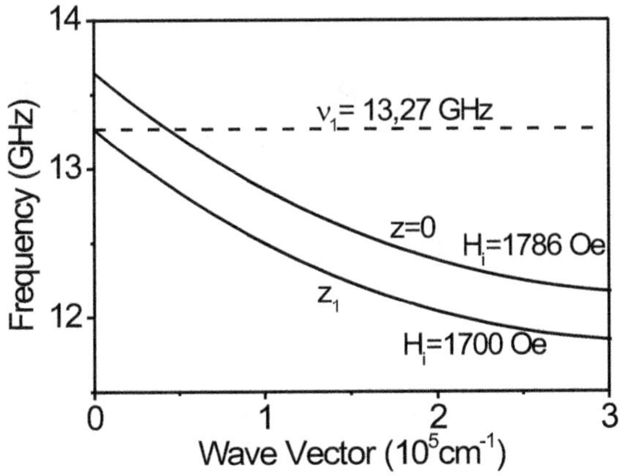

Fig. 21. Spin-wave dispersion relations (*solid lines*) for an external applied field $H_e = 2000$ Oe at different points of a 1 μm wide and 33 nm thick stripe calculated with (37). The *dashed line* shows the frequency of the first mode $n' = 1$ from Fig. 18 localized in the center. z_1 indicates the turning point of the first dipolar localized mode and is shown also in Fig. 22

$z = z_1$ inside the stripe. With the increase of z the internal field decreases and, starting from the coordinate $z = z_1$, there are no allowed real values of the plane-wave wavevector for the mode. Thus, the mode will be localized in the interval $-z_1 < z < z_1$ near the center of the stripe.

The dependence of the plane-wave wavevector on the coordinate z calculated for dipole-dominated localized spin-wave modes $n' = 1$ and $n' = 2$ is shown in Fig. 22. Analogously to the case of exchange-dominated localized modes we introduce the localization length $\Delta z_1^{\mathrm{dip}} = 2z_1$ for dipolar modes localized near the center of the stripe. Mode profiles similar to the ones shown in Fig. 18 have been calculated in [37] for magnetic elements with larger *in-plane* dimensions, while the effect of dipolar localization of spin-wave modes has been discussed in detail in [38, 39]. Note that, using the dependences $q = q(z)$ presented in Fig. 22 in the quantization integral (39), we find that the approximate phase differences between successive spin-wave modes are close to 2π. Assuming that the phase jumps $\Delta\phi_l, \Delta\phi_r$ are nearly identical for successive modes this supports again the assumption of the applicability of the WKB approach.

It should be mentioned that a clear separation of the "exchange-dominated" modes ($n = 1, 2, 3, 4$) and "dipolar-dominated" modes ($n' = 1, 2, 3, 4, \ldots$) exists only in large fields $H_e > 1000$ Oe. At smaller values of the applied field the modes have a mixed character. This is illustrated by Fig. 23 where the profiles of the lowest modes of a stripe ($l = 1$ μm) are calculated from (31) for different values of the field $H_e = 300$ Oe, 450 Oe and 950 Oe. As is seen, for a field $H_e = 300$ Oe these modes have no pronounced "exchange" character

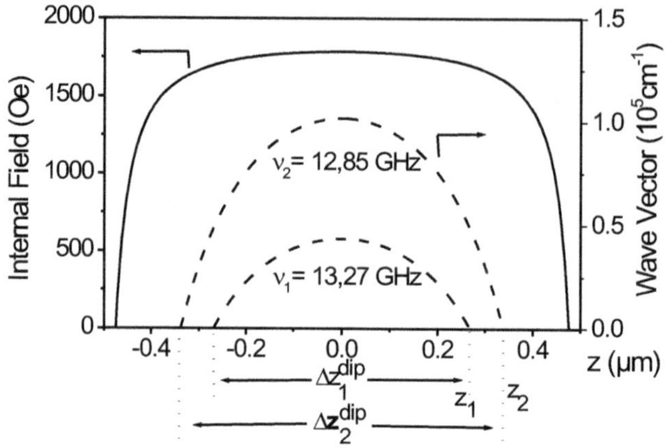

Fig. 22. Internal field (*solid line*) and wavevector of the localized modes in the center (*dashed lines*) calculated as functions of z in a transversely applied magnetic field of $H_e = 2000\,\mathrm{Oe}$. z_1 and z_2 are the turning points of the dipolar localized modes with numbers $n' = 1$ and $n' = 2$ with the frequencies as given in the figure. $\Delta z_1^{\mathrm{dip}} = 2z_1$ and $\Delta z_2^{\mathrm{dip}} = 2z_2$ indicate the localization lengths of the two modes. (From [11])

and are not localized near the edges of the stripe. Therefore n is not a good quantum number in the low-field range.

Numerically calculated frequencies of the spin-wave modes of a transversely magnetized stripe as functions of the field H_e are shown in Fig. 24 by solid lines. Symbols in the same figure show the experimentally measured frequencies from Fig. 12. The triangular and circular symbols show the localized modes, while the position of the band of nonresolved MSBVW modes is shown as a gray region. It is evident that the approximative numerical solution of (31) gives a reasonably good qualitative description of the experiment. We also note that the agreement between the experiment and the calculation for the frequencies of the localized modes improves with increasing external magnetic field.

We would like to stress that all our calculations of the spin-wave modes in a transversely magnetized stripe were made under the assumption that everywhere in the stripe the magnetic material is magnetized to saturation and the direction of static magnetization is parallel to the direction of the applied field. This assumption, strictly speaking, is correct only for very strong external fields comparable in magnitude to the static magnetization of the material. In a small applied field, comparable to the demagnetizing field in the center of the stripe, the magnetization will rotate across the whole width of the stripe [36]. For moderate external fields below the demagnetizing field there will be "edge domains" at the edges of the stripe, or regions where the effective internal field vanishes and the static magnetization rotates (see

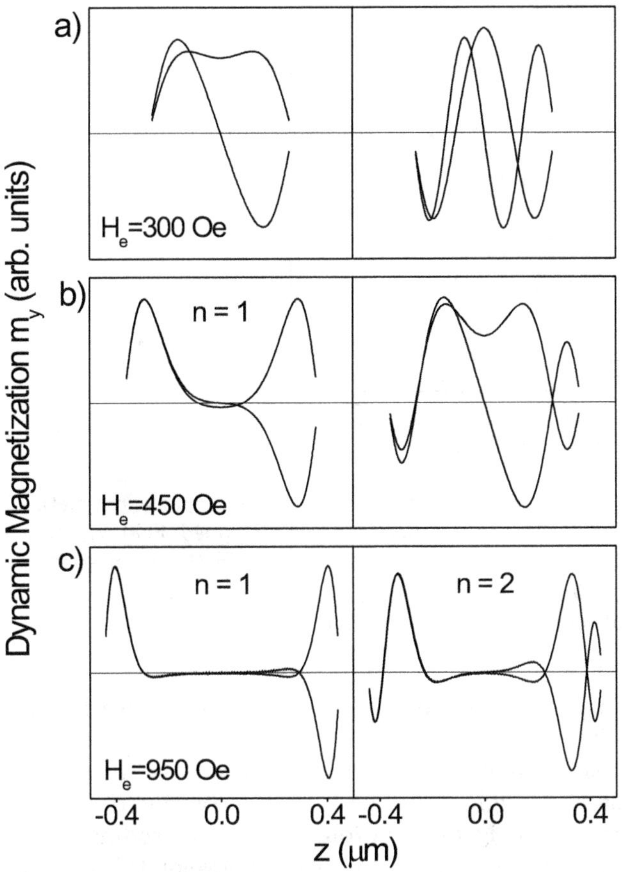

Fig. 23. Numerically calculated eigenfunctions $m_y(z)$ of the dynamic magnetization for the two lowest doublets of modes and for different applied fields (a) $H_e = 300$ Oe, (b) $H_e = 450$ Oe, (c) $H_e = 950$ Oe. The mode profiles are calculated in that region where the internal field is nonzero. The stripe is 1 μm wide and 33 nm thick

Fig. 9). The spin-wave modes in a transversely magnetized stripe in the case of the inhomogeneous magnetization were discussed in [36, 40–42].

4.3 Rectangular Magnetic Elements

As discussed above, the eigenfrequencies and two-dimensional precession amplitude profiles, i.e., the dynamic part of the magnetization in a rectangular magnetic element (see Fig. 16) can be obtained from the numerical solution of (7). However, technically, this solution is not much simpler than a direct micromagnetic numerical simulation using, for example, Langevin dynamics or zero-temperature simulations of the magnetization relaxation in a pulsed

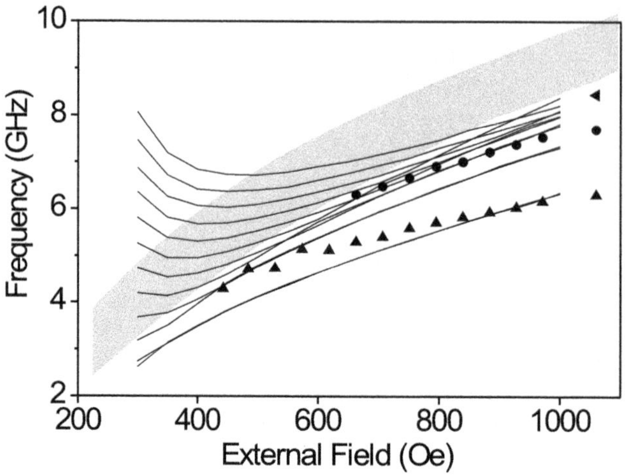

Fig. 24. Numerically calculated eigenfrequencies for a transversely magnetized 1 μm wide and 33 nm thick stripe as a function of the applied field H_e (*solid lines*). The experimental frequencies of the modes observed in the 1 μm stripe at $q = 0.47 \times 10^5\,\mathrm{cm}^{-1}$ are indicated by the *triangles* and *circles*. The *gray region* illustrates the experimental band of nonresolved excitations

field. Examples of such simulations were presented in a recent publication [10] and will be discussed in Sect. 5.

A different approach is to create an approximative analytic theory describing such modes using the insight obtained from the solution of the two above-described one-dimensional eigenvalue problems for longitudinally and transversely magnetized magnetic stripes. Below we will develop this qualitative theory and will compare its results with the results of BLS experiments and with the results of micromagnetic simulations based on Langevin and on zero-temperature dynamics.

First, we make the strong assumption that the two-dimensional distribution of the dynamic magnetization $m_{m,n}(y,z)$ in the spin-wave eigenmodes of a rectangular element can be approximately factorized as

$$m_{mn}(y,z) = \varphi_m(\kappa_{my} y) \times \mu_n(\kappa_{nz} z), \tag{40}$$

where φ_m are similar to the eigenfuntions of a longitudinally magnetized infinitely long stripe shown in Fig. 17, while the functions μ_n are similar to the eigenfunctions of a transversely magnetized infinitely long stripe shown in Figs. 18 and 23. Here, κ_{my} and κ_{nz} are the quantized values of the corresponding projections of spin-wave wavevectors, and $m, n = 1, 2, 3, \ldots$. This assumption is justified by the fact that the inhomogeneity of the internal magnetic field in the rectangular element occurs mainly along the direction of this field (z-direction), while the internal field is nearly constant along the y-direction, as is shown in Fig. 13.

The quantized values of the wavevector projection κ_{my} perpendicular to the applied field can be found from the approximation (27). The situation with the discrete projections of the spin-wave wave number κ_{nz} along the applied field is more complicated. As was shown in Figs. 18 and 23, in a transversely magnetized stripe one has two different types of modes along this direction: exchange-dominated modes localized near the element edges (indexed by n in Fig. 18) and dipole-dominated modes localized near the stripe center (indexed by n' in Fig. 18). The quantization condition in this case has the integral form (39), and the evaluation of the discrete values of the wavevector for a particular mode is nontrivial. However, using (27) and taking the mean value of the wavevector of each mode, we can approximately evaluate the characteristic wavevector of a localized mode as

$$\kappa_{nz} = \frac{n\pi}{\Delta z_{\mathrm{n}}}\,, \tag{41}$$

where $\Delta z_n = z_l - z_r$ is the localization length for a particular mode that can be estimated using the method illustrated in Figs. 19–22. As was shown both theoretically and experimentally [10–12], the localization length of an exchange-dominated spin-wave mode can be less than $\frac{1}{10}$ of a stripe width $\Delta z_1^{\mathrm{ex}} < \frac{l}{10}$, so that the characteristic wavevector of the mode along the direction of magnetization can be rather high, e.g., for $l = 1\,\mu\mathrm{m}$

$$\kappa_{n=1,z} = \frac{\pi}{\Delta z_1^{\mathrm{ex}}}\,, \tag{42}$$

easily exceeding the maximum value of the wavevector accessible by the BLS experimental technique (usually $2.5 \times 10^5\,\mathrm{cm}^{-1}$). An alternative approximate analytic treatment of exchange-dominated localized spin-wave modes that approximates the internal field profile by a cosine and localized modes by Mathieu functions is given in [35].

Analogously, we can calculate the quantized wavevectors for dipole-dominated localized modes:

$$\kappa_{n'=1,z} = \frac{\pi}{\Delta z_1^{\mathrm{dip}}}\,, \tag{43}$$

where $\Delta z_{n'}^{\mathrm{dip}}$ is the localization length for a dipolar localized mode. This dipolar localization length for the lowest dipolar mode $\Delta z_{n'=1}^{\mathrm{dip}}$ is between l and $\frac{l}{2}$. For the modes in the transitional regime, where the influence of dipolar and exchange interactions are comparable, this approximation, of course, cannot be applied (see, e.g., Fig. 23 for $H_{\mathrm{e}} = 300\,\mathrm{Oe}$).

Using approximate eigenfunctions (40) we can represent the variable magnetization of a rectangular element in the form of a series of eigenfunctions

$$\boldsymbol{m}(y,z) = \sum_{m,n} \boldsymbol{A}_{mn}\varphi_m(\kappa_{my}y) \times \mu_n(\kappa_{nz}z)\,. \tag{44}$$

Substituting (44) for $\boldsymbol{m}(y,z)$ in (7) and evaluating the matrix elements of the Green's function $\hat{\boldsymbol{G}}$ on the spin-wave eigenmodes (43), we can, in principle, obtain an approximative dispersion equation for the discrete spin-wave modes of the rectangular element.

There is, however, an even simpler way to approximately evaluate these frequencies as suggested in [38]. Using the fact that all the matrix elements of the dynamic dipole–dipole interaction in a thin magnetic element ($\frac{L}{l} \ll 1$, $\frac{L}{w} \ll 1$) contain small aspect ratios of the element $\frac{L}{l}$ or $\frac{L}{w}$, one can simply use the dispersion equation for an infinite magnetic film where projections of the spin-wave wavevector κ_{my} and κ_{nz} have discrete values defined by (27) and (42) or (43).

Before doing so we have to consider an important qualitative feature that distinguishes a rectangular element from an infinite film, namely the inhomogeneity of the internal field inside the element. This inhomogeneity will lead to different effective values of the internal field for different spin-wave modes. These effective values can be calculated using the mode distributions (40) and the coordinate-dependent effective demagnetizing factors (10).

Taking this into account, approximate equations can be written as [38]

$$\omega_{mn}^2 = \left(\omega_H^{mn} + \alpha\omega_M\kappa_{mn}^2\right)\left[\omega_H^{mn} + \alpha\omega_M\kappa_{mn}^2 + \omega_M F_{mn}(\kappa_{mn}L)\right] , \qquad (45)$$

where the quantity ω_H^{mn} (proportional to the effective value of the internal field for a particular spin-wave mode H_{mn}) is defined as

$$\omega_H^{mn} = \omega_H - \omega_M N_{mn} , \qquad (46)$$

where

$$N_{mn} = \frac{4}{wl}\int \mathrm{d}\boldsymbol{\rho}\, m_{mn}^2(\boldsymbol{\rho}) N_{zz}(\boldsymbol{\rho}) . \qquad (47)$$

The quantity $F_{mn}(\kappa_{mn}L)$ plays the role of a quantized matrix element of the dipole–dipole interaction

$$F_{mn}(\kappa_{mn}L) = 1 + P(\kappa_{mn}L)\left[1 - P(\kappa_{mn}L)\right] \times \left(\frac{\omega_M}{\omega_H^{mn} + \alpha\omega_M\kappa_{mn}^2}\right)\left(\frac{\kappa_{my}^2}{\kappa_{mn}^2}\right) - P(\kappa_{mn}L)\left(\frac{\kappa_{nz}^2}{\kappa_{mn}^2}\right) , \qquad (48)$$

with $P(\kappa_{mn}L)$ defined by (38) and $\kappa_{mn}^2 = \kappa_{my}^2 + \kappa_{nz}^2$.

The eigenfrequencies of a rectangular Permalloy element as used in the experiment (see Fig. 14) have been calculated using this approach. The following parameters of the element were used: thickness $L = 33\,\mathrm{nm}$, lateral sizes $l \times w = 1.75\,\mu\mathrm{m} \times 1\,\mu\mathrm{m}$, $4\pi M_\mathrm{S} = 10.2\,\mathrm{kOe}$, exchange constant $\alpha = 2.5 \times 10^{-13}\,\mathrm{cm}^2$, or $A \simeq 1 \times 10^6\,\mathrm{erg/cm}^3$, $H_\mathrm{e} = 600\,\mathrm{Oe}$, gyromagnetic ratio for the electron $\gamma/2\pi = 2.97\,\mathrm{MHz/Oe}$. The magnetic parameters were determined independently from experiments on unstructured films. The results of the calculation are presented in Table 1. The lowest mode having the frequency of 4.9 GHz was identified as an exchange-dominated mode localized along the direction of the magnetic field (z-axis), the localization length for $n = 1$ being evaluated as $\Delta z_{n=1} = 0.3\,\mu\text{m}$. The spatial distribution of the exchange-dominated localized mode within the region of its localization was assumed to be cosinusoidal

$$\mu_{n=1}(\kappa_{1z}^{\text{ex}} z) = \cos\left(\frac{\pi}{\Delta z_1^{\text{ex}}}(z - l^*)\right) \tag{49}$$

in the intervals $l^* - \Delta z_1^{\text{ex}} \leq |z| \leq l^*$ and zero elsewhere ($|z| < l^* - \Delta z_1^{\text{ex}}$). $z = \pm l^*$ are here the points where the internal field becomes zero.

For all the other experimentally observed modes it was assumed that the distribution along the z-axis is similar to the lowest dipole-dominated localized mode (mode $n' = 1$ in Fig. 18). For the experimental conditions ($H_\text{e} = 600\,\text{Oe}$) the localization length of this distribution was evaluated as $\Delta z_1^{\text{dip}} = 0.85\,\mu\text{m}$. The spatial distribution of the dipole-dominated modes along the z-axis was also assumed to be cosinusoidal

$$\mu_{n'=1}(\kappa_{1z}^{\text{dip}} z) = \cos\left(\frac{\pi}{\Delta z_1^{\text{dip}}} z\right) \tag{50}$$

in the interval $|z| \leq \Delta z_1^{\text{dip}}/2$ and zero elsewhere (i.e., in the interval $\Delta z_1^{\text{dip}}/2 < |z| < l/2$).

The dynamic magnetization distributions of both exchange-dominated and dipole-dominated modes along the y-axis were assumed to be described by (26). The two-dimensional spatial distributions of the variable magnetization in these analytically calculated modes are shown in the right panel of Fig. 27.

It should be mentioned that further numerical micromagnetic calculations [43–45] give results that quantitatively coincide with the results of the approximate analytic formalism presented in Sect. 4 and with the micromagnetic simulations of this section (see Fig. 8 in [43]).

5 Micromagnetic Simulations

To check the validity of the approximate results described in the preceding section, corresponding dynamical micromagnetic simulations have been performed. The commercially available micromagnetic packages MicroMagus [46] and OOMMF [21] have been employed to produce raw simulation data, i.e., spin precessions in different points of the sample. To calculate the excitation spectra further processing of the data was performed, as described below.

In the MicroMagus package a system of magnetization dynamics is simulated using the standard extension of the Landau–Lifshitz (LL) equation of motion for magnetic moments that includes finite-temperature random fluctuations:

$$\frac{\mathrm{d}\boldsymbol{M}_i}{\mathrm{d}t} = -\gamma \cdot [\boldsymbol{M}_i \times (\boldsymbol{H}_i^{\mathrm{det}} + \boldsymbol{H}_i^{\mathrm{fl}})] - \gamma \cdot \frac{\lambda}{M_{\mathrm{S}}} \cdot [\boldsymbol{M}_i \times [\boldsymbol{M}_i \times (\boldsymbol{H}_i^{\mathrm{det}} + \boldsymbol{H}_i^{\mathrm{fl}})]] , \tag{51}$$

where the precession constant $\gamma(>0)$ is equal to the gyromagnetic ratio of the spins for small dissipation $\lambda \ll 1$, and the deterministic field $\boldsymbol{H}_i^{\mathrm{det}}$ acting on each magnetic moment $\boldsymbol{M}_i$ includes all standard micromagnetic interactions: external, anisotropy, exchange and dipolar fields.

The fluctuation (or Langevin) field $\boldsymbol{H}_i^{\mathrm{fl}}$ is introduced to take into account random thermal noise responsible for spontaneous fluctuations of the system magnetization for finite temperatures. The question concerning the statistical properties of this random field is crucially important for obtaining the correct physical results from numerical simulations. Hence we will briefly address this question here leaving a detailed methodological explanation to a separate publication.

To our knowledge, in virtually all micromagnetic simulations the simplest statistical properties, written in Cartesian coordinates, of the fluctuation field

$$\langle H_{\xi,i}^{\mathrm{fl}}(t)\rangle = 0 \tag{52}$$

$$\langle H_{\xi,i}^{\mathrm{fl}}(0) \cdot H_{\psi,j}^{\mathrm{fl}}(t)\rangle = 2D \, , \, \delta(t) \, , \, \delta_{ij} \, , \, \delta_{\xi\psi} \tag{53}$$

were used (here indices i, j refer to discretization cells, $\xi, \psi = x, y, z$ and the noise power $D = \lambda/(1+\lambda^2)(kT/M_{\mathrm{S}})$ is proportional to the system temperature T). The relation (53) was rigorously derived from the fluctuation-dissipation theorem in [47] for a system of noninteracting single-domain particles (actually for point dipoles with a uniaxial onsite anisotropy), so that the usage of the same simple correlation properties for a much more complicated micromagnetic system requires careful justification.

First, we point out that the sole presence of interactions between elementary system moments (representing the magnetization vectors in finite element discretization cells) does *not* require the introduction of nontrivial correlations between the random field components, because these interactions are taken into account via the deterministic part of the effective field $\boldsymbol{H}_i^{\mathrm{det}}$ in (51). The corresponding generalization of the fluctuation-dissipation theorem for a system of interacting magnetic moments can be found in [48].

However, in micromagnetic simulations there exist in principle at least three reasons to use more complicated random field correlations. The first – the physical one – is the existence of correlations of thermal-bath phonon fluctuations (supposed to cause the magnetization fluctuations). Their exact correlation properties are not known; however, since the system is considered at room temperature, i.e., at a temperature comparable to the Debye temperature of the lattice, it is reasonable to suggest that these correlations are short ranged (in space and time) and with an exponentially decaying correlation function. Hence the δ-functional expression (53) can be used as a good first approximation for these correlations, provided that we are interested in processes with characteristic time and space scales much larger than the corresponding decay time (picosecond, i.e., $2\pi/\omega_D$, where ω_D is the Debye frequency) and the correlation length (several interatomic distances) of the heat-bath fluctuations.

The other two reasons are due to the finite-difference approximation of the continuous micromagnetic problem. First of all, such an approximation does not allow one to take into account magnetic excitations with a wavelength smaller than the grid cell size Δx. However, due to magnon–magnon interactions such magnons can excite a precession of the cell magnetic moment. Since they can have a mean free path much larger than Δx thus causing substantial correlations of the effective fields acting on the magnetic moments of the neighboring cells (see [49] for details). Whether these correlations are important for the particular magnetization dynamics under study can be determined by comparing simulation results for grids with various cell sizes. We have checked that in our case these correlations do not play any significant role as soon as the grid cell is smaller than the exchange correlation length: $\Delta x < l_{\mathrm{ex}} = \sqrt{A/2\pi M_S^2}$ (where A denotes the exchange constant).

The *qualitative* change in the density of magnon states (m-DOS) due to the transition from the formally continuous (discrete on the atomic scale) to a finite element micromagnetic system is the third reason why more complicated correlations should be used in principle. However, it is shown elsewhere [50] that for frequencies much less than the frequency of the van-Hove singularity in the m-DOS for the discretized (not initial!) system this effect can also be neglected. This is the case for the current study.

Based on the discussion above, we used in our micromagnetic simulations the δ-correlated noise $\boldsymbol{H}_i^{\mathrm{fl}}$.

The spectra of magnetic eigenmodes in Permalloy elements were simulated in the following way. The rectangular Permalloy element with lateral sizes $l \times w = 1.75\,\mu\mathrm{m} \times 1\,\mu\mathrm{m}$ and thickness $L = 33\,\mathrm{nm}$ (as used in the experimental setup) was discretized using a grid with $N_x \times N_z \times N_y = 4 \times 180 \times 100$ cells (mesh size $\simeq 10\,\mathrm{nm}$); it was checked on a smaller system that further grid refinement did not lead to any noticeable change in the magnon spectrum for the frequency range studied here (except for the lowest mode – see below). Periodic boundary conditions were applied with the distance between the Permalloy elements equal to the experimental spacing $\Delta = 100\,\mathrm{nm}$.

For simulations the same material parameters for Permalloy as in the analytical study were used. The cubic crystallographic anisotropy of Permalloy ($K_{\mathrm{cub}} = 5 \times 10^3\,\mathrm{erg/cm^3}$) was proven to be too low to have any influence on the simulation results and hence was neglected. The dissipation constant was set to $\lambda = 0.01$.

By choosing the initial state for dynamical simulations we have found that in the external field $H_e = 600\,\mathrm{Oe}$ used for experimental measurements three types of magnetization configuration for $T = 0$ are possible: the so called S-, C- and flower states. Among them we have used the S-type magnetization state as the initial state, because the flower state was found to be unstable with respect to thermal fluctuations: By the equilibration of the system at room temperature this state switches to either a S- or C-state. Magnon spectra for the C-state were identical to the S-state spectra up to the accuracy of small frequency shifts of the lowest mode near the domain walls between the closure and central domains.

After the initial magnetization configuration was obtained, dynamical simulations using (51) were started. The system temperature was set to room temperature $T = 300\,\text{K}$. To obtain the magnon spectrum in thermodynamical equilibrium, the system was first equilibrated by integrating (51) until the system energy has stopped to increase in frames of statistical errors (the "heating" phase). Afterwards, the time dependences of all grid cell moments were recorded with the sampling interval $\Delta t_{\text{min}} = 1\,\text{ps}$, which provided sufficiently good statistics to determine spectra up to the maximum frequency $\nu_{\text{max}} \sim 0.1/\Delta t_{\text{min}} = 100\,\text{GHz}$. To achieve the required lowest frequency $\omega_{\text{min}}/2\pi \sim 5\,\text{GHz}$ simulations were performed for the total time $\Delta t_{\text{min}} = 5 \times 2\pi/\omega_{\text{min}} = 1\,\text{ns}$. The resulting spectra (see Fig. 25) were obtained using Fourier analysis of the recording time dependences by averaging over at least $N_{\text{run}} = 8$ independent simulation runs. The magnon spectra obtained are displayed in Fig. 25. The eigenmodes manifest themselves as sequences of peaks with the same frequency for several subsequent wavevectors.

For the computation of the *spatial* distribution of the oscillation power of a magnetization component $m_{(x,y,z)}$ for the given frequency $\omega/2\pi$ we have first performed the time Fourier transforms of $m_{(x,y,z)}$-time dependences for all discretization cells thus obtaining the spectral power as a function of frequency for all cells. Then we used the spectral power values for the required frequency $\omega_0/2\pi$ (extracted from all these spectra) to build up the corresponding spatial maps (see Fig. 26).

The comparison of the eigenmode frequencies is shown in Table 1. One can see the excellent agreement between experimental and simulated frequencies for all modes except the lowest one (we point out once more that our simulations do not use any adjustable parameters). This mode with the smallest frequency corresponds to the oscillations in the region of inhomogeneous internal field near the domain walls of the closure domains in our S-structure. The difference between the simulated and experimentally measured frequencies for this mode can be attributed to the fact that the discretization into the $N_{\perp} = 4$ layers in the direction perpendicular to the layer plane is insufficient to resolve the structure of the domain wall sufficiently fine in order to reproduce its exact oscillation behavior.

Also, for all oscillation modes the regions where the oscillation power is concentrated are extended along the short side of the magnetic element (perpendicular to the applied field direction). This feature explains the experimental observation that the corresponding spectra in the q-space show a

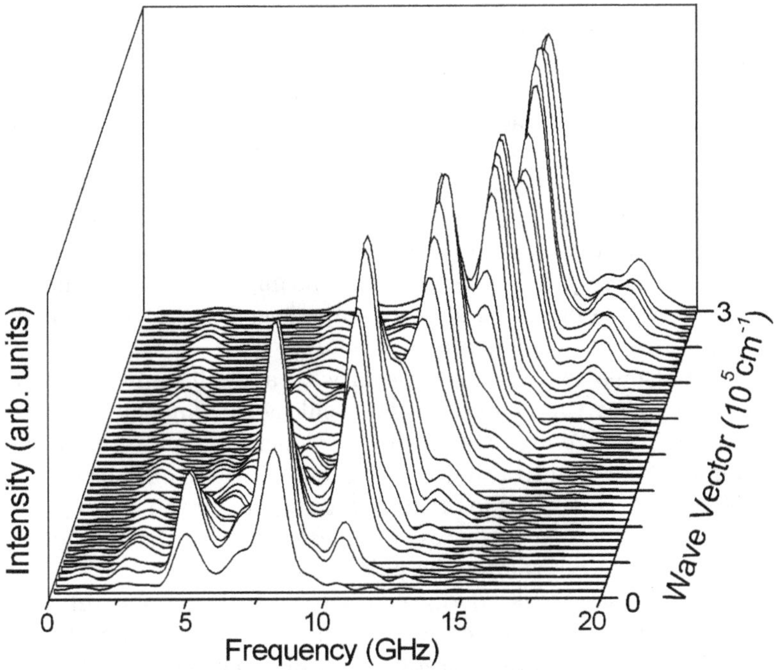

Fig. 25. Power spectra of the m_x-oscillations for a sequence of inplane wavevectors $q_\perp$ perpendicular to the field direction (and to the long side of the Py element) obtained using Langevin dynamics simulations [46]

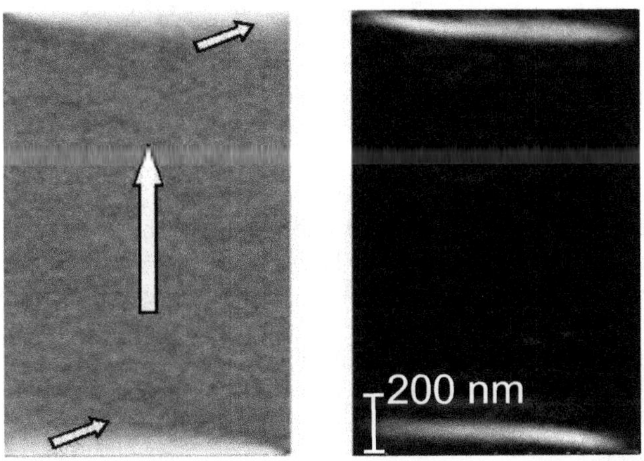

Fig. 26. *Left panel:* Distribution of static magnetization used for Langevin dynamics (S-state). *Right panel:* Spatial distribution of the lowest exchange-localized mode having a frequency of 4.94 GHz

q-dependence with a pronounced peak along the q_y-direction but extend over the whole measured q_z-region.

For the simulation with the object-oriented micromagnetic framework (OOMMF) [21] the rectangle was discretized along the y- and z-directions with a cell size of 5 nm × 5 nm × 33 nm. The initial condition for the dynamic simulation was obtained by relaxing from a state of homogeneous magnetization $M_S \mathbf{e}_x$ with a large Gilbert damping parameter $\lambda = 0.5$. Contrary to MicroMagus, the flower state was found to be stable in OOMMF and so it was used as an initial condition. The reasons for this might be 1. that simulations with OOMMF were performed at zero temperature and 2. the element was not discretized in sublayers when using OOMMF.

The dynamical calculation was undertaken with a small damping parameter of $\lambda = 0.01$ so that the magnetization oscillations could be observed over several periods. The eigenfrequencies did not depend significantly on the damping. A uniform Gaussian field pulse with a full-width at half-maximum (FWHM) of 150 ps and an amplitude of 10 Oe was applied in the x-direction, and the response over the next 10 ns was calculated using 10 ps time steps during the pulse and 20 ps steps after the pulse.

A local Fourier transform was performed with these time-domain data and power-spectral images as a function of frequency were computed. All images are normalized by the total spectral weight at each position. A more detailed discussion of this procedure can be found in [12, 36]. Adding all the local Fourier transform data one obtains a so-called integrated Fourier transform. There, several distinct peaks corresponding to spin-wave modes can be observed.

The left panel in Fig. 27 shows some power spectral images for several frequencies corresponding to maxima in the integrated Fourier transform in comparison with the results of the analytical theory. Note that this approach accounts only for modes whose mean value of the dynamical magnetization is nonzero, i.e., modes with an even number of nodes ($m = 1, 3, 5, \dots$), since the precession was excited by an uniform pulse field. The small wavelength noise is due to the fact that the cell size is of the order of the exchange correlation length and not significantly below. The mode ($m = 1, n' = 1$) in Fig. 27 corresponds rather to a band of nonresolved modes with a broad frequency peak, while the modes with higher $m > 1$ are well defined in frequency. The values for the frequencies are compared with the other methods in Table 1. One can see that the results obtained with OOMMF agree well with the results of the other methods (spatial mode distribution in Fig. 27 and frequencies in Table 1).

6 Outlook

In this Chapter we have discussed dynamic magnetic excitations in homogeneously magnetized rectangular elements. For simplicity we have assumed

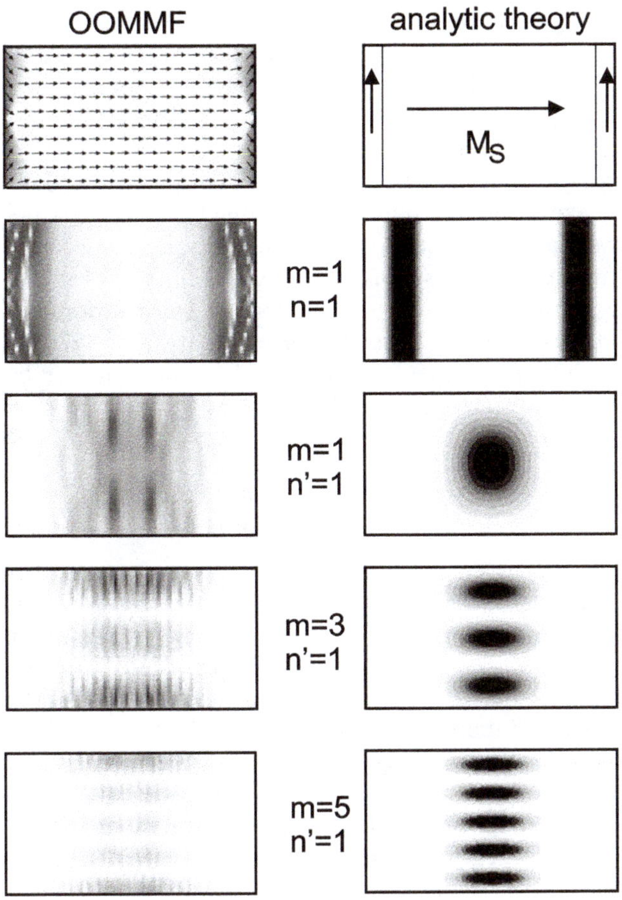

Fig. 27. Comparison of OOMMF with analytical theory for several modes. The upper graphs show the static equilibrium magnetization used in the two models. The corresponding effective field is shown in Fig. 13

Table 1. Comparison of frequencies (in gigahertz) of eigenmodes in a $1 \times 1.75\,\mu m^2$ rectangle obtained from experiment and different theoretical techniques

Experiment	Mode index	Analytic theory	Langevin dynamics	OOMMF
5.2– 5.6	$(m = 1, n = 1)$	4.9	4.94	4.3
7.1– 8.2	$(m = 1, n' = 1)$	8.6	7.98– 8.11	8.2
10.0–10.4	$(m = 2, n' = 1)$	10.6	10.52–10.65	
12.0–12.7	$(m = 3, n' = 1)$	12.1	12.55–12.67	11.7
13.2–13.9	$(m = 4, n' = 1)$	13.1	not observed	
14.5–14.8	$(m = 5, n' = 1)$	14.0	13.94	13.6
	$(m = 6, n' = 1)$	14.7	14.57–14.70	

single-layer films and no magnetic anisotropy apart from shape anisotropy. Several initiatives have recently started to address more complex scenarios and have produced impressive results so far: the influence of confinement on spin waves in layered stripes comprised by two magnetic metallic layers separated by a nonmagnetic layer under the dipolar as well as the exchange coupling between the magnetic layers has been studied in detail [51–54]. Furthermore, collective spin-wave modes in an array of parallel Permalloy stripes were also studied [55] as well as the spectrum of inhomogeneously magnetized single stripes [36]. Substantial effort has also been made in the investigation of the excitation spectrum of domain structures like closure domains in squares [56] or magnetic rings with various spin configurations [57–59]. The spectrum of a magnetic vortex [60–65] is discussed in the Chapter by *Back* et al. of this book.

7 Conclusion

Summarizing, we have presented a detailed description of the spin-wave eigenmodes in homogeneously magnetized magnetic rectangles. We showed that lateral confinement effects drastically affect the properties of the magnetic eigenmodes. Plane spin waves, which are eigenmodes of unconfined films are substituted by quantized and localized spin-wave modes, the properties of the latter are mainly determined by inhomogeneous static and dynamic fields.

For longitudinally magnetized stripes the quantization effect determined by the dipolar pinning is theoretically discussed and experimentally proven. For a transversely magnetized stripe we further developed the model to describe exchange and dipolar localization of spin-wave modes in an inhomogeneous field. For rectangles, both cases are combined to evaluate the mode spectrum of a rectangle.

References

[1] B. Hillebrands, K. Ounadjela (Eds.): *Spin Dynamics in Confined Magnetic Structures*, vol. I (Springer, Berlin, Heidelberg 2002)

[2] M. L. Plumer, J. van Ek, D. Weller (Eds.): *The Physics of Ultra-High-Density Magnetic Recording* (Springer, Berlin, Heidelberg 2001)

[3] B. Hillebrands, K. Ounadjela (Eds.): *Spin Dynamics in Confined Magnetic Structures*, vol. II (Springer, Berlin, Heidelberg 2003)

[4] B. A. Gurney, P. Baumgart, V. S. Speriosu, R. E. Fontana, A. Patlac, A. Logan, P. Humbert: in *Thin Films and Surfaces, Glasgow, Proc. of the 13th Int'l Colloq. on Mag.*, vol. 474 (1991) pp. 7–12

[5] C. Mathieu, C. Hartmann, M. Bauer, S. R. O. Büttner, B. Roos, S. O. Demokritov, B. Hillebrands, B. Bartenlian, C. Chappert, D. Decanini, F. Rousseaux, E. Cambril, B. H. A. Müller, U. Hartmann: Appl. Phys. Lett. **70**, 2912 (1997)

[6] C. Mathieu, J. Jorzick, A. Frank, S. O. Demokritov, B. Hillebrands, A. N. Slavin, B. Bartenlian, C. Chappert, D. Decanini, F. Rousseaux, E. Cambril: Phys. Rev. Lett. **81**, 3968 (1998)
[7] J. Jorzick, S. O. Demokritov, C. Mathieu, B. Hillebrands, B. Bartenlian, C. Chappert, F. Rousseaux, A. N. Slavin: Phys. Rev. B **60**, 15194 (1999)
[8] Y. Roussigné, S. M. Chérif, C. Dugautier, P. Moch: Phys. Rev. B **63**, 134429 (2001)
[9] Z. K. Wang, M. H. Kuok, S. C. Ng, D. J. Lockwood, M. G. Cottam, K. Nielsch, R. B. Wehrspohn, U. Gösele: Phys. Rev. Lett. **89**, 027201 (2002)
[10] J. Jorzick, S. O. Demokritov, D. B. B. Hillebrands, N. L. Gorn, K. Guslienko, A. N. Slavin: Phys. Rev. Lett. **88**, 047204 (2002)
[11] C. Bayer, S. O. Demokritov, B. Hillebrands, A. N. Slavin: Appl. Phys. Lett. **82**, 607 (2003)
[12] J. P. Park, P. Eames, D. M. Engebretson, J. Berezovsky, P. A. Crowell: Phys. Rev. Lett. **89**, 277201 (2002)
[13] S. O. Demokritov, B. Hillebrands: Spinwaves in laterally confined magnetic structures, in B. Hillebrands, K. Ounadjela (Eds.): *Spin Dynamics in Confined Magnetic Structures*, vol. I (Springer, Berlin, Heidelberg 2002) pp. 65–92
[14] N. Smith, P. Arnett: Appl. Phys. Lett. **78**, 1448 (2001)
[15] J.-G. Zhu: J. Appl. Phys. **91**, 7273 (2002)
[16] Y. Zhou, A. Roesler, J.-G. Zhu: J. Appl. Phys. **91**, 7276 (2002)
[17] C. Kittel: Phys. Rev. **73**, 2 (1948)
[18] L. R. Walker: Phys. Rev. **105**, 390 (1957)
[19] R. I. Joseph, E. Schlömann: J. Appl. Phys. **36**, 1579 (1965)
[20] K. Y. Guslienko, S. O. Demokritov, B. Hillebrands, A. N. Slavin: Phys. Rev. B **66**, 132402 (2002)
[21] M. J. Donahue, D. G. Porter: *OOMMF User's Guide, Version 1.0, in Interagency Report No. NISTIR 6376*, National Institute of Standards and Technology, Gaithersburg, Maryland (1999)
[22] B. Heinrich, R. Urban, G. Woltersdorf: J. Appl. Phys. **91**, 7523 (2002)
[23] W. K. Hiebert, A. Stankiewicz, M. R. Freeman: Phys. Rev. Lett. **79**, 1134 (1997)
[24] A. Barman, V. V. Kruglyak, R. J. Hicken, J. M. Rowe, A. Kundrotaite, J. Scott, M. Rahman: Phys. Rev. B **69**, 174426 (2004)
[25] Y. Acremann, C. H. Back, M. Buess, A. V. O. Portmann, D. Pescia, H. Melchior: Science **290**, 492 (2000)
[26] B. Hillebrands: Rev. Sci. Instrum. **70**, 1589 (1999)
[27] R. W. Damon, J. R. Eshbach: J. Phys. Chem. Solids **19**, 308 (1961)
[28] B. A. Kalinikos, A. N. Slavin: J. Phys. C **19**, 7013 (1986)
[29] G. T. Rado, J. R. Weertman: J. Phys. Chem. Solids **11**, 315 (1959)
[30] S. O. Demokritov: J. Phys.: Condens. Matter **15**, 2575 (2003)
[31] P. Bryant, H. Suhl: Appl. Phys. Lett. **54**, 2224 (1989)
[32] M. J. Donahue, D. G. Porter, R. D. McMichael, J. Eicke: J. Appl. Phys. **87**, 5520 (2000)
[33] E. Schlömann: J. Appl. Phys. **159**, 35 (1964)
[34] J. Jorzick, C. Krämer, S. O. Demokritov, B. Hillebrands, B. Bartenlian, C. Chappert, D. Decanini, F. Rousseaux, E. Cambril, E. Søndergard, M. Bailleul, C. Fermon, A. N. Slavin: J. Appl. Phys. **89**, 7091 (2001)

[35] G. Gubbiotti, M. Conti, G. Carlotti, E. P. Candeloro, Di Fabrizio, K. Y. Guslienko, A. Andre, A. N. Slavin, C. Bayer: J. Phys.: Condens. Matter **16**, 7709–7721 (2004)
[36] C. Bayer, J. P. Park, H. Wang, M. Yan, C. E. Campbell, P. A. Crowel: Phys. Rev. B. **69**, 134401 (2004)
[37] P. H. Bryant, J. F. Smyth, S. Schultz, D. R. Fredkin: Phys. Rev. B **47**, 11255 (1993)
[38] K. Y. Guslienko, R. W. Chantrell, A. N. Slavin: Phys. Rev. B **68**, 024422 (2003)
[39] S. O. Demokritov, A. A. Serga, A. Andre, V. E. Demidov, M. P. Kostylev, B. Hillebrands, A. N. Slavin: Phys. Rev. Lett. **93**, 047201 (2004)
[40] M. Bailleul, D. Olligs, C. Fermon: Phys. Rev. Lett. **91**, 137204 (2003)
[41] Y. Roussigné, S. M. Chérif, P. Moch: J. Magn. Magn. Mater. **263**, 289 (2003)
[42] Y. Roussigné, S. M. Chérif, P. Moch: J. Magn. Magn. Mater. **268**, 89 (2004)
[43] M. Grimsditch, G. K. Leaf, H. G. Kaper, D. A. Kapreev, R. E. Camley: Phys. Rev. B **69**, 174428 (2004)
[44] M. Grimsditch, L. Giovannini, F. Montoncello, F. Nizzoli, G. K. Leaf, H. G. Kaper: Phys. Rev. B **70**, 054409 (2004)
[45] L. Giovannini, F. Montoncello, F. Nizzoli, G. Gubbiotti, G. Carlotti, T. Okuno, T. Shinjo, M. Grimsditch: Phys. Rev. B **70**, 172404 (2004)
[46] D. V. Berkov, N. L. Gorn:
URL `http://www.micromagus.de`
[47] W. F. Brown, Jr.: Phys. Rev. **130**, 1677 (1963)
[48] H. B. Braun: Stochastic magnetization dynamics in magnetic nano-structures: from Neel–Brown to soliton-antisoliton creation, in P. Entel, D. Wolf (Eds.): *Structure and Dynamics of Heterogeneous Systems* (World Scientific, Singapur 2000)
[49] D. V. Berkov, N. L. Gorn: J. Magn. Magn. Mater. **272–276P1**, 687–689 (2004)
[50] D. V. Berkov, N. L. Gorn: Stochastic dynamic simulations of fast remagnetization processes: recent advances and applications, J. Magn. Magn. Mater. **290–291P1**, 442 (2005)
[51] G. Gubbiotti, M. Kostylev, N. Sergeeva, M. Conti, G. Carlotti, T. Ono, A. Stashkevich: Brillouin light scattering investigation of magnetostatic modes in symmetric and asymmetric NiFe/Cu/NiFe trilayered wires, Phys. Rev. B **70**, 224422 (2004)
[52] N. A. Sergeeva, S. M. Cherif, M. P. K. A. A. Stachkevitch, J. B. Youssef: Spin waves quantization in patterned exchange-coupled double-layers, J. Magn. Magn. Mater. **288C**, 250–258 (2005)
[53] M. P. Kostylev, A. A. Stashkevich, Y. R. N. A. Sergeeva: Spin wave modes localized on a nano-stripe with two dipole coupled layers, J. Magn. Magn. Mater. **278**, 397–406 (2004)
[54] N. A. Sergeeva, S.-M. Chérif, M. K. A. Stachkevitch, Y. Roussigné: Spin-waves in ferromagnetic double layers: effect of a lateral patterning, phys. stat. sol. (C) **1**, 1587–1590 (2004)
[55] M. P. Kostylev, A. A. Stashkevich, N. A. Sergeeva: Collective magnetostatic modes on a one-dimensional array of ferromagnetic stripes, Phys. Rev. B **69**, 064408 (2004)
[56] K. Perzlmaier, M. Buess, C. H. Back, V. E. Demidov, B. Hillebrands, S. O. Demokritov: Phys. Rev. Lett. **94**, 057202 (2005)

[57] F. Giesen, J. Podbielski, T. Korn, D. Grundler: J. Appl. Phys. **97**, 10A712 (2005)
[58] F. Giesen, J. Podbielski, T. Korn, M. Steiner, A. van Staa, D. Grundler: Appl. Phys. Lett. **86**, 112510 (2005)
[59] J. Podbielski, F. Giesen, M. Berginski, N. Hoyer, D. Grundler: Superlatt. Microstruct. **37**, 341–348 (2005)
[60] J. P. Park, P. Eames, D. M. Engebretson, J. Berezovsky, P. A. Crowell: Phys. Rev. B **67**, 020403(R) (2003)
[61] M. Buess, R. Höllinger, T. Haug, K. Perzlmaier, U. Krey, D. Pescia, D. W. M. R. Scheinfein, C. H. Back: Phys. Rev. Lett. **93**, 077207 (2004)
[62] M. Buess, T. P. J. Knowles, R. Höllinger, T. Haug, U. Krey, D. Weiss, D. Pescia, M. R. Scheinfein, C. H. Back: Phys. Rev. B **71**, 104415 (2005)
[63] K. Y. Guslienko, W. Scholz, R. W. Chantrell, V. Novosad: Phys. Rev. B **71**, 144407 (2005)
[64] B. A. Ivanov, C. E. Zaspel: Phys. Rev. Lett. **94**, 027205 (2005)
[65] R. Zivieri, F. Nizzoli: Phys. Rev. B **71**, 014411 (2005)

Index

Ferromagnetic Resonance Force Microscopy

Philip E. Wigen[1], Michael L. Roukes[2], and Peter C. Hammel[1]

[1] Ohio State University, 191 West Woodruff Ave., Columbus, OH 43210-1117, USA
wigen,hammel@mps.ohio-state.edu
[2] California Insitute of Technology, Pasadena, CA 91125, USA
roukes@caltech.edu

Abstract. The magnetic resonance force microscope (MRFM) is a novel scanning probe instrument that combines the three-dimensional imaging capabilities of magnetic resonance imaging (MRI) with the high sensitivity and resolution of atomic force microscopy (AFM). In the nuclear magnetic resonance (NMR) mode or the electron spin resonance (ESR) mode it will enable nondestructive, chemical-specific, high-resolution microscopic studies and imaging of subsurface properties of a broad range of materials. In its most successful application to date, MRFM has been used to study microscopic ferromagnets. In ferromagnets the long-range spin–spin couplings preclude localized excitation of individual spins. Rather, the excitations employed in ferromagnetic resonance (FMR) are the normal magnetostatic wave (or spin-wave) modes determined by the geometry of the sample. In this case the response of the cantilever will be a measure of the amplitude of the FMR signal integrated over the volume where the magnetic field gradient of the tip magnet is significant. Thus, as the magnetic tip is scanned across the material under study, the signal intensity will be proportional to the local amplitude of the normal modes. In addition, the MRFM technique has proven useful for the observation of relaxation processes in microscopic samples. The MRFM will also enable the microscopic investigations of the nonequilibrium spin polarization resulting from spin injection. Microscopic MRFM studies will provide unprecedented insight into the physics of magnetic and spin-based materials at micrometer and submicrometer dimensions.

1 Introduction

Recent years have seen an extraordinary increase in the density of magnetic storage media and access speeds of read/write heads as well as great interest in the field of spin electronics. With such devices having dimensions at the nanometer scale, there is a need for the development of new techniques to characterize these materials. Two such examples include the near-field microwave microscopes [1] and the ferromagnetic resonance microscope [2]. The near-field microwave microscope uses a loop probe to measure local magnetic properties in ferromagnetic samples measures local magnetic properties on the length scale of 200 μm. The ferromagnetic resonance microscope uses a small hole, ≈ 1 mm diameter, in a thinned side wall of a microwave cavity to couple the microwaves to a sample. By moving the sample with respect to the opening a resolution on the order of the hole diameter is obtained. Scanning

B. Hillebrands, A. Thiaville (Eds.): Spin Dynamics in Confined Magnetic Structures III, Topics Appl. Physics **101**, 105–136 (2006)

optical microscopes have recently been developed to observe phenomena at submicrometer dimensions. Using "solid immersion lenses" and stroboscopic techniques, the measurement of the magnetization of a thin-film recording head was measured obtaining a resolution on the order of 200 nm [3], In another experiment using microfocus Brillouin light scattering the magnetic dynamics of magnetic elements that are submicrometer in size were investigated [4, 5].

The magnetic resonance force microscope (MRFM) provides another technique that has the sensitivity to detect a single electronic spin [6] and thus to obtain atomic resolution [7, 8]. MRFM achieves such resolution by sensitively detecting the force between a small probe magnet mounted on a compliant cantilever and the magnetic moment of spins in the sample that are driven to vary their z-component at the cantilever resonance frequency [9].

High-resolution imaging techniques such as atomic force microscopy (AFM) and magnetic resonance imaging (MRI) have had a substantial impact in the fields of electonic and structual materials and medical science. AFM provides atomic-scale resolution but is essentially limited to surface studies. MRI is a fully three-dimensional, noninvasive imaging technology that employs an applied field gradient, $\nabla \boldsymbol{B}$, to distinguish magnetic resonance signals arising from different locations in the sample. The high sensitivity of MRFM offers the possibility of shrinking the linear dimension of resolved volumes into the submicrometer regime with the potential of achieving spatial resolution comparable to that obtained from scanning tunneling and atomic force microscopy, while obtaining detailed information such as that obtained from optical spectroscopies [10, 11]. MRFM offers several unique advantages over other scanned probes including:

- A 3-D imaging field with the extent of scanning below the surface being determined by the spatial dependence of the field gradient.
- Because each nucleus has a unique gyromagnetic ratio, NMR imaging is chemical-species specific.
- The well-developed and validated theory of magnetic resonance interactions provides a reliable basis for the design and operation of imaging instruments.

The MRFM is proving to be a versatile instrument that has been demonstrated in a variety of magnetic resonance experiments:

- Electron spin resonance [8, 12];
- Nuclear spin resonance [13, 14];
- Ferromagnetic resonance [15].

A schematic diagram of an MRFM apparatus is shown in Fig. 1. The time dependence of $\boldsymbol{m}(\boldsymbol{r}, t)$ is driven by modulating the bias magnetic field $\boldsymbol{B}_0$ at some amplitude B_{M} and at some frequency f_{M} and simultaneously modulating the amplitude of the RF field B_1 at some frequency f_1 such that the beat frequency, $f_\mathrm{b} = f_\mathrm{M} - f_1$, produces a time variation in the transfers component of the magnetization $\boldsymbol{m}(\boldsymbol{r}, t)$ at the resonance frequency of the mechanical resonator, $f_\mathrm{c} = f_\mathrm{b}$. This produces a small variation in the z-component of the magnetization at the resonance frequency of the mechanical resonator [16, 17]. This time-dependent variation in $M_\mathbf{z}$ is coupled to a mechanical resonator via the gradient in the magnetic field due to a small magnet on its tip. The oscillation amplitude of the mechanical resonator is detected by an optical-fiber interferometer. By scanning the cantilever across the surface of a sample, a spatially resolved evaluation of the amplitude of $\boldsymbol{m}(\boldsymbol{r}, t)$ can be obtained.

The RF microstrip excites the spin resonance in the sample, while the probe magnet mounted on the tip of a compliant micromechanical resonantor produces a bowl-shaped region in the sample ("resonant slice") in which the resonant conditions for the spins in the sample are satisfied (Fig. 1). The field gradient produced by the tip magnet provides the conditions for establishing a force on the compliant cantilever necessary for the detection of the resonance. A time variation of the spin moments in the sample, $\boldsymbol{m}(\boldsymbol{r}, t)$, will couple to the probe magnet mounted on a resonantor, producing a force that is given by the relation

$$\boldsymbol{F}(\boldsymbol{r}, t) = -[\boldsymbol{m}(\boldsymbol{r}, t) \cdot \nabla]\boldsymbol{B}_\mathrm{probe}(\boldsymbol{r}) . \tag{1}$$

2 MRFM Detection of Weakly Interacting Spins (ESR and NMR)

The MRFM technique relies on the coupling between a time-dependent resonating moment $\boldsymbol{m}(\boldsymbol{r}, t)$ and a probe magnet mounted on a compliant micromechanical resonator via the force interaction given in (1) with the strength of the interaction being proportional to the gradient of the inhomogeneous magnetic field of the probe magnet. This force is measured through the detection of the displacement of the resonator that is deflected by the force $\boldsymbol{F}(\boldsymbol{r}, t)$, or by the change in the resonant frequency of the cantilever.

The sensitivity of force detection is ultimately limited by the thermomechanical noise F_n of the detector

$$F_\mathrm{n} = \sqrt{(2kk_\mathrm{B}T\Delta\nu/\pi Q f_\mathrm{c})} . \tag{2}$$

This noise depends on the temperature T (k_B is Boltzmann's constant) and the detection bandwidth $\Delta\nu$ as well as the mechanical characteristics of the resonantor, such as its spring constant k, resonance frequency f_c, and quality factor Q [9, 12, 19].

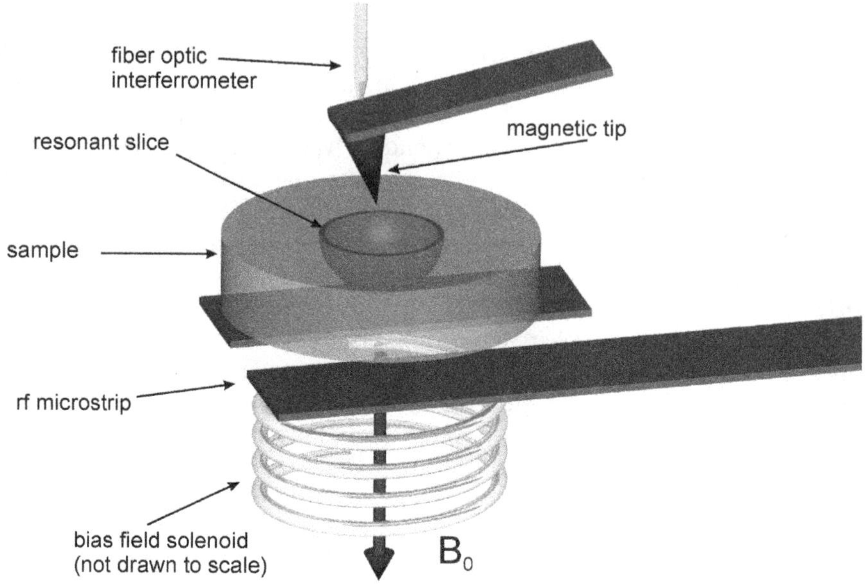

Fig. 1. A shematic view of the MRFM apparatus. In addition to the external field $\boldsymbol{B}_0$ the sample is also exposed to an alternating field applied by the field solenoid, the field gradient of the magnetic tip on the mechanical resonator and the RF field produced by the RF microstrip. The resonant slice is the region in which the bias field plus the gradient field are sufficient to satisfy the resonant conditions for spins in that region of the sample [18]

A second key function of the magnetic field gradient is the definition of the volume of the spin magnetization that will be coupled to the force detector. The electron spin precesses at the Larmor frequency f_L

$$f_\mathrm{L} = g\mu_\mathrm{B} B_\mathrm{Tot} \,, \tag{3}$$

where g is the electron **g**-factor of the given paramagnetic species in the host material and B_Tot is the total magnetic field at the site of the resonating species and is the sum of the applied magnetic field $\boldsymbol{B}_0$ and the magnetic field of the tip, $\boldsymbol{B}_\mathrm{probe}(\boldsymbol{r})$,

$$\boldsymbol{B} = \boldsymbol{B}_0 + \boldsymbol{B}_\mathrm{probe}(\boldsymbol{r}) . \tag{4}$$

If the applied field, $\boldsymbol{B}_0$, is set just below the value sufficient to establish resonance of the paramagnetic spins in the sample, then spins that are too close to the micromagnetic probe will have a resonance frequency that is too high to couple to the RF field, b_RF, and will not resonate. Similarly, those too far from the probe magnet will have a resonance frequency that is too low to couple to the RF field. The region in the sample in which the resonant frequency of the paramagnetic spins satisfy the resonant condition determined by the frequency of the RF field, f_{RF}, defines a bowl-shaped "resonance slice" as shown in Fig. 1. The width of the sensitive slice will be determined by the magnitude of the field gradient, $|\nabla_z \boldsymbol{B}|$ established by the probe magnet and the resonance linewidth ΔH_{lw}. The width of the resonant slice, z_{sl}, is given by

$$z_{\mathrm{sl}} \approx \Delta H_{\mathrm{lw}} / |\nabla_z \boldsymbol{B}| \,. \tag{5}$$

This resonance slice can be scanned in the z-direction by changing the value of the applied field, $\boldsymbol{B}_0$, and in the xy-plane by scanning the position of the cantilever over the sample. Field gradients sufficiently large to obtain angstrom-scale resolution can be obtained using probe magnets with submicrometer tip radii. Reliable interpretation of MRFM signals requires a thorough and detailed understanding of the interaction between the micromagnetic probe and the sample [18, 20]. The deconvolution of the signal to obtain sample images is more complicated for the bowl-shaped resonance slice but imaging has been demonstrated [21] and this is a topic of active research [22–24]. Further details of the ESR and NMR application are reviewed in the literature [18].

3 MRFM in Ferromagnetic Systems

Ferromagnetic resonance force microscopy (FMRFM) is a variation of MRFM that enables the characterization of the dynamic magnetic properties of magnetic structures at the micrometer scale. Ferromagnetically coupled systems pose unique challenges for magnetic resonance imaging due to the strong exchange coupling of the spins. The resulting magnetostatic/exchange resonance modes involve spins occupying the entire sample. The detector then monitors the amplitude of the oscillating magnetization within the range of the resonance slice of the detector probe and enables the characterization of the ferromagnetic resonance in three regimes defined by the degree to which the field of the micromagnetic probe perturbs the resonance modes [25, 26]:

- If the detector is scanned sufficiently far above the sample, the perturbation field of the tip magnet B_{tip} is small, the amplitude of the various excited intrinsic resonance modes can be spatially resolved.
- At intermediate heights, the magnetic field of the probe magnet is sufficiently strong to alter the spatial "shape" of the resonance mode near the region of the probe. These perturbations will alter the detected *amplitudes* of the modes [27] and when scanned across the sample, this perturbation breaks the symmetry of the magnetic field within the sample and the normally "hidden" modes having odd symmetry in the unperturbed case will have a net dipole moment and will be observed.
- As the probe magnet is moved very close to the sample surface the strong perturbations in the local magnetic field results in those modes having a half-wavelength approximately the size of the "resonant slice" becoming strongly excited [26].

3.1 Magnetostatic Modes

The ability of FMRFM to detect resonance in thin magnetic yttrium iron garnet (YIG) films was first reported by *Zhang* et al. [15]. While this initial report clearly showed a spectrum of magnetostatic modes, the irregular shape of the sample made it impossible to consider the details of the modes. Since that time a number of additional groups have observed the ferromagnetic resonance spectra in a variety of magnetic materials having well-structured and characterized geometries. *Midzor* et al. [25] investigated the ability of the spectrometer to scan the resonant modes in a series of well-structured microscopic YIG films having rectangular shapes. *Klein* et al. have investigated the magnetostatic mode spectra and relaxation processes in a 160 μm diameter YIG disc [28–33]. *Rugar* and coworkers [34] have reported the observation of magnetostatic modes in YIG discs. *Zhang* et al. [19] have also reported the observation of the force detected resonance in thin cobalt microscopic films.

The observations by the group of *Klein* in thin YIG discs reported the resonances ascribed to magnetostatic modes having axial symmetry across the sample. The two-dimensional Bessel-function-like modes are labeled by (n, m), the number of nodes, respectively, in the radial and circumferential directions. The various modes resonate at different fields as a consequence of the dependence of their excitation energies on the dipolar interactions between spins in the cylinder, and hence on its aspect ratio [35].

Figure 2b shows a sample fabricated by ion milling from a single-crystal YIG film having a thickness of 4.75 μm with the [111] direction, the easy axis, oriented normal to the film plane. The disk has a radius $R = 80\,\mu\mathrm{m}$. The dimensions are large enough so that standard FMR experiments can be carried out on the sample. Figure 2c shows the microwave susceptibility of the disk as a function of the DC magnetic field applied parallel to the disk axis (perpendicular resonance). The absorption spectrum was measured at 10.46 GHz. Four magnetostatic modes are resolved corresponding to the longest-wavelength magnetostatic modes.

The same disk measured by standard FMR methods was then tested by the mechanical force detection. The sample temperature was fixed at $T = 285\,\mathrm{K}$, where the saturation magnetization is $4\pi\mathbf{M}_s(T) = 1815\,\mathrm{G}$. A probe magnet (shown in Fig. 2a) 18 μm in diameter and 40 μm in length is glued to the end of a cantilever having a spring constant $k = 0.5\,\mathrm{N/m}$ in a bias field of 5.3 kG.

The probe magnet is set at a distance of 110 μm above the YIG sample. This large separation is required so that the magnetic field of the probe magnet at the sample is sufficiently weak so as not to perturb the "shape" of the magnetostatic modes. At this height the magnetic field gradient produced on the sample is less than 0.16 G/μm. The MRFM is then used not for sensitive detection of FMR, but to allow measurement of the longitudinal sample magnetization that can be compared with data obtained by traditional FMR techniques. The MRFM signal is proportional to variations in

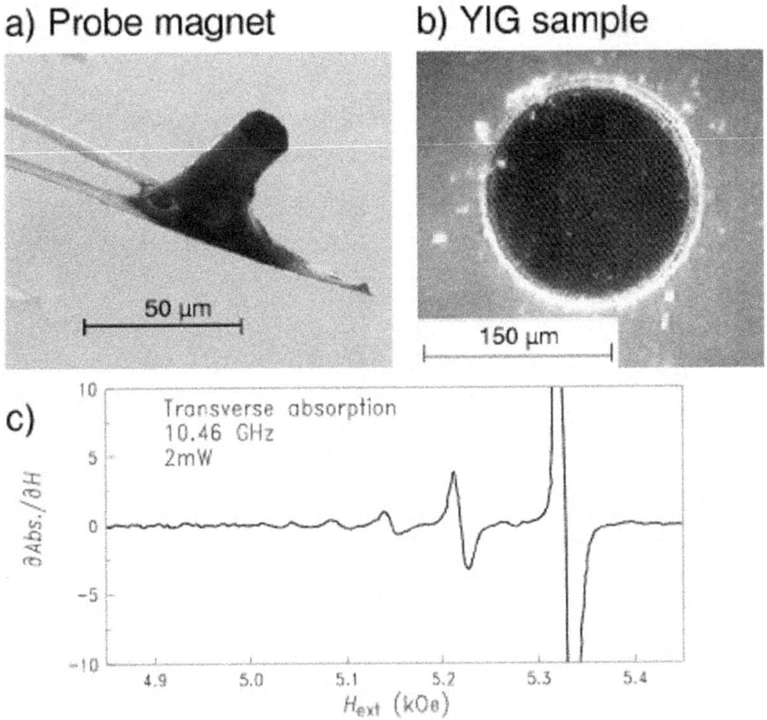

Fig. 2. Images of both the cylindrical probe magnet (**a**) and the YIG disk (**b**). (**c**) The derivative of the imaginary part of the microwave susceptibility of the disk obtained from a microwave cavity [28]

the longitudinal magnetization ΔM_z and thus it increases linearly with microwave power ($\propto b_{\mathrm{RF}}^2$) below saturation where the transverse component of the magnetization $\boldsymbol{m}(\boldsymbol{r},t)$ is proportional to b_{RF}.

Figure 3 shows the field dependence of the FMRFM signal when the probe magnet is placed on the symmetry axis of the disk and the amplitude of b_{RF} is fully modulated at the resonance frequency of the cantilever $f = 2.8\,\mathrm{kHz}$. The microwave peak power is increased gradually during the sweep, from $25\,\mu\mathrm{W}$ for the longest wavelength modes up to $2.5\,\mathrm{mW}$ at $B_0 = 4.7\,\mathrm{kOe}$. The normalized result is shown on a logarithmic scale. A series of 100 absorption peaks is resolved, demonstrating the sensitivity of mechanical detection.

A fit of the magnetostatic modes to the dispersion relation is shown in Fig. 4. Although exact modeling can be carried out numerically, an analytical expression that assumes that the disk is uniformly magnetized has been used. In practice, the rotation of the magnetization at the outer edge of the sample results in an effective decrease of the radial wavevector k, which is equivalent to an increase of the effective disc radius. To fit the data with a uniformly magnetized disk model at $n = 30$, a radius of $85\,\mu\mathrm{m}$ (compared to the actual

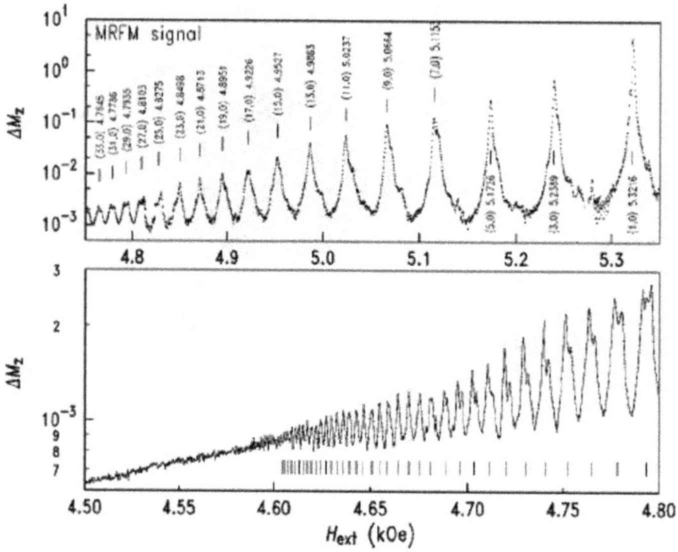

Fig. 3. Mechanically detected FMR spectrum of the normally magnetized YIG disk. The signal is proportional to the changes of the longitudinal component of the magnetization ΔM_z. The absence of even absorption peaks, $(2n, 0)$, is a signature that the probe is placed precisely on the symmetry axis of the disk [28]

radius of 80 µm) had to be assumed. In addition, for $n > 50$, a better fit was obtained by including exchange effects into the dispersion relation. The FMRFM data agree quantitatively with the model for the entire range of observed modes.

3.2 Linewidths

In addition to determining the internal fields of ferromagnetic materials, another important application of FMR is to understand magnetization dynamics; measurement of relaxation of magnetization provides insight into the dissipation of magnetic energy. A recent report [31] describes the application of MRFM to the measurement of the transverse relaxation time T_2 and the longitudinal relaxation time T_1 in micrometer-size ferromagnetic films.

By measuring the linewidth of the ferromagnetic system at different RF-frequencies three separate contributions to the linewidth can be extracted from the data:

- A radiation damping term ΔH_{rd} of 0.62 G due to relaxation of the magnetization through coupling to the microstrip resonator.
- A second contribution ΔH_{lin} with a linear dependence on frequency having a slope of 0.043 G/GHz.

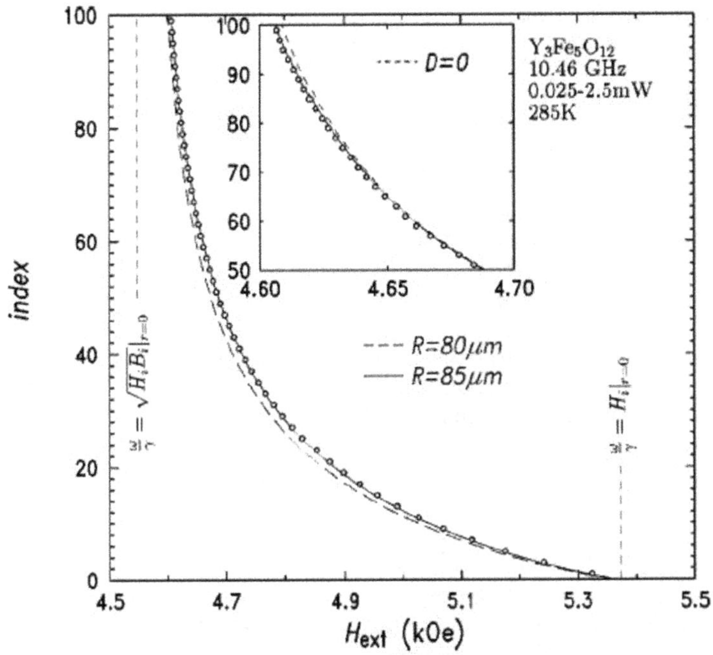

Fig. 4. Mode number n as a function of the external field, where n is the number of standing waves in the radial direction. The *open circles* are the field position of each absorption peak measured in Fig. 3. The *solid line* is the theoretical prediction for a uniformly magnetized disk of radius 85 mm. The *long dashed* line is the same calculation for $R = 80$ mm (approximately the physical dimension of the sample). The *short dashed line* in the *inset* shows the behavior when exchange effects are omitted ($D = 0$) [28]

– A frequency-independent term $\Delta H_{\rm cst} = 0.50$ G, which is due to inhomogeneous broadening and to scattering inside the magnon manifold.

The results are shown in Fig. 5. These measurements are consistent with the early reports of *LeCraw* and *Spencer* [36] .

The contribution of the homogeneous and inhomogeneous broadening was obtained by performing a series of experiments in which the amplitude of the longitudinal and transverse components of the magnetization are independently observed for various modulation frequencies. As the resonant frequency of the cantilever $\omega_{\rm c}$ is fixed at 3 kHz, the broadband modulation experiments were performed by using anharmonic techniques [16, 17]. The amplitude of the RF field was fully modulated at a frequency $\omega_{\rm s}$, while the frequency output of the generator was modulated at a frequency $\omega_{\rm f}$ such that $\omega_{\rm f} = \omega_{\rm s} + \omega_{\rm c}$. Figure 6 shows the decrease of $\langle M_z \rangle$ and $\langle m_{\rm tr} \rangle$ with increasing modulation frequency $\omega_{\rm f}$. From this data the homogeneous

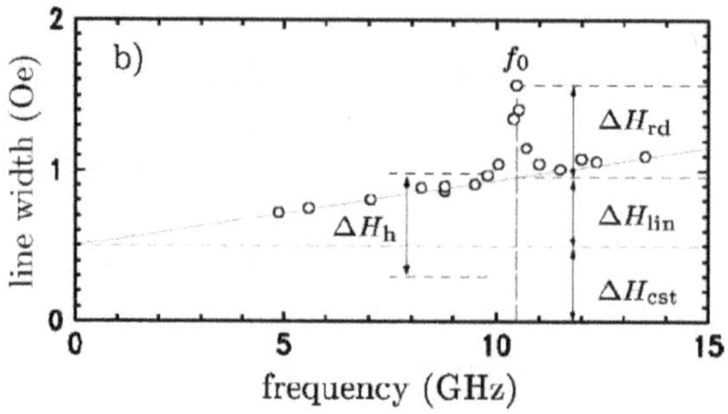

Fig. 5. The frequency dependence of the linewidth measured mechanically. The magnetic field strength is 5324.5 Oe. Contributions to the linewidth are separated into linear, ΔH_{lin}, and frequency-independent, ΔH_{cst}, relaxation channels. Homogeneous broadening, ΔH_h and radiation damping effects, ΔH_{rd}, are indicated by *arrows* [31]

broadening of $\Delta H_h = 0.70 \pm 0.05$ G and a longitudinal relaxation time of $T_2 = 2/(\gamma \Delta H_h) = 162 \pm 10$ ns was determined. From the measurement of the power dependence of the longitudinal magnetization a value of $T_1 = 106 \pm 10$ ns is obtained for the transverse relaxation time.

A consistency check was obtained by repeating the measurements at higher power when foldover effects take place [31]. The locus of the resonance, observed during sweeps of H_{ext}, decreases quadratically with the precession angle θ. When the shift is greater than the linewidth, the response becomes hysteretic. Figure 7 shows the lineshape asymmetry of the mechanical signal when the disk is excited at different powers.

More recently the group has reported the first measurements of M_z, the time-average magnetization at the saturation of the main resonance [33]. They find that M_z decreases rapidly when saturation effects set in. This decrease results from a rapid growth of the nonequilibrium degenerate magnons. The sample is large enough that they are able to simultaneously observe the reflected signal from the microwave stripline as in conventional FMR and observe the longitudinal signal with the MRFM. Figure 8 is a plot of both the transverse susceptibility, χ'' and ΔM_z as a function of the driving field h. The results are in qualitative agreement with the Suhl model describing the saturation of the main resonance in the presence of two-magnon scattering [37].

3.3 Scanning Mode

The most versatile application of FMRFM has been reported by *Midzor* [25, 26]. The probe magnet was a 170 nm thick Permalloy film deposited on an ultrasharp conical AFM tip as shown in Fig. 9. The calculated strength of the

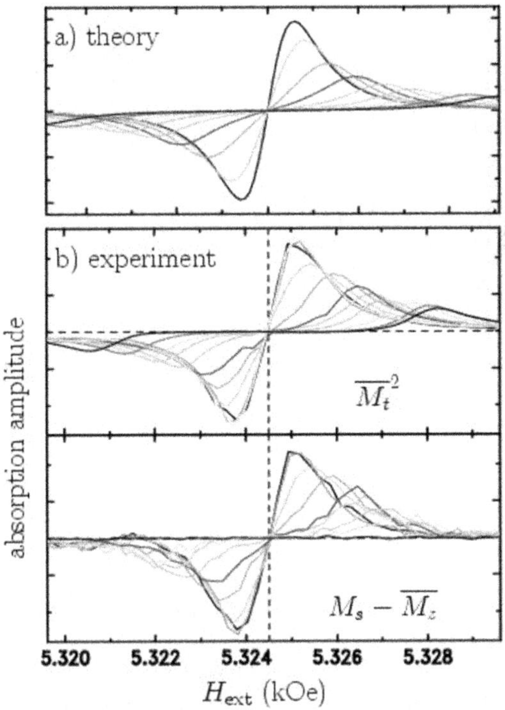

Fig. 6. (**a**) Theoretical and (**b**) experimental distortion of the anharmonic absorption line (longitudinal and transverse) for different modulation frequencies between 0.1 MHz and 10 MHz in steps of 1 MHz. The amplitude of the frequency modulation corresponds to 10 % of the linewidth [31]

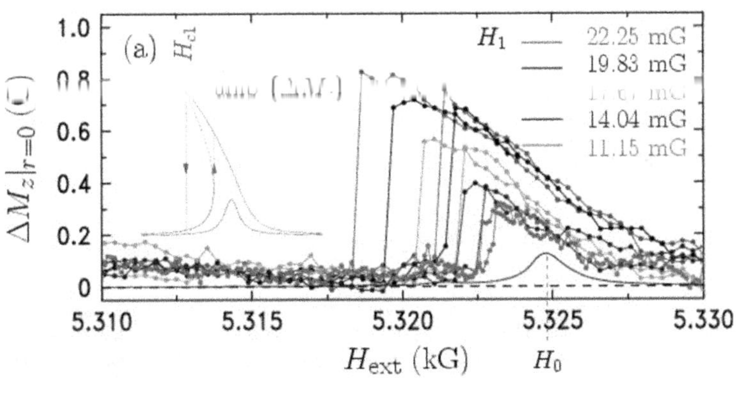

Fig. 7. The upsweep and downsweep profile of the resonance peak at high power [31]

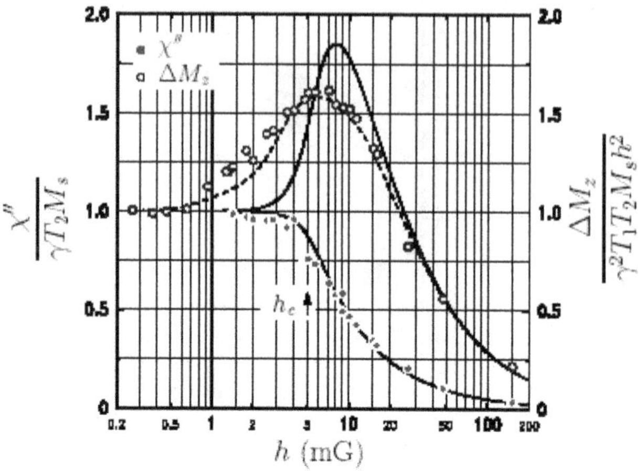

Fig. 8. The microwave field strength dependence of the transverse and longitudinal components of the magnetization at 10.47 GHz [33]

magnetic field and the magnetic field gradient as a function of the distance from the tip of the probe magnet are shown in Fig. 10.

In these experiments a 3 μm thick single-crystal YIG film was patterned into a geometrical series of rectangular samples by optical lithography and ion beam milling. Two sets of rectangular samples of different widths were fabricated: (a) $w = 10$ μm having lengths $L = 10, 20, 40, 80$ and 160 μm and (b) $w = 20$ μm having lengths $L = 20, 40, 80, 160$ and 320 μm. A typical spectrum observed for the 20 μm × 80 μm sample is shown in Fig. 11.

In the measurements represented in Fig. 11, the probe magnet is approximately 10 μm above the sample surface and produces a negligible additional field, H_{probe}, at the sample (calculated to be about 15 G). In this weak-field perturbation limit the modes are identified by the values of n_x (half-wavelength modes across the width of the sample) and n_y (half-wavelength modes along the length of the sample).

3.3.1 Dependence of the Fundamental Mode on Sample Dimensions

The positions of the peaks in the magnetostatic mode spectra depend on the size of the sample. For the series of 20 μm wide samples, the spectra are shown in Fig. 12.

For an ellipsoidal sample the resonance condition is given by the relation:

$$\left(\frac{\omega}{\gamma}\right)^2 = [B_0 - 4\pi (N_z - N_x) M]\,[B_0 - 4\pi (N_z - N_y) M] \ , \tag{6}$$

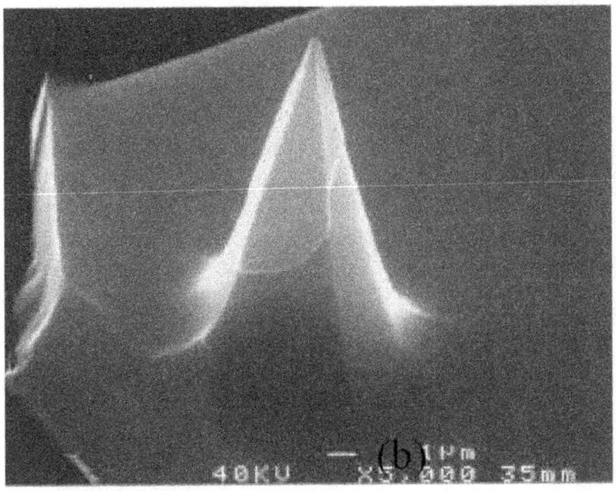

Fig. 9. Permalloy film selectively deposited on an ultrasharp conical AFM tip [26]

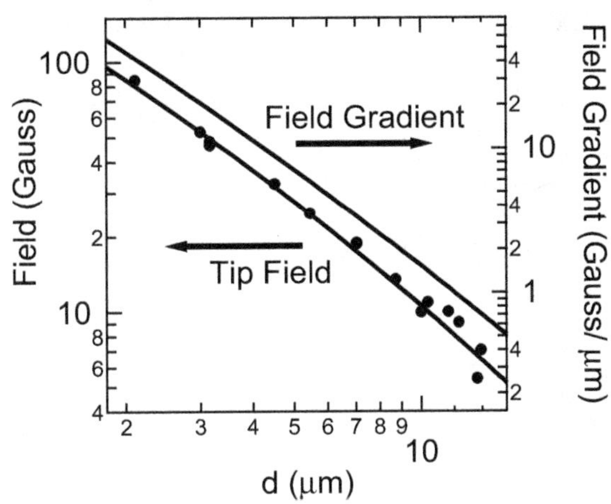

Fig. 10. Magnetic field and the magnetic field gradient of the coated tip shown in Fig. 9 vs. distance from the tip is shown. The upper line is from a micromagnetic calculation of the tip field and the lower line is obtained by differentiating that curve [26]

where ω is the radial frequency of the RF field, γ is the gyromagnetic ratio, H_0 is the applied field, M is the magnetic moment of the media and the N_i are the demagnetization factors along the three principle axes of the ellipsoid.

Because the samples used in this study are not ellipsoidal in shape, the internal fields vary within the sample; this can be calculated as a function of the position within the material. Assuming that the magnetization is saturated along a given axis, the magnetic "surface charge" at the surfaces will produce a demagnetization field that varies as a function of position between the two surfaces normal to the direction of the magnetization and having

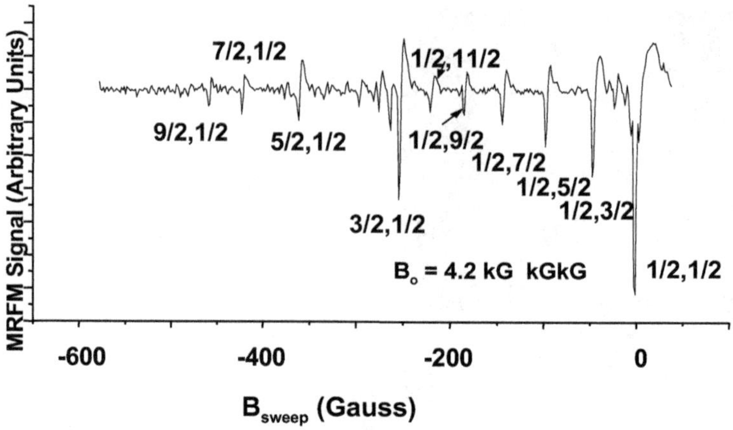

Fig. 11. The spectrum obtained from the 20 μm × 80 μm sample with the magnetostatic modes identified by the numbers (n_x, n_y). B_0 is the field position of the fundamental mode and B_{sweep} is the field separation from the fundamental mode [26]

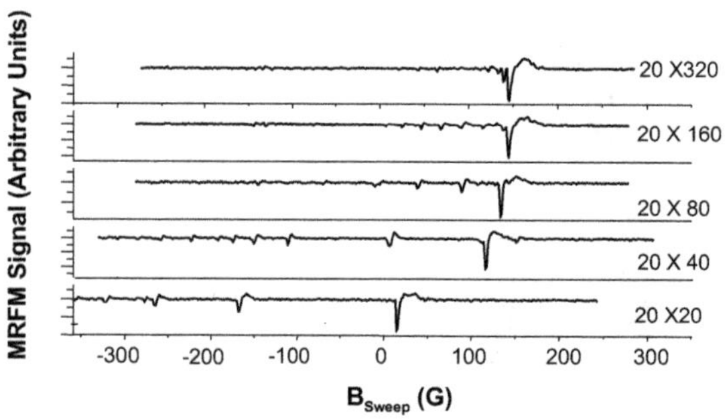

Fig. 12. The observed spectra for a family of samples having lengths of 20, 40, 80, 160 and 320 μm. The positions of the fundamental modes are compared with the theory in Fig. 14 and the mode spacings are compared with theory in Fig. 15 [26]

a minimum value at the center of the film. Typical results are shown for the 20 μm × 80 μm sample in Fig. 13.

The value of the internal field at the center of the sample (zero per cent in Fig. 13) for each orientation is used for the demagnetization field in the dispersion relation for the case of ellipsoidal shapes of (6) to give an approximate dispersion relation of the form:

$$\left(\frac{\omega}{\gamma}\right)^2 = [H_0 - (H_{z,\text{int}} - H_{x,\text{int}})]\,[H_0 - (H_{z,\text{int}} - H_{y,\text{int}})]\ . \tag{7}$$

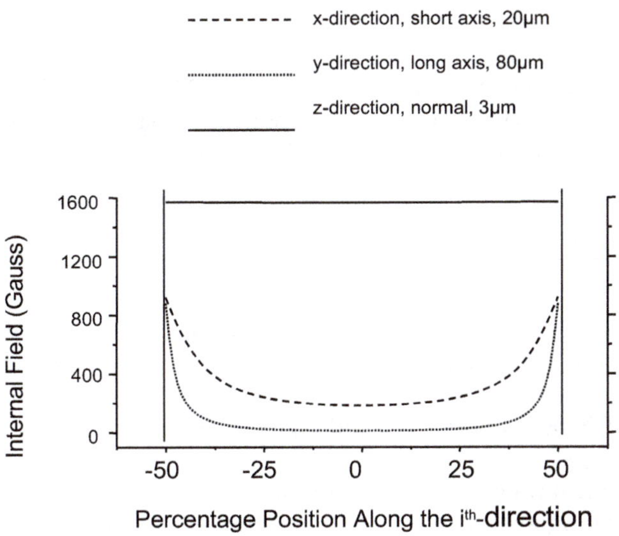

Fig. 13. The internal field along the various axes of the 20 μm by 80 μm sample. The value of the internal field at the center of the sample (zero per cent) was taken as the value of $H_{i,\mathrm{int}}$ in (4) above

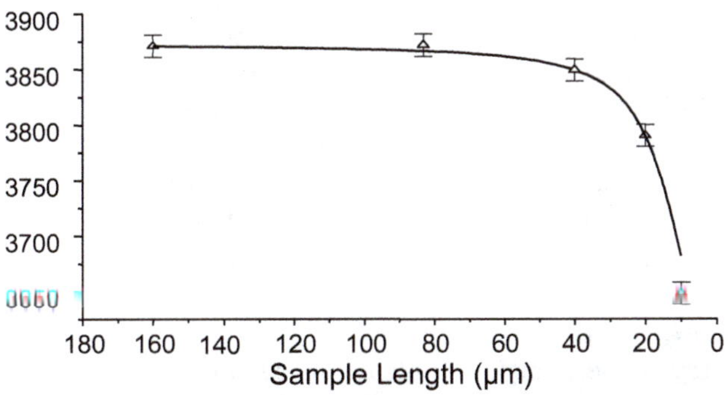

Fig. 14. The magnetic field at which the fundamental modes were excited as a function of the length of the series of 10 μm wide samples. The fit to the experimental data is represented by the *curve* that results if the best fit values $\omega/\gamma = 2.7\,\mathrm{kG}$ and $4\pi M = 1.6\,\mathrm{kG}$ are used [26]

Using a value of $4\pi M = 1730\,\mathrm{G}$, the values of the calculated resonances compare quantitatively with the observed resonances for the fundamental modes of the 20 μm wide series in Fig. 14.

3.3.2 Dispersion Relation

The dispersion of the higher-order modes shown in Fig. 11 is plotted in Fig. 15. At these dimensions the linear approximation to the Damon–Eshbach (DE) theory is not applicable and the dispersion curve must be solved explicitly [36, 37]:

$$\omega^2 = \omega_\mathrm{i}\left[\omega_\mathrm{i} + \omega_M\left(1 - \frac{1-\exp(-k_\mathrm{t}d)}{k_\mathrm{t}d}\right)\right], \tag{8}$$

where $\omega_\mathrm{i} = \gamma H_\mathrm{i}$, $\omega_M = 4\pi\gamma M_\mathrm{s}$ and $H_\mathrm{i} = H_\mathrm{res} - H_\mathrm{demag}$. H_i is the internal field required to support the i-th magnetostatic mode having the transverse wave number k_t, H_demag is the demagnetization field at the center of the sample, ω is the applied RF radial frequency and H_res is the external field at which resonance occurs.

By imposing the natural physical boundary conditions established by the lateral dimensions of a rectangular sample, k_t is approximated as plane waves having the allowed values

$$k_\mathrm{t} = \left(k_x^2 + k_y^2\right)^{1/2} = \left(\frac{n_x^2\pi^2}{w^2} + \frac{n_y^2\pi^2}{L^2}\right)^{1/2}. \tag{9}$$

The mode numbers n_x and n_y are positive integers equal to the number of half-wavelengths along the width, w, and the length, L, respectively. Since the samples are not ellipsoidal the internal fields are not uniform across the sample, yet it is observed that the complicated amplitude dependence of the modes can be reasonably represented in the plane-wave approximation. The modes having even values of n will have odd symmetry and thus a zero net dipole moment. As a result they will not be excited by the uniform RF field (hidden modes). The following parameters were used in the calculation: $\omega =$ 7.6 GHz, $d = 3.15\,\mu\mathrm{m}$, $4\pi M_\mathrm{s} = 1760\,\mathrm{G}$ and $g = 2$. H_demag is calculated for the sample of this geometry to be 1660 G. The calculated positions are also plotted in Fig. 15.

3.3.3 Spatial Mapping of Magnetostatic Modes

A series of measurements showing the variation in the amplitude of the time-dependent magnetization associated with various magnetostatic modes are shown in Fig. 16. The probe magnet is positioned about 5 µm vertically above the sample surface. The spectra are then obtained with the probe positioned at nine different lateral locations along the long axis of the 20 µm × 80 µm sample.

The lateral spatial resolution of FMRFM is demonstrated by plotting the lateral position dependence (along the 80 µm axis) of several mode amplitudes as shown in Fig. 17. The signal amplitude measured at the detector depends

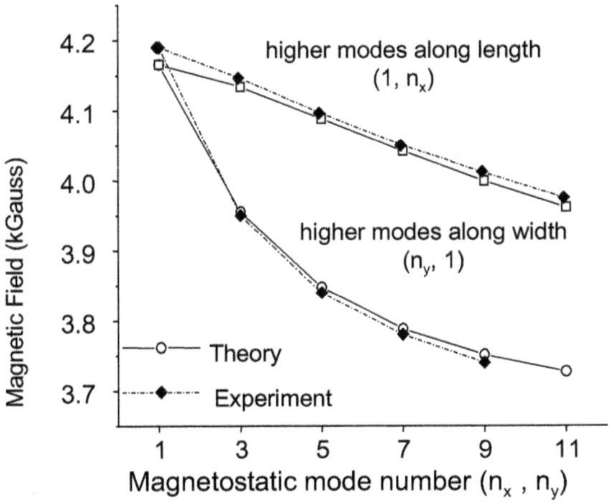

Fig. 15. A comparison of the magnetic field at which magnetostatic modes are observed in experiment and predicted by theory (6) [26]

on the spatial variation of the amplitude of the magnetostatic modes $m_t(r, t)$. The amplitude of the fundamental mode, $n_x = n_y = 1$, has a maximum amplitude at the center of the film and falls off as expected for the cosine dependence of a mode having a wavelength equal to twice the length of the sample. The wavelength of the first higher-order mode, $n_y = 3$, is 2/3 the sample length, hence 3/2 of the spatial period is contained in the long axis of the film, so the amplitude shows a minimum near the 13 μm position and a maximum near the 26 μm one as expected. The scan of the $n_y = 5$ mode ($\lambda/2 = 16$ μm) evidenced no detectable variation in the mode amplitude. This suggests a resolution limit of about 20 μm, a diameter established by the extent of the dipole field of the probe magnet, approximately twice the beam height.

3.3.4 Hidden Modes

A second interesting feature of the spectra in Fig. 16 is the demonstration of the ability of the probe magnet field to break the field symmetry in the sample and enable observation of "hidden modes". At a distance of 5 μm above the film the strength of the probe magnetic field at the surface is approximately 20 G. When the probe is located at the center of the film, the internal field remains symmetric and the hidden modes having even n-values are not excited by the RF field. However, as the probe magnet is scanned along the long axis, the 20 G magnetic field due to the probe magnet is sufficient to break the symmetry of the internal field in the sample. The normally hidden mode $n_y = 2$, $n_x = 1$ is then no longer antisymmetric and has a weak maximum in its intensity at $y = \pm 20$ μm, as shown in Fig. 16.

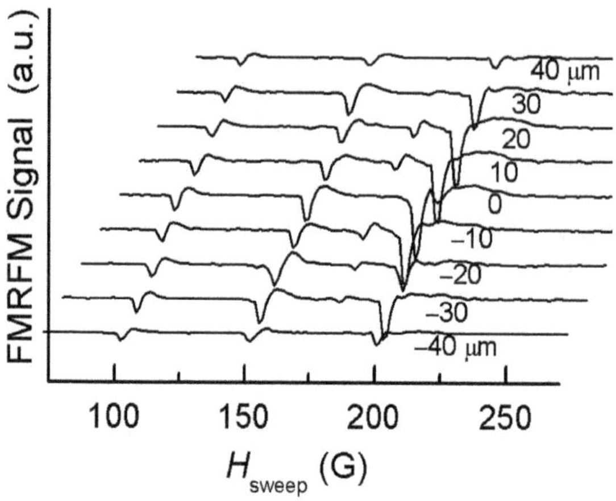

Fig. 16. The spectra obtained from the 20 μm × 80 μm as the probe is scanned along the length from one end of the sample to the other [26]

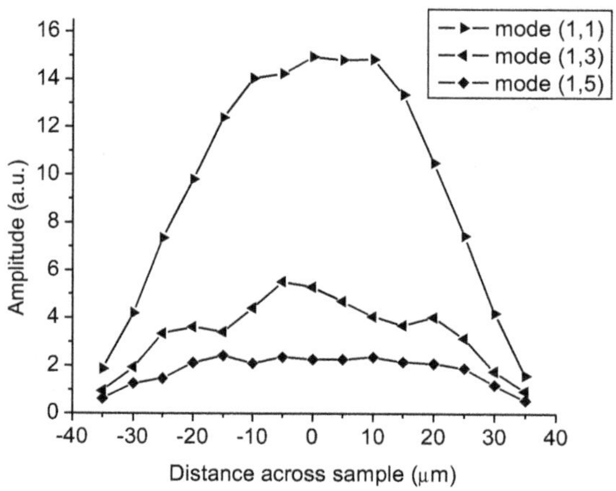

Fig. 17. The amplitude of the peaks as a function of the position of the probe magnet along the length of the 20 μm × 80 μm sample

3.3.5 Mapping RF Force Fields

FMRFM also has the potential to determine the strength and the shape of the RF field associated with the precessing ferromagnetic moment in the volume surrounding the resonant sample. Figure 18a shows the spectra observed when the probe magnet is scanned across the narrow dimension of the 20 μm × 80 μm sample at a probe height of 10 μm. Note that the amplitude of the fundamental mode goes to zero at $x = 8$ μm, as expected, and the phase rotates by π for larger values of x. In Fig. 18b the data is repeated for a sweep height of 5 μm showing the intensity of the modes going to zero at

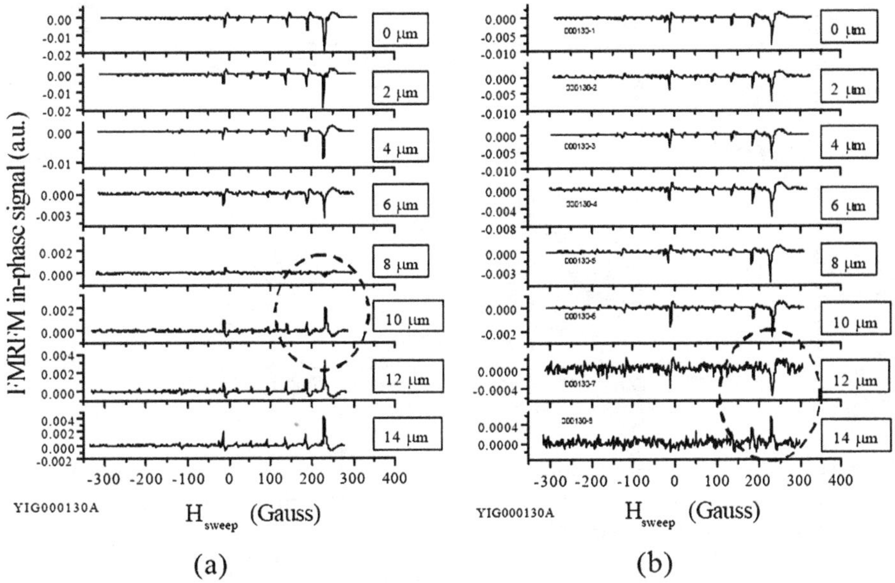

Fig. 18. Spectra obtained from the 20 µm × 80 µm sample as the probe magnet is scanned perpendicular to the long axis of the sample; the traces are labeled by the distance of the probe from the center of the film at a height of 10 µm (**a**) and 5 µm above the sample (**b**) [26]

$x = 12\,\mu\text{m}$ and again a phase rotation at larger distances. When the spectrum is observed as a function of the height of the probe magnet above the sample at the position $x = 10\,\mu\text{m}$, the modes have a zero intensity at $z = 8\,\mu\text{m}$ with the phase of the signal changing by π above and below that height.

The origin of this effect is the spatial dependence of the interaction between the gradient field of the probe magnet and the precessing moment of the sample [16]. The gradient of the dipole field generated by the probe magnet is negative directly below the probe and becomes positive in the plane away from the tip. As the probe is moved laterally beyond the edge of the sample the gradient of the dipole field decreases and eventually changes sign. As the sample is moved under the probe magnet there will be a position at which the integrated sum of the positive and the negative forces will cancel and beyond that position the negative force, observed as a reversal of the signal phase, will dominate. Calculated positions of expected phase reversals as a function of the height of the probe magnet are consistent with the data shown in Fig. 18.

3.4 FMRFM in Metal Films

FMRFM in metal films have been reported [38, 39] for metallic Co films deposited on the cantilever with the magnetic field applied in the film plane.

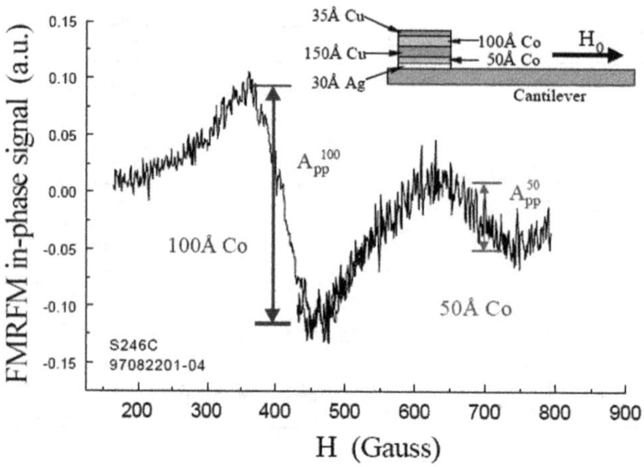

Fig. 19. FMRFM spectrum obtained for the multilayer composite indicated in the inset. The two Co films have different resonance fields due to the different volume and surface anisotropy energies [26]

Midzor [22] investigated a bimagnetic layer structure of Ag 30 Å \ Co 50 Å \ Cu 150 Å \ Co 100 Å \ Cu 35 Å as shown in Fig. 19. A 40 μm by 40 μm film composite was sputter deposited directly on the cantilever. The relatively thick 150 Å Cu layer was used in order to ensure negligible exchange coupling between the Co layers. The gradient field was $\approx 0.15\,\mathrm{G/\mu m}$ so the field variation of 6 G across the sample is small in comparison to the ≈ 80 G linewidth of the samples. The amplitude ratio of the two signals is $A_{\mathrm{pp}}^{50}/A_{\mathrm{pp}}^{100} \approx 0.3$. The discrepancy with the ratio of layer thicknesses is within the accuracy of the ability to determine the thicknesses of films of this area depositied by shadow-masking techniques. In the analysis below, it was assumed that the thinner Co layer has a thickness of 30 Å.

The volume and the surface anisotropy energies of the Co layers can be estimated by fitting their resonance fields to the empirical formula

$$H_{\mathrm{U}}^{\mathrm{eff}} = \frac{2}{M_{\mathrm{S}}}\left(K_{\mathrm{V}} + \frac{2K_{\mathrm{S}}}{t_{\mathrm{film}}}\right), \tag{10}$$

where K_{V} and K_{S} are the volume and surface anisotropies for the films, t is the film thickness and $H_{\mathrm{U}}^{\mathrm{eff}}$ is the effective uniaxial anisotropy field. The calculated values of K_{V} and K_{S} are given in Table 1, where it is assumed that the Co films have bulk values of $4\pi M_{\mathrm{s}}$. The experimental values agree well with typical results reported in the literature [40–42].

4 Torque Measurements in a Uniform Field

Moreland and colleagues have demonstrated micromechanical detection of ferromagnetic moments and ferromagnetic resonance in thin magnetic films

Table 1. Experimentally determined values of the volume and surface anisotropy energies for the Co/Cu interfaces in the bilayer film [26]

$H_{\mathrm{U}}^{\mathrm{eff}} = \frac{2}{M_{\mathrm{S}}}\left(K_{\mathrm{V}} + \frac{2K_{\mathrm{S}}}{M_{\mathrm{S}} t_{\mathrm{film}}}\right)$	Fit to data Co/Cu interface	Literature values Co/Cu interface
t_{Co} (Å)	100	10–50
K_{V} ($\times 10^6$ erg/cm^3)	1.1 ± 0.2	0.9–2.0
K_{S} (erg/cm^2)	0.15 ± 0.05	0.1–0.35

For the Co/Ag interface

$H_{\mathrm{U}}^{\mathrm{eff}} = \frac{2}{M_{\mathrm{S}}}\left(K_{\mathrm{V}} + \frac{2K_{\mathrm{S}}}{M_{\mathrm{S}} t_{\mathrm{film}}}\right)$	Fitted Co/Ag interface	Typical values Co/Ag interface
t_{Co} (Å)	30	15–30
K_{V} ($\times 10^6$ erg/cm^3)	1.7 ± 0.3	1.0–1.4
K_{S} (erg/cm^2)	0.45 ± 0.05	0.2–0.4

using torque deflection of the cantilever in a homogeneous magnetic field [43–47]. The detection scheme monitors the deflection of the cantilever with a laser beam-bounce method with the laser beam focused on the cantilever and reflected onto a split four-quadrant photodiode detector as shown in Fig. 20. The cantilever deflection signal corresponds to the $(\mathrm{C}+\mathrm{D})-(\mathrm{A}+\mathrm{B})$ signal, whereas the cantilever torque signal corresponds to the $(\mathrm{A}+\mathrm{C})-(\mathrm{B}+\mathrm{D})$ signal. This configuration enables the detection of both the deflection and the torque signals with the same apparatus. The Si cantilever has a deflection spring constant of 0.35 N/m with a resonant frequency of 17 kHz and a torsion spring constant of 3.0×10^{-20} N · m/ rad with a torsional resonance frequency of 250.3 kHz. The system is capable of detecting 10 pm amplitudes of vibration under ambient conditions.

4.1 Mechanical Torque on a Thin Film

In the presence of an applied torque field, H_{T}, the magnetization M of a thin film will generate a mechanical torque T. In many cases the shape anisotropy is sufficient to generate the mechanical torques that can be measured with micromechanical detectors. Considering the geometry in Fig. 21 the torque is given by

$$T = |\boldsymbol{M}_{\mathrm{s}} \times \boldsymbol{H}_{\mathrm{T}}| V = M_{\mathrm{s}} H_{\mathrm{T}} V \,, \tag{11}$$

where M_{s} is the sample magnetization, H_{T} is the total magnetic field and V is the volume of the sample.

4.2 Magnetization versus Field (M–H) Loops

Figure 22 shows the experimental configuration for measuring M–H loops [46]. The torque field, H_{T}, was applied by a solenoid and kept constant, while the

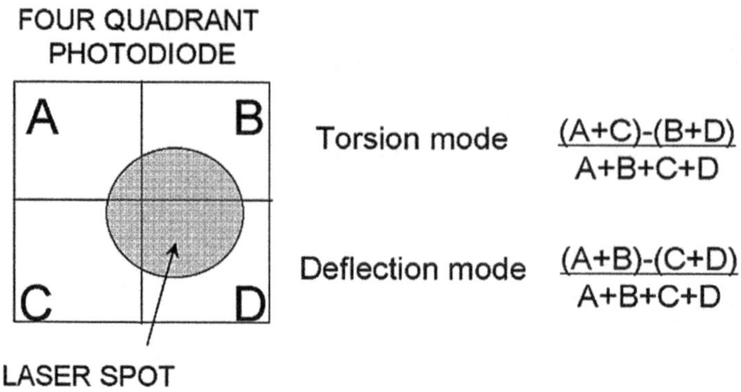

Fig. 20. Reflected laser spot on photodiode detector [43]

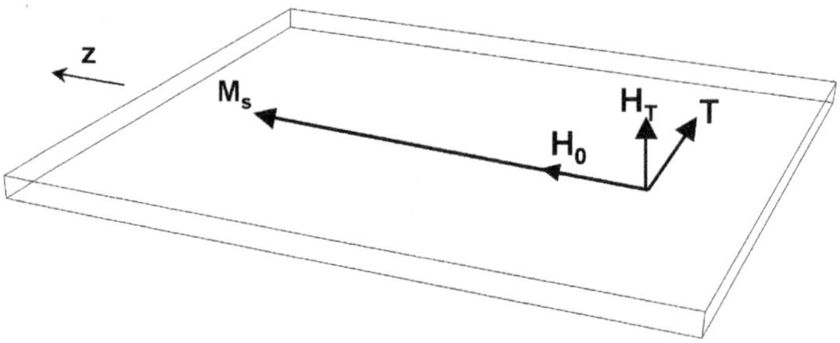

Fig. 21. Vector diagram showing the orientations of the magnetic fields and torque on a thin film magnetized inplane along the z-direction [43]

applied field, H_0, is applied in the film plane perpendicular to the long axis of the cantilever and cycled over the range of the observation. A typical result is shown in Fig. 23. Fe films having thicknesses as small as 1 nm and a total volume of $2.2 \times 10^{-11}\,\mathrm{cm}^3$ could be measured.

4.3 Microresonating Torque Magnetometer (μRTM)

In the FMR mode [43, 45, 46] the change in the mechanical torque in FMR is proportional to the change in the longitudinal component of the magnetization as shown in Fig. 24

$$\Delta T_{\mathrm{FMR}} = \Delta M_z H_{\mathrm{T}} V \,, \tag{12}$$

where ΔM_z is the change in the magnetization due to the FMR precession

$$M_Z = \left[M_{\mathrm{S}}^2 - m_{\mathrm{in}}^2 - m_{\mathrm{out}}^2\right]^{1/2} \approx M_{\mathrm{S}} - \left(\frac{m_{\mathrm{in}}^2 + m_{\mathrm{out}}^2}{2M_{\mathrm{S}}}\right) , \tag{13}$$

where M_{S} is the change in the magnetization, m_{in} is the inplane component of the magnetization and m_{out} is the out-of-plane component.

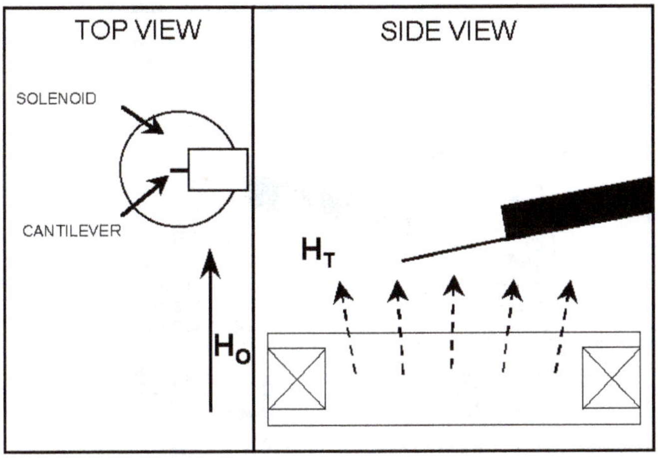

Fig. 22. Experimental configuration for magnetic-torque measurements with a cantilever [43]

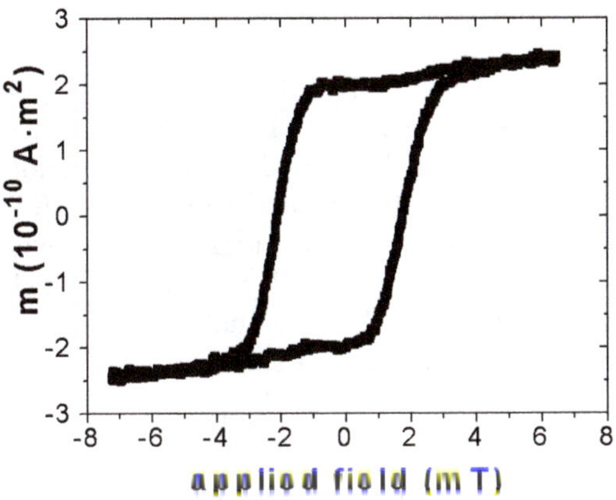

Fig. 23. M–H loops of an Fe film measured by MTM [43]

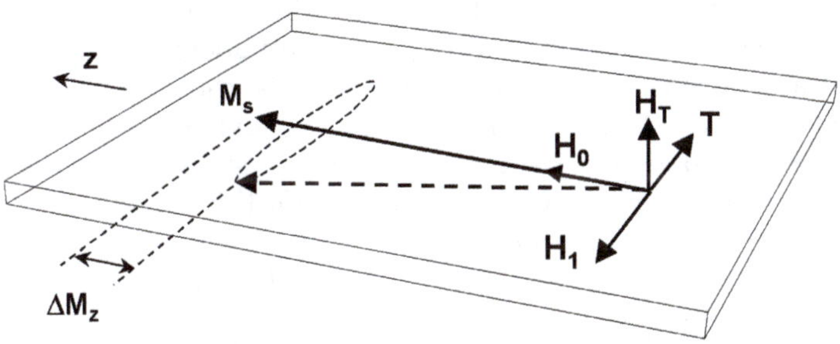

Fig. 24. Vector diagram showing the orientation of the applied fields and mechanical torque generated in an FMR experiment [43]

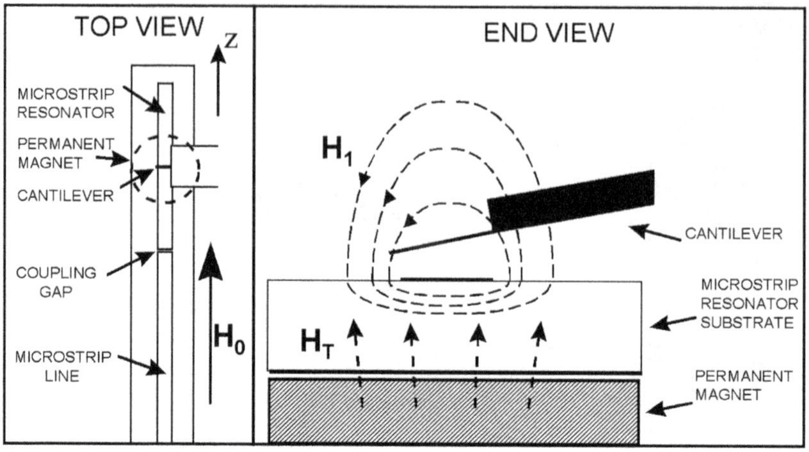

Fig. 25. Experimental configuration for FMR with a μRTM [43]

Figure 25 shows the experimental configuration. The torque on the cantilever is measured as a function of the magnetic field applied along the axis of the cantilever and swept over the desired range. The Si cantilever was positioned 200 μm to 300 μm above a microstrip resonator having a resonance frequency of 9.17 GHz. At resonance, the torque developed by the precession is coupled to the cantilever and by modulating the amplitude of the microwave field at the torsional resonance frequency of 250.3 kHz the cantilever is excited in its torsional mode. Figure 26 shows the result for a 30 nm thick Permalloy film having a volume of $1.1 \times 10^{-10}\,\mathrm{cm}^3$. Note that the direction of the torque is reversed upon reversing the direction of the sweep magnetic field, H_0.

4.4 Bimaterial Micromechanical Calorimeter Sensor for FMR

Ferromagnetic resonance in magnetic metal films was also detected by using calorimetric detection of the microwave absorption using a micromechanical bimaterial sensor [47]. The detection method can be understood within the mathematical framework developed for other bimaterial thermal sensors. Consider the silicon cantilever, layer 1, as a rectangular beam fixed at one end with its metallic magnetic coating, layer 2, as a two-layer system each having different thermal properties. Solving the heat equation for this configuration the deflection at the free end of the beam will be

$$z = a\frac{E_1}{E_2}\frac{t_1^2 l^3}{t_2^3 w}\left(\frac{\gamma_1 - \gamma_2}{\lambda_1 t_1 + \lambda_2 t_2}\right)P\,, \tag{14}$$

where γ, λ, t, w, l and E are, respectively, the thermal expansion coefficient, thermal conductivity, thickness, width, length, and Young's modulus of the beam layers (subscripts refer to the different materials) and P is the absorbed

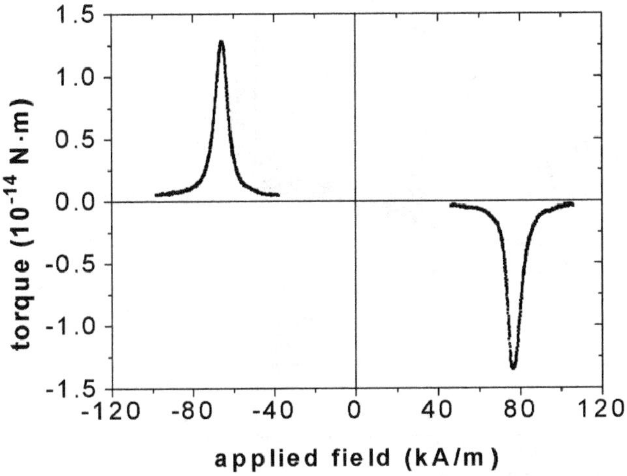

Fig. 26. Torque versus applied field measured with the μRTM for a 30 nm thick NiFe film [43]

power. Equation (14) applies only in the limit $t_1 \ll t_2$ (t_1 is the thickness of the magnetic film and t_2 is the thickness of the silicon cantilever). In addition, it is assumed that the temperature is constant over any cross section along the axis of the cantilever – this is a good approximation if $t_1, t_2 \ll l$. The constant a ranges from a value of 2, if power is absorbed near the end of the beam, to a value of 1.25, if power is absorbed uniformly along the beam.

A schematic of the experimental arrangement is shown in Fig. 27. The microwave output is amplitude modulated by a 1 kHz square wave and the deflection of the cantilever is determined as a function of the magnetic field applied in the plane of the film. The results for the Co, Ni and $Ni_{81}Fe_{19}$ films together with the response observed from a nonmagnetic Au film are shown in Fig. 28. The constant offset is due to the nonresonant absorption of microwaves due to the eddy currents in the metallic films. The calculated magnetic-moment sensitivity of this technique is $\Delta m = 1.3 \times 10^{-12}\,\mathrm{A \cdot m^2}$. The microwave power used in the experiment was 75 mW.

5 Conclusion

The MRFM employs a micromechanical resonator to detect the force between a micromagnetic probe tip and the time-dependent spin magnetization of a well-defined resonant slice within the sample. One-, two- and three-dimensional imaging capabilities of MRFM have been demonstrated using ESR and NMR techniques. The recent demonstration of the detection of a single electron spin by MRFM will strongly stimulate additional interest in the field [6].

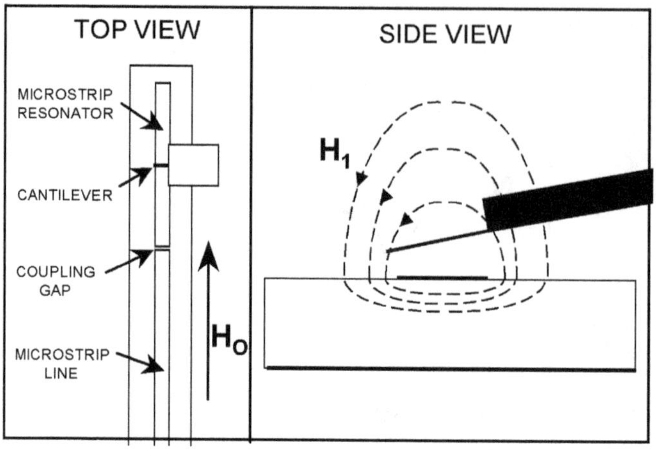

Fig. 27. Top view and side view of the experimental configuration for FMR

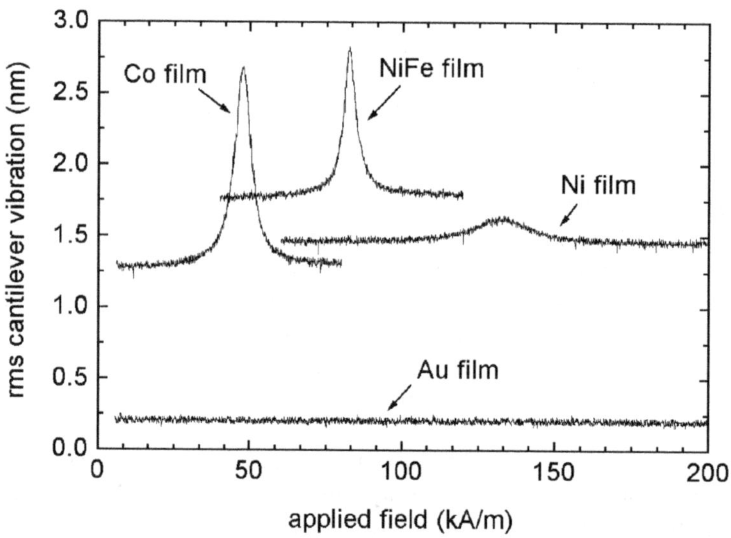

Fig. 28. Cantilever vibration vs. applied field showing microwave absorption in Co, NiFe, Ni and Au thin-film samples [47]

The high sensitivity of MRFM takes advantage of the high Q of a mechanical cantilever. The amplitude of the magnetic resonance signal is modulated at a frequency that matches the cantilever resonance frequency f_c thus generating a large-amplitude cantilever oscillation. This time-dependent drive is generated by modulating the uniform magnetic field and/or the RF field. The oscillation amplitude of the cantilever depends sensitively on the modulation amplitude, the RF field strength and the external field gradient.

In ferromagnetic systems the resonance conditions are strongly influenced by exchange coupling and long-range dipole–dipole effects so that the dispersion relation depends upon the total magnetic field resulting from external, effective internal anisotropy fields and upon the sample geometry. With the application of the probe magnet field, the wavevector k of magnetostatic modes will be modified by the magnetic field near the probe as the sample accommodates the localized inhomogeneous fields of the micromagnetic tip in order to satisfy the resonance condition in the entire sample. The volume of sample studied in an FMRFM experiment is thus not determined solely by the magnetic field gradient and the sample linewidth as in the case of the electron spin or the nuclear spin resonant versions of MRFM. The challenge of achieving spatially localized FMR within an extended sample remains a topic of active research.

FMRFM opens the possibility of conducting spatially resolved, subsurface studies of many solid-state materials. Examples include:

- Magnetostatic modes have been detected in YIG films having microscopic dimensions. The observed dispersion is in quantitative agreement with the Damon–Eshbach theory [26, 28].
- In a scanning mode, the spatial variations in the amplitudes of the magnetostatic modes have been observed in microscopic samples of YIG films [26, 27].
- When the perturbation of the tip field is sufficient, the variation of the magnetic field in the media will modify the resonance condition of the magnetostatic modes that will produce modification of the signal amplitude and when the tip is moved off the center of the sample the symmetry of the internal field is broken and the normally "hidden modes" can be excited [26–28].
- Mapping of the force fields reflecting the dipole nature of the magnetic interaction between the sample and the probe magnet has been observed [26].
- Subsurface studies have been demonstrated in magnetic layer structures. FMRFM signals from microscopic Co/Cu/Co trilayer films demonstrate that MRFM is sensitive enough to perform microscopic evaluation of local magnetic environments that can affect the performance of magnetic layered devices [40, 48].
- Bulk and surface anisotropy energies in microscopic metal films have been evaluated [26].
- Measurement of both the transverse susceptibility χ'' and ΔM_z as a function of the driving field H and evaluation of the relaxation times T_1 and T_2 in ferromagnetic resonance [31, 33].
- Investigation of the dramatic decrease in M_z when the driving field reaches the threshold for 2-magnon excitation [33].

Using a mechanical torque effect on the cantilever:

- From the torque acting on the cantilever, hysteresis loops have been observed in microscopic samples [43].
- The torque-induced rotation of the cantilever due to the precession of spins has been used to detect the ferromagnetic resonance in thin metal films mounted on the cantilever [45, 46].
- Treating the cantilever/film structure as a bilayer material, the heating of the ferromagnetic metal layer at resonance induces a thermal stress on the cantilever giving rise to a calorimeter sensor for FMR [47].

The extremely strong FMR signals obtained from microscopic samples of magnetic thin films indicates that MRFM has the potential to study a large variety of magnetic materials with very high sensitivity. By increasing the magnetic-field gradients associated with the probe magnets it is expected that it will be possible to conduct microscopic FMR experiments with micrometer to submicrometer resolution.

Future generations of MRFM instruments will operate at lower temperatures, apply larger magnetic field gradients and employ advanced micromechanical resonators. Such instruments would enable unprecedented insight into topics of scientific and technological interest in the fields of electronic and magnetic materials.

As the size and magnetic moment of the probe magnet is reduced, the resolution limit of FMRFM will approach that of the magnetic correlation length, 100 nm, the limit of the resolution of the magnetic properties in ferromagnetic materials.

In addition to its application to ferromagnetic resonance phenomena, magnetic resonance force microscopy holds significant promise for applications in spin-injection devices and in magnetic semiconductor devices where a spin-polarized electron current is employed to enhance information processing capabilities.

Acknowledgements

The authors wish to acknowledge the assistance received from their students and postdoctoral fellows who have contributed to this program and assisted with the preparation of this manuscript, Z. Zhang, D. Pelekov, M. Midzor, A. Putilin and R. Urban and to Prof. M. Cross. PEW acknowledges the support of the R. J. Yeh fund during his visits at California Institute of Technology.

References

[1] S.-C. Lee, C. P. Vlahacos, B. J. Feenstra, A. Schwartz, D. E. Steinhauer, F. C. Wellstood, S. M. Anlage: Magnetic permeability imaging of metals with a scanning near-field microwave microscope, Appl. Phys. Lett. **77**, 4404 (2000)

[2] S. E. Lofland, S. M. Bhagat, Q. Q. Shu, M. C. Robsen, R. Ramesh: Magnetic imaging of perovskite thin films by ferromagnetic resonance microscopy-$La_{0.7}Sr_{0.3}MnO_3$, Appl. Phys. Lett. **75**, 1947 (1999)

[3] J. A. H. Stotz, M. R. Freeman: A stroboscopic scanning solid emersion lens microscope, Rev. Sci. Instrum. **68**, 4468 (1997)

[4] K. Perzlmaier, M. Buess, C. H. Back, V. E. Demidov, B. Hillebrands, S. O. Demokritov: Spin-wave eigenmodes of permalloy squares with closure domain structure, Phys. Rev. Lett. **94**, 057202 (2005)

[5] V. E. Demidov, S. O. Demokritov, B. Hillebrands, M. Laufenberg: Radiation of spin waves by a single micrometer-sized magnetic element, Appl. Phys. Lett. **85**, 2866 (2004)

[6] D. Rugar, R. Budakian, H. J. Mamin, B. W. Chui: Single spin detection by magnetic resonance force microscopy, Nature **430**, 329 (2004)

[7] J. A. Sidles: Noninductive detection of single-proton magnetic resonance, Appl. Phys. Lett. **58**, 2854 (1991)

[8] D. Rugar, C. S. Yannoni, J. A. Sidles: Mechanical detection of magnetic resonance, Nature **360**, 563 (1992)

[9] J. A. Sidles, d. Rugar: Signal-to-noise ratios in inductive and merchanical detection of magnetic resonance, Phys. Rev. Lett. **70**, 3506 (1993)

[10] B. C. Choi, M. Belov, W. K. Heibert, G. E. Ballentine, M. R. Freeman: Ultrafast reversal magnetization dynamics investigated by time domain imaging, Phys. Rev. Lett. **86**, 728 (2001)

[11] J. P. Park, P. Eames, D. M. Engebretson, J. Berezovsky, P. A. Crowell: Spatially resolved spin dynamics of localized spin-wave modes in ferromagnetic wires, Phys. Rev. Lett. **89**, 277201 (2002)

[12] P. C. Hammel, Z. Zhang, G. J. Moore, M. L. Roukes: Subsurface imaging with the magnetic resonance force microscope, J. Low Temp. Phys. **101**, 59 (1995)

[13] D. Rugar, O. Zuger, S. T. Hoen, C. S. Yannoni, H. M. Vieth, R. D. Kendrick: Force detection of nuclear magnetic resonance, Science **264**, 1560 (1994)

[14] T. A. Barrett, C. R. Miers, H. A. Sommer, K. Mochizuki, J. T. Merkert: Design and construction of a sensitive nuclear magnetic resonance force microscope, J. Appl. Phys. **83**, 6235 (1999)

[15] Z. Zhang, P. C. Hammel, P. E. Wigen: Observation of ferromagnetic resonance using magnetic resonance force microscopy, Appl. Phys. Lett. **68**, 2005 (1996)

[16] D. Rugar, C. S. Yannoni, J. A. Sidles: Mechanical detection of magnetic resonance, Nature **360**, 563 (1992)

[17] K. J. Bruland, J. Krzystek, J. L. Garbini, J. A. Sidles: Anharmonic modulation for noise reduction in magnetic resonance force microscopy, Rev. Sci. Instrum. **66**, 2853 (1994)

[18] P. C. Hammel, D. V. Pelekov, P. E. Wigen, T. R. Gosnell, M. M. Midzor, M. L. Roukes: The magnetic-resonance force microscope: A new tool for high-resolution, 3-D, subsurface scanned probe imaging, Proc. IEEE **91**, 789 (2003)

[19] Z. Zhang, M. L. Roukes, P. C. Hammel: Sensitivity and spatial resolution for electron-spin-resonance detection by magnetic resonance force microscopy, J. Appl. Phys. **80**, 6931 (1996)
[20] A. Suter, D. V. Pelekov, M. L. Roukes, P. C. Hammel: Probe sample coupling in the magnetic resonance force microscope, J. Magn. Reson. **154**, 210 (2002)
[21] O. Zugar, D. Rugar: First images from a magntic resonance force microscope, Appl. Phys. Lett. **63**, 2496 (1993)
[22] R. C. Puetter: Pixon-based multiresolution image reconstruction and the quantification of picture information content, Int. J. Syst. Tech. **6**, 314 (1995)
[23] R. K. Pina, R. C. Puetter: Bayesian image reconstruction: The pixon method and optimal image modeling, Publ. Astron. Soc. Pac. **105**, 630 (1993)
[24] R. C. Ruetter, A. Yahil: Astronomical data analysis software and systems VIII, in D. M. Mehringer, R. L. Plante, D. A. Roberts (Eds.): *Astron. Soc. Pac. Conf. Ser.*, vol. 172 (San Francisco, CA 1999) p. 307
[25] M. M. Midzor, P. E. Wigen, D. Pelekov, W. Chen, P. C. Hammel, M. L. Roukes: Imaging mechanisms of force detected FMR microscopy, J. Appl. Phys. **87**, 6493 (2000)
[26] M. M. Midzor: *Ferromagnetic Resonance Force Microscopy*, Ph.D. dissertation, California Institute of Technology, Pasadena, CA (2000)
[27] R. Urban, A. Putilin, P. E. Wigen, M. Cross, M. L. Roukes: Perturbation of the magnetostatic modes observed by FMRFM, Phys. Rev. B **73**, 212410 (2006)
[28] V. Charbois, V. V. Naletov, J. Ben Joussef, O. Klein: Mechanical detection of ferromagnetic resonance spectrum in a normally magnetized yttrium-iron-garnet disk, J. Appl. Phys. **91**, 7337 (2002)
[29] V. Charbois, V. V. Naletov, J. Ben Joussef, O. Klein: Influence of the magnetic tip in ferromgnetic resonance force microscopy, Appl. Phys. Lett. **80**, 4795 (2002)
[30] V. V. Naletov, V. Charbois, O. Klein, C. Fermon: Quantitative measurement of the ferromagnetic resonance signal by force detection, Appl. Phys. Lett. **83**, 3132 (2003)
[31] O. Klein, V. Cahrbois, V. V. Naletov, C. Fermon: Measurement of the ferromagnetic relaxation in a micron-size sample, Phys. Rev. B Rap. Commun. **67**, 220407 (2003)
[32] V. V. Naletov, V. Charbois, O. Klein, C. Fermon: Quantitative measurement of the ferromagnetic resonance signal by force detection, Appl. Phys. Lett. **83**, 3132 (2003)
[33] V. V. Naletov, G. deLoudens, O. Klein: Magnetization reduction induced by non-linear effects, Phys. Rev. B **71**, 180411 (2005)
[34] K. Wago, D. Botkin, C. S. Yannoni, D. Rugar: Paramagnetic and ferromagnetic resonance imaging with a tip-on-cantilever magnetic force microscope, Appl. Phys. Lett. **72**, 2757 (1998)
[35] J. F. Dillon: J. Appl. Phys. **31**, 1605 (1960)
[36] R. C. LeCraw, E. G. Spencer: J. Phys. Soc. Jpn., Suppl. (B1) **17**, 401 (1962)
[37] H. Suhl: J. Phys. Chem. Solids **1**, 209 (1957)
[38] R. W. Damon, J. R. Eshbach: Magnetostatic modes of a ferromagnetic slab, J. Phys. Chem. Solids **19**, 308 (1961)
[39] B. A. Kalinikos: Excitation of propagating spin waves in ferromagnetic films, IEE Proc. **127 H**, 4 (1980)

[40] B. J. Suh, P. C. Hammel, Z. Zhang, M. M. Midzor, M. L. Roukes, J. R. Childress: Ferromagnetic resonance imaging of Co films using magnetic resonance force microscopy, J. Vac. Sci. Technol. B **16**, 2275 (1998)

[41] J. W. Feng, S. S. Kang, F. M. Pan, G. J. Jin, A. Hu, S. S. Jiang, D. Feng: Magnetic anisotropy and interlayer exchange coupling in the sputtered Co/Ag multilayers, J. Appl. Phys. **78**, 5549 (1995)

[42] F. J. A. Den Broeder, W. Having, P. J. H. Bloeman: Magnetic anisotropy of multilayers, J. Magn. Mater. **93**, 562 (1994)

[43] J. Moreland, P. Kabos, A. Jander, M. Lohndorf, M. R. McMichael, C.-G. Lee: Micromechanical detectors for ferromagnetic resonance spectroscopy, in E. Peters, O. Paul (Eds.): *Micromachined Devices and Components VI*, vol. 4176, Proc. SPIE (2000) pp. 84–95

[44] M. Lohndorf, J. Moreland, P. Kabos, N. Rizzo: Microcantilever torque magnetometery of thin magnetic films, J. Appl. Phys. **87**, 5995 (2000)

[45] M. Lohndorf, J. Moreland, P. Kabos: Ferromagnetic resonance detection with a torsion-mode atomic-force microscope, Appl. Phys. Lett. **76**, 1176 (2000)

[46] A. Jander, J. Moreland, P. Kabos: Angular momentum and energy transferred through ferromagnetic resonce, Appl. Phys. Lett. **78**, 2348 (2001)

[47] J. Moreland, M. Lohndorf, P. Kabos, R. D. McMichael: Ferromagnetic resonance spectroscopy with a micromechanical calorimeter sensor, Rev. Sci. Instrum. **71**, 3099 (2000)

[48] Z. Zhang, P. C. Hammel, M. M. Midzor, M. L. Roukes, J. R. Childress: Ferromagnetic resonance force microscopy an microscopic Co single layer films, Appl. Phys. Lett. **73**, 2036 (1998)

Index

Vortex Dynamics

Christian H. Back[1], Danilo Pescia[2], and Matthias Buess[1,2]

[1] Institut für Experimentelle und Angewandte Physik, Universität Regensburg, Universitätsstr. 31, 93040 Regensburg, Germany
christian.back@physik.uni-regensburg.de

[2] Laboratorium für Festkörperphysik, Eidgenössische Technische Hochschule Zürich, CH-8093 Zürich, Switzerland
pescia@solid.phys.ethz.ch

Abstract. In this Chapter we will report on progress that has been made in recent years in the investigation of magnetization dynamics in confined magnetic structures of micrometer size. Emphasis is laid on two issues: 1. on time-resolved microscopy of the dynamic mode structure of small ferromagnetic elements and 2. the investigation of magnetic elements in a domain state. We will restrict ourselves to two simple domain states, the magnetic vortex state of thin ferromagnetic disks and the Landau configuration in flat ferromagnetic squares. Both states may be summarized as flux-closure states. The topology of these micromagnetic configurations brings the magnetic vortex at the center of the elements naturally into play.

1 Introduction

The growing importance of high-speed magnetoelectronics has triggered intensive research in the field of magnetization dynamics [1–12]. In recent years substantial effort has been put into the identification of the excitation spectrum of small ferromagnetic elements [2,9,13–19] and into the investigation of their switching behavior in the precessional regime [20–23]. The response of small magnetic elements to ultrashort magnetic field pulses and the accurate modeling of the spin excitations following a magnetic field pulse is a central question in advanced magnetic recording technology as switching times are further reduced and pushed well into the gyromagnetic regime. While these problems can be addressed in the frequency domain using Brillouin light scattering or ferromagnetic resonance methods [18, 19, 24–26], time-resolved scanning Kerr microscopy experiments offer the opportunity to observe magnetic excitations on the picosecond time scale directly [2, 9, 13, 14].

The dynamics of magnetic elements of micrometer- and submicrometer-size is of special interest from the fundamental point of view, since both exchange and magnetic dipolar interactions contribute to magnetic confinement phenomena in such elements. In the last several years high-frequency confined spin-wave eigenmodes of micrometer-sized magnetic elements have been systematically studied for the straightforward case of elements possessing an almost monodomain state [2, 18, 25, 27–29]. Only in recent years has the experimental analysis of modes confined to inhomogeneously magnetized

B. Hillebrands, A. Thiaville (Eds.): Spin Dynamics in Confined Magnetic Structures III, Topics Appl. Physics 101, 137–160 (2006)

ferromagnetic elements been tackled. In this Chapter we will review some recent progress made in the analysis and understanding of spin-wave modes in magnetic elements in a flux-closure state. This magnetic state is characterized by a curling magnetization with a magnetic vortex at the center of the element. Fundamental experiments aim at the detailed understanding of the modal structure and – intimately connected – the boundary conditions of the small-angle excitations of these ferromagnetic elements [9, 13, 14, 19, 30–33].

2 Experimental Techniques

For the investigation of magnetization dynamics in small magnetic elements several experimental techniques are available. In this experimental section not all techniques will be introduced, in particular we will not review methods based on inductive techniques nor ferromagnetic resonance methods. These techniques – the most widely used one being "pulsed inductive magnetometry" – were developed in the 1960s [34], but have become increasingly important with the availability of modern pulse generators and oscilloscopes [35,36]. A second important inductive technique is ferromagnetic resonance (FMR) using vector network analysis (VNA-FMR) [37–39]. This method has become successful in the investigations of confined magnetic structure due to its distinct difference to conventional FMR: using a VNA-FMR one can sweep the frequency at a fixed external bias field while in conventional FMR the exciting RF frequency is fixed and the external field is swept through the resonance. For micrometer-sized magnetic structures in a domain configuration this means that by using VNA-FMR a spin configuration may be conserved in a fixed bias field and the resonance of this configuration is found by sweeping the frequency. However, in all inductive techniques a detailed understanding of the dynamic response of the ferromagnetic elements under investigation can only be obtained when comparison with micromagnetic simulations or – in simple cases such as the magnetic vortex state – with simple theories can be made. In the following we will introduce optical methods that allow observation of the modal structure of small magnetic elements directly.

2.1 Time-Resolved Magneto-Optical Microscopy

Time-resolved Kerr microscopy has been a successful technique since the early 1970s. In its simplest form as used by *Argyle* and coworkers in an early experiment [40] a continuous-wave light source illuminates a ferromagnetic sample and magnetic contrast is obtained by various contrast mechanisms including the Kerr effect, Voigt effect, Faraday effect or Cotton–Mouton effect [41]. When a given domain structure is excited by a microwave magnetic-field excitation, the domain image is blurred by the motion of the domains. The blurring becomes maximum when the resonance of the mode is reached. In this case the response of the system can only be imaged qualitatively. Stroboscopic techniques have been pioneered by *Humphrey* [42] and have become more important with the availability of modern pulsed laser sources [43, 44]. This experimental approach has been reviewed in detail by *Freeman* and *Hiebert* in [45] and will only be reviewed briefly here. In a time-resolved Kerr (TR-Kerr) microscopy experiment the equilibrium state of the magnetic elements is disturbed by a short, but weak – typical field strengths do not exceed 50 Oe to 100 Oe – magnetic tipping pulse $\boldsymbol{H}_{\mathrm{p}}$ in the plane of the magnetic element or perpendicular to it. The magnetic tipping field pulse is synchronous to a probing short laser pulse. Synchronization can be achieved by electronic delay lines when using electronic trigger signals or by optical delay lines when part of the laser pulse is used to launch a current pulse via an optical switch [45, 46]. The precessional motion of the spins is excited by the tipping-field pulse and one is left to observe the response on a macroscopic length scale given by the optical experiment with typically 300 nm spatial resolution [9]. In most cases, polar Kerr images are obtained as a function of the time delay between application of the tipping pulse and arrival of the probe pulse in a stroboscopic experiment. Two methods are possible: scanning microscopy where the laser spot is focused onto the sample and the sample is scanned using a piezostage [2, 9, 13, 22, 47], or wide-field imaging where the pulsed laser system replaces the lamp of a conventional microscope and a full image is obtained without scanning [48].

The combination of a sinusoidal excitation field with a pulsed laser system is also possible, as has been demonstrated by *Tamaru* et al. [15]. Replacing the tipping-field pulse used in the TR-Kerr setup by a sinusoidal high-frequency (hf) field created by a hf source with a single frequency, one is able to directly investigate the eigenmode spectrum of a magnetic system in the frequency domain still benefitting from the spatial resolution of the Kerr setup. In order to ensure correlation between the phase of the excitation and the probe beam, a Ti:sapphire mode locked laser is phase locked to exactly 80 MHz by introducing an additional piezomirror that tunes the cavity of the laser. As a source of the hf signal one uses a frequency source that is phase locked to the laser at 80/8 MHz = 10 MHz. The frequency resolution is given by the repetition rate of the laser. In order to investigate the magnetic response at any specific phase of the excitation, the hf signal is guided to the coplanar waveguide via an electric delay. A pin modulator acts as a switch for the hf signal to allow detection of the magnetic response using lock-in techniques.

2.2 Brillouin Light Scattering

The Brillouin light scattering (BLS) technique has been reviewed in detail in [25, 49, 50]. Together with FMR and TR-Kerr it is one of the most commonly used methods to study magnetization dynamics. Its merits lie in its large frequency range (up to 100 GHz and beyond) and in its extreme sensitivity allowing detection of thermally excited modes. Its working principle will be reviewed only briefly. A monoenergetic photon beam with fixed wavelength (typically at 514.5 nm wavelength) and wavevector interacts with a magnetic sample. It can interact with a magnon, a quasiparticle described by $\boldsymbol{q}$ and ω. After a scattering process the scattered photon may gain or lose energy and momentum (corresponding to ω and $\boldsymbol{q}$ of the involved quasiparticle) depending upon if a magnon has been annihilated or created, respectively. Changing the scattering geometry one can sweep the value of $\boldsymbol{q}$ and measure the frequency dependence $\omega(\boldsymbol{q})$. In this way the spin-wave dispersion can be investigated with a frequency resolution of 0.1 GHz to 0.2 GHz. In modern BLS spectrometers the frequency shift is analyzed using a tandem Fabry–Perot spectrometer [18, 19, 50, 51]. In a recent extension of the BLS technique it has become possible to detect spectra of thermally excited quantized spin waves using a microfocus Brillouin light scattering setup [52]. The setup is characterized by a diameter of the probing laser spot smaller than 300 nm and allows for simultaneous detection of spin-wave excitations in a wide range of wavevectors of up to $2 \times 10^5\,\mathrm{cm}^{-1}$. In addition to the standard detection of thermally excited spin waves, spin waves can also be detected in this setup when the magnetic element is excited by microwave magnetic fields either via a microcoil [53] or via coplanar waveguides. Microwave excitation allows enhancement of the amplitude of those quantized spinwaves that couple efficiently to the external microwave magnetic field.

2.3 Synchrotron-Based Techniques

Synchrotron-based techniques to image the magnetization dynamics of small magnetic elements may be viewed as an extension of standard magneto-optical microscopies. Here, the probing laser pulse is replaced by the X-ray pulse of a third-generation synchrotron source. Due to its pulse width, the time resolution is usually limited to about 50 ps. One exception is the low alpha mode of BESSY II that allows for a time resolution of about 10 ps. The magnetic contrast method in all imaging techniques is the X-ray magnetic circular dichroism (XMCD) [54, 55]. X-ray magnetic circular dichroism (XMCD) in core-level absorption detects the dependence of the X-ray absorption coefficient of circularly polarized radiation on the magnetization in a ferromagnetic material. The magnetic contrast that is obtained by comparing the absorption of left circular polarized light to right circular polarized light at, e.g., the spin-orbit coupled L2,3 edges in transition metals can reach values up to 50 %. This large contrast allows the combination of XMCD with high-resolution imaging techniques. Imaging techniques may make use of the difference of the absorption directly, as for example in magnetic transmission X-ray microscopy (MTXM) or may rely on the difference in secondary electron yield due to the XMCD effect as first used in a photoemission electron microscope (PEEM) [56]. Currently, the full-field MTXM at the Advanced

Light Source in Berkeley CA, has a lateral resolution down to 25 nm provided by Fresnel zone plates used as optical elements. The typical magnetic resolution of a commercial PEEM is 50 nm. Both techniques may be extended for the study of spin dynamics in stroboscopic pump and probe experiments where the magnetization is excited by short magnetic field pulses (pump) and probed by short (< 70 ps) X-ray flashes of the synchrotron source delayed with respect to the pump probe to obtain the spatiotemporal evolution of the ultrafast magnetization. The short magnetic field pulse must be synchronized to the revolution clock of the electron bunches in the storage ring. In most cases the RF frequency of the electron accelerator is used as a reference signal to phase lock either an electronic delay generator in combination with an electronic pulse generator [57, 58] or a laser system. If a laser system is used, it serves solely to trigger an optical switch that in turn launches a current pulse in a waveguide where the sample is located [59, 60].

3 Excitation Schemes

The field pulse excitation may be in the plane of the ferromagnetic element or perpendicular to it, depending on the lithographically prepared samples. Figure 1 shows two typical schemes. For perpendicular excitation microcoil geometries are used [2, 9, 30, 61, 62] as depicted in Fig. 1a, while for inplane excitations coplanar waveguides are chosen, see Fig. 1b. For both geometries the field pulse strength as a function of injected current can be calculated using the Biot–Savart law and is usually in the range of 10 Oe to 50 Oe. Exact information about the temporal evolution of the magnetic field pulse is harder to obtain. Two experimental methods are used: 1. *Freeman* and *Elezzabi* [63] proposed to use a biased ferromagnetic garnet film to image the temporal response directly via the fast response of this particular material. 2. Ringing the ferromagnetic element in a strong external magnetic field may give a similar estimate of rise and fall times. Unfortunately, lithographically prepared magnetic field sources such as microcoils or coplanar waveguides are often accompanied by unwanted damping and reflections.

3.1 Pulsed Precessional Motion

Figure 1 shows two schematic examples of the possible excitation direction used for pulsed precessional motion experiments. Here, a short field pulse disturbs the equilibrium state of a magnetic element. The element may be placed inside the microcoil or on top of the coplanar waveguide. In pulsed precessional motion, the field pulse may be viewed as a broadband excitation and a Fourier transformation reveals the frequency spectrum available in the experiment. Typical magnetic field pulse rise times are 10 ps to 20 ps if Auston switches are used as voltage-pulse sources triggered by a laser pulse [46], but may increase to 50 ps to 100 ps if photodiodes or commercial electronic

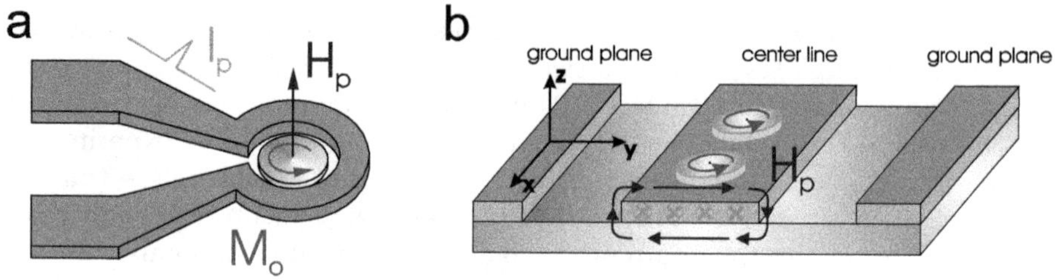

Fig. 1. Illustration of geometries used for magnetic-field generation. (**a**) Lithographically produced microcoils provide perpendicular pulse fields H_p at the sample region (*gray*), while coplanar waveguides (**b**) are used for fields in the plane of the sample. The latter geometry has the advantage that many different structures can be placed on the same sample

pulse generators are used as voltage-pulse sources. Fall times are typically longer than 100 ps. In a recent experiment *Gerrits* et al. [64] have shown that by superposition of two laser-triggered current pulses narrow magnetic field pulses with pulsewidths of the order of the rise time can be produced.

3.2 Microwave Excitation

The same structures (microcoils and coplanar waveguides) can be used to excite a sample with a sinusoidal single-frequency magnetic field. In this case a RF generator is connected to the waveguides. *Argyle* et al. [40] used a continuous-wave illumination in a Kerr microscope to observe the blurring of domain walls after excitation with oscillating external fields up to 40 MHz. In more advanced experiments, the excitation source is synchronized to a pulsed laser system and the response to the excitation is imaged as a function of position or phase between the probing laser pulse and the sinusoidal field [15]. Here, the frequency resolution is given by the repetition rate of the laser.

4 The Energy Landscape of Confined Magnetic Structures

The excitation spectrum of thin magnetic films has been studied intensely in the last decade and their spin-wave spectra seem to be understood well, both, experimentally and theoretically. Confined magnetic objects add more complexity to the spectrum of spin waves. Quantization of modes caused by the inplane confinement may be observed. The exact behavior of the confined modes is governed by the dynamic boundary conditions of the elements [30,32]. Early studies have revealed confined modes in disks or stripes, but in most cases experiments have been performed in large external fields, ensuring saturation of the elements along the field direction [13,24]. In some cases [13, 16, 18, 49] localized edge modes have been found. These modes illustrate the inhomogeneous internal field distribution even in almost monodomain elements. Significantly less is known about the magnetization dynamics in mesoscopic systems with an inhomogeneous distribution of the static magnetization. The simplest of these systems is a magnetic disk in the vortex state characterized by an axial symmetry. The vortex configuration is characterized by a smooth variation of the static magnetization around the vortex core. The core may be viewed as a topological defect and as a consequence the spins in the central region turn out of the plane either up or down [65–68]. The direction of the spins in the core and the direction of flux closure in the disk give the system a handedness or chirality. The basic variable is the magnetization $\boldsymbol{M}(\boldsymbol{r}) = M_\mathrm{s}\boldsymbol{m}(\boldsymbol{r})$, its norm being considered constant, only the direction of the reduced magnetization $\boldsymbol{m}(\boldsymbol{r})$ changing with the constraint $\boldsymbol{m}^2 = 1$, and M_s being the constant saturation magnetization. From the derivative of the total internal energy density ϵ_tot with respect to the magnetization one obtains the total internal field $\boldsymbol{H}_\mathrm{eff} = -\frac{1}{\mu_0 M_\mathrm{s}}\frac{\delta\varepsilon_\mathrm{tot}}{\delta\boldsymbol{m}}$ that enters the Landau–Lifshitz equation and thus governs the precessional motion of the magnetization and the modal structure

$$\frac{\mathrm{d}}{\mathrm{d}t}\boldsymbol{M} = -\gamma_0\boldsymbol{M} \times \boldsymbol{H}_\mathrm{eff}\,. \tag{1}$$

Here, γ_0 is the gyromagnetic ratio. We will ignore the energy dissipation term that describes at what rate the magnetization relaxes back into equilibrium in a pulsed precessional motion experiment as we will only describe the mode spectrum of simple mesoscopic ferromagnetic elements in the following paragraphs. As long as the decay times are long compared to the precessional frequencies we expect no influence of damping on the modal structures. What is left to do is to determine the relevant contributions to the total internal energy density. In the case of Permalloy ($Ni_{80}Fe_{20}$), the material studied most frequently, one needs to take dipole–dipole and exchange contributions into account. The corresponding confined spin-wave modes are now being intensively studied both theoretically and experimentally and seem to be well understood [9, 18, 29, 30, 61, 62, 69].

5 Experimental Results

In the following we will review some experimental result obtained for confined magnetic elements in a flux-closure state. As a consequence of flux closure, the elements exhibit a vortex in the central part. To address the dynamic behavior of this topological defect, experiments will be reviewed for square-shaped samples as well as for disks, while for the normal modes of these elements we will restrict ourselves to disk-shaped elements.

5.1 Gyroscopic Vortex Motion in Micrometer Sized Ferromagnetic Squares and Disks

Flat circular disks made out of soft magnetic material (for example $Ni_{80}Fe_{20}$) exhibit a circulating magnetization distribution for large enough inplane sizes [65, 66, 70]. As a consequence of the circulating magnetization a vortex at the center of the element appears where the spins point perpendicular to the disk plane. The appearance of this singularity leads to a modification of the normal-mode spectrum of circular magnetic elements. For large sizes, e.g., in a size range where magnetostatic modes dominate, the normal mode spectrum of quasiuniformly magnetized disks or cylinders has been investigated by *Dillon* [71] for the cases of an external magnetic field perpendicular to the disk plane and in the plane of the disk. However, when a vortex enters the magnetic disk, the normal-mode spectrum is altered considerably. In particular, low-frequency modes appear that are associated with the motion of the vortex as a whole. The motion of the vortex has been described theoretically by *Thiele* [72] and *Huber* [73]. Essentially, they conclude that a magnus force perpendicular to the velocity of the moving vortex acts on the magnetic structure leading to a damped circular orbit for the translation mode described by *Thiele's* equation [72]. The frequency of this mode depends on the details of the magnetic structure, while the direction of the motion (clockwise or counterclockwise) depends on the direction of the spins in the vortex core (up or down). The direction of flux closure and the polarization of the vortex core impose a handedness or chirality to this particular spin structure. *Ivanov* and *Zaspel* [74] have calculated the frequencies of this mode for a finite spin lattice, albeit taking into account only exchange interactions. In a recent article *Guslienko* et al. [75] have extended the investigation of vortex oscillations to more realistic ferromagnetic system taking the magnetostatic energy into account. They could show that this energy term determines the vortex translation eigenfrequency. They also show by comparison to micromagnetic simulations that a rigid vortex model where the spin structure is displaced as a whole cannot describe the eigenfrequencies accurately, while their so-called "two-vortex model", where the spin structure is deformed to avoid surface charges at the edges of the disks, can.

The appearance of a singular spin structure at the center of a magnetic element is not limited to ferromagnetic disks, but must also occur for other flux-closed spin structures, a prominent one being the so-called Landau state in rectangles or squares. Here, a magnetic vortex is located at the intersection of four domain walls, illustrated in Fig. 2a for the case of a square and a "Swiss cross" geometry with Néel walls in Fig. 2b and Fig. 2c. The "Swiss cross" geometry was investigated by *Argyle* and coworkers [40] in 1984 using magneto-optical methods. They used a continuous-wave laser system to image the magnetic structures in an optical microscope with about 300 nm spatial resolution. Their setup leads to three consequences: 1. they cannot resolve the vortex directly (the spatial extent of the vortex region for the ferrimagnetic

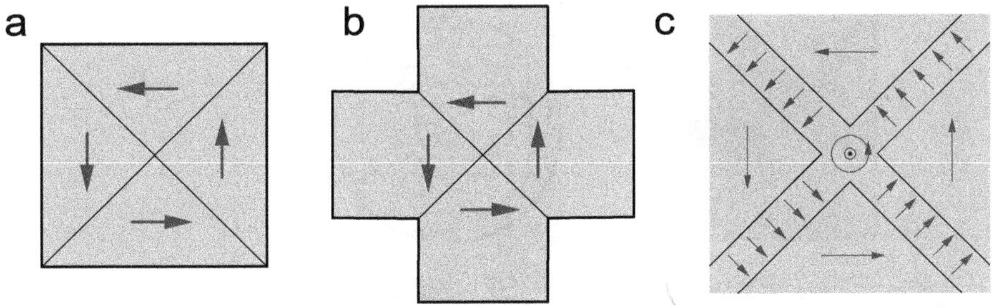

Fig. 2. (**a**) The flux-closed Landau state formed by the four intersecting domains is very similar to the center region of the "Swiss cross" structure shown in (**b**). (**c**) is the central detail of the continuous magnetization distribution (*arrows*). Generally, $\boldsymbol{M}$ lies in the plane of the film plane, except within the vortex region where it turns normal to the plane

garnet film used in their experiment is about 20 nm), but they can follow its motion indirectly by monitoring the domain-wall motion. 2. When using a cw-laser source in connection with a microwave excitation that brings the vortex into a circular orbit they can only observe blurring of the domain walls. When sweeping the microwave frequency, they record the size of the blurred region as a function of frequency and can locate the resonant frequency for the vortex-displacement mode. 3. They cannot observe the direction of motion of the spin structure. In their analysis they show that only the central region of the spins leading to the formation of the vortex is responsible for the resonant frequency, while the Néel walls contribute only negligibly.

In 2003 *Park* and coworkers [14] investigated the vortex motion in circular $Ni_{80}Fe_{20}$ disks in a stroboscopic time-resolved Kerr microscopy experiment using a pulsed laser source. In their approach, the vortex motion is triggered by a 150 ps inplane magnetic field pulse that displaces the position of the vortex region. They subsequently probe the dynamic response of the polar magnetization component in the central region of the magnetic disks with 540 nm spatial resolution. The result is shown in Fig. 3. Besides the high-frequency response of the magnetization caused by the oscillation of $\boldsymbol{M}$ about the local internal magnetic field, a low-frequency component can clearly be observed. This component is associated with the spiraling motion of the vortex back into its equilibrium position. Comparison to micromagnetic simulations (Fig. 3 right side) and to the calculated frequencies by *Guslienko* et al. [75] confirms that the "pole-free" model is best suited to describe the spiraling motion.

Again, the optical resolution of Park's experiment does not allow for the determination of the sense of rotation of the vortex core, nor is it possible to determine the direction of flux closure.

In 2004 *Choe* et al. [59] addressed this problem. They used a synchrotron-based technique to image the vortex motion in soft CoFe alloy thin-film ele-

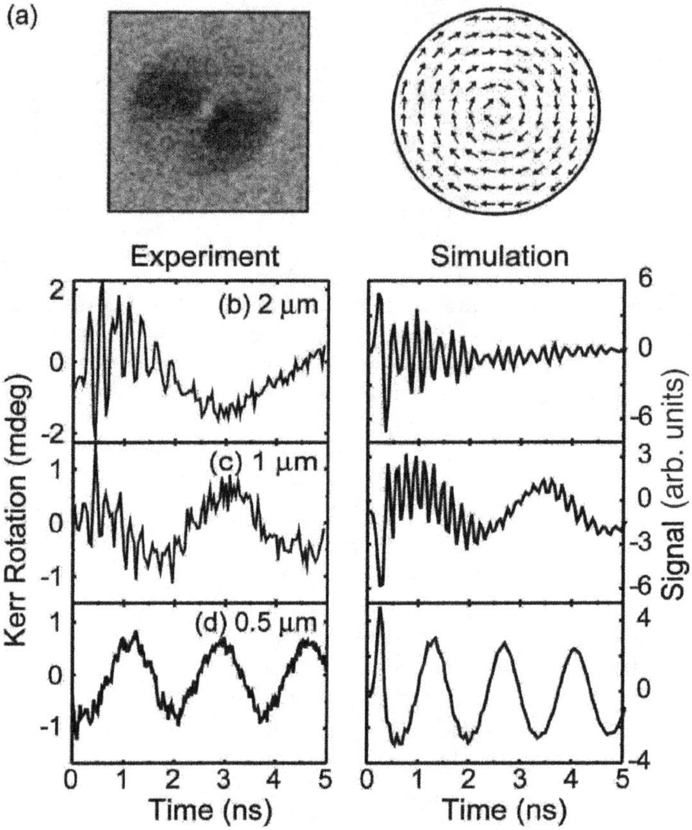

Fig. 3. Figure 3 from [14]: Magnetic force microscope image (*left*) and schematic of the vortex structure (*right*) of a 500 nm disk. The bright spot at the center of the disk in the image is due to the large z-component of the magnetization. (**b**), (**c**), (**d**) Experimental (*left*) and simulated (*right*) time-domain polar Kerr signals for vortex structures of diameters 2 mm, 1 mm, and 500 nm near the center of each disk. The low-frequency signal that is particularly prominent in the case of the 500 nm disk is the gyrotropic mode discussed in the text

ments. Their squares and rectangles (typical sizes are 1 µm × 1 µm, 1.5 µm × 1 µm, and 2 µm × 1 µm) showed the flux-closed Landau-spin configuration. These elements were imaged with high resolution ($\leq$ 100 nm) in a photoemission electron microscope exploiting the XMCD effect. Using XMCD the direction of flux closure (clockwise or counterclockwise) could be imaged directly. The magnetic structures were placed on top of a coplanar stripline that was connected to an optical switch. Using a ps laser pulse, which was synchronized to the revolution clock of the electron bunches in the storage ring, and focused onto the optical switch, 300 ps current pulses were launched into the stripline and excited the spin motion of the elements similar to the experiment of *Park* et al. [14]. The spin motion was then imaged in a time-resolved version of a regular PEEM experiment with about 70 ps temporal resolution. This time resolution was sufficient to image the spiraling vortex motion. While in this experiment the spatial resolution still was not sufficient to resolve the vortex directly, the authors inferred the direction of the central spins by observing the sense of rotation, as the direction of flux closure was known. As predicted by *Thiele* [72] the sense of rotation depends on the polarization of the magnetic vortex structure. In their experiment the magnetic elements were excited every 8 ns due to the repetition rate of the laser system. This may result in a "quasiresonant" experiment as the vortex still possesses a finite velocity when the next exciting laser pulse arrives.

5.2 The Modal Spectrum of Permalloy Disks

In the following we will review results obtained for the modal spectrum of ferromagnetic disks in the flux-closure state, but we will limit ourselves to the normal-mode spectrum of disks in the size range of about 1 μm diameter to several micrometers. In this size regime the normal-mode spectrum is dominated by the magnetostatic energy. First experiments have been performed by *Hiebert* et al. [2] using an 8 μm diameter, 100 nm thick $Ni_{80}Fe_{20}$ disk. They report on the frequency dependence of the oscillations of the perpendicular magnetization component of the disks after excitation with a perpendicular magnetic field pulse in various external inplane fields. Due to the inplane fields the vortex has been pushed out of the sample and the disks are in a "quasiuniform state". Similar experiments for saturated disks with thicknesses $d = 10$ nm, 20 nm, and 40 nm and diameters 1 μm and 2 μm have been performed with BLS by *Jorzick* et al. [27]. Quantized spin-wave modes have been found and are reviewed in detail in [24, 25]. Similar results are obtained for a variety of smaller disks by *Gubbiotti* et al. [18]. This group found two families of spin-wave modes with wavevector parallel and perpendicular to the magnetization direction similar to results obtained in stripes [13, 16, 33]. In [19] spin excitations in disks in the vortex as well as in the saturated states are measured using BLS and compared to calculations. Excitations with positive and negative spin-wave dispersion are found in the calculations. Experimental data for the normal-mode spectrum of disks in the vortex state were obtained by *Novosad* et al. [29] using BLS. They observed two modes at zero bias field that merge into a Kittel-like mode for saturated disks when an externally applied field is increased. In 2002, *Ivanov* and *Zaspel* [74] gave some insight into the normal-mode spectrum of disks in the vortex state.

In 2000, *Acremann* et al. [9] imaged the response of a Co disk in a flux-closure state as a function of time elapsed after the excitation with a perpendicular field pulse with 300 nm resolution in a time-resolved Kerr microscopy experiment. The measured spatially inhomogenous response has been analyzed by *Buess* et al. [30] and has been attributed to a superposition of magneto-static modes based on a simple dipolar model. In 2003, *Park* et al. [14] showed that modes can be imaged directly in micrometer-sized ferromagnetic elements by Fourier transforming a whole image sequence obtained by time-resolved Kerr microscopy. Their experiments show two modes in micrometer-sized square samples in the flux-closed Landau configuration located in the domains and in the domain walls.

In this section the multimode excitation spectrum measured by time-resolved Kerr microscopy is analyzed. The two key components of our experimental setup are a Ti:sapphire pulsed laser system and a sturdy polarization conserving optical microscope. The laser operates at 800 nm wavelength with pulses of 120 fs at 80 MHz repetition rate. The fundamental wavelength of the laser is frequency doubled to improve the optical resolution in our microscope to about 300 nm.

Single Permalloy disks with diameters of 3 µm, 4 µm and 6 µm are studied that were produced by e-beam evaporation on a Si substrate and capped with 2 nm Al for corrosion protection. Each disk was placed inside a single-turn Au loop of 300 nm thickness connected to a 50 Ω coplanar transmission line. The loop has inner and outer diameters of 8 µm and 12 µm, respectively. For a sketch of the sample see Fig. 1. Using nonpatterned films evaporated in the same batch as the samples under investigation, the sample thickness, d and its saturation magnetization, M_s were measured by means of X-ray fluorescence and a superconducting quantum interference device magnetometer. The obtained values are 15 nm and 800 G, respectively.

Let us now consider applying a short magnetic field pulse to a magnetic element. At positions where the field pulse is not collinear with the equilibrium magnetization direction, a torque will act on the magnetization and $\boldsymbol{M}$ will start a precessional motion, according to (1). However, the dynamic response is strongly affected by the finite size of the element and the existence of boundaries resulting in various eigenmodes whose spatial distribution and frequency depends on the exact shape and size of the element.

In our experiment a current pulse of about 100 ps rise time launched into the microcoil provides the perpendicularly oriented magnetic tipping field pulse exciting the spin precessional motion [2, 9, 30] of the ferromagnetic microstructures under examination. The magnetic response of the individual disks is examined by recording a time-resolved movie with a high spatial resolution of 300 nm. The maximum strength of the field pulse is less than $\approx$ 50 Oe and leads to a small deviation from the magnetic equilibrium.

Figure 4 shows a stroboscopic sequence of polar Kerr microscopy images obtained during and after application of the field pulse for a disk with 3 µm radius. The time interval between each image is 30 ps. The image contrast is produced by the z-component of the magnetization vector undergoing a spatially nonuniform motion. Notice that our experiments measure changes of M_z with respect to the equilibrium configuration. Therefore the image of the initial state (the first image) shows no contrast. Despite the complex-

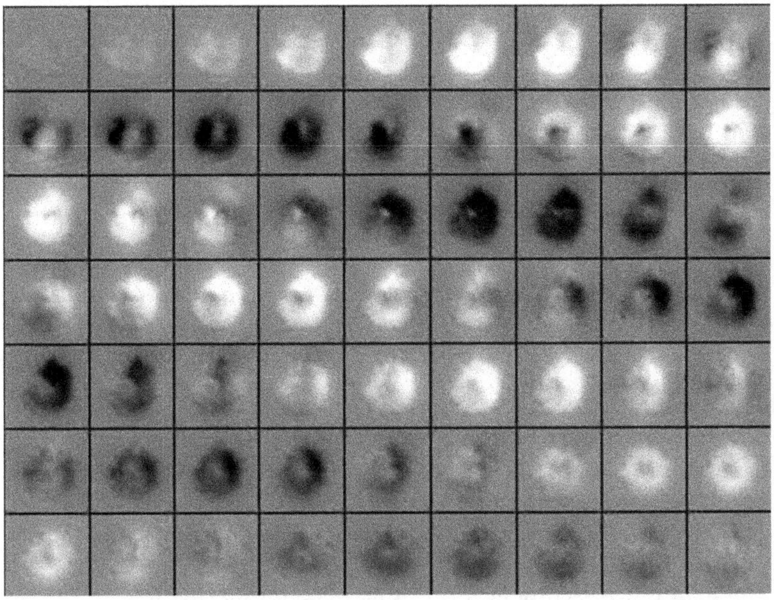

Fig. 4. Temporal evolution of the precessional motion. $M_z(t) - M_z(t=0)$ – i.e., the difference between the value of the z-component of the magnetization vector before application of the pulse and after application of the pulse – is imaged as a function of the time elapsed after the magnetic field pulse. Notice that some images show a central spot that is clearly distinguishable from the surroundings. This feature is discussed in [61]. From such a sequence one may extract an overall periodicity corresponding to the strongest excited mode. The spin motion, however, is not uniform, but is the superposition of a number of modes (see also [13, 33])

ity of the sequence, one recognizes an overall periodicity of the motion with maxima (bright) and minima (dark) of M_z recurring after a characteristic time of the order of 350 ps. The modal structure of the motion is not immediately apparent from Fig. 4 but will emerge from a Fourier analysis. Notice that the regime of small deviations from equilibrium the inplane tangential component of the magnetization vector is, to first order, unaffected by the motion. The radial component M_r (not measured in the present experiment) is connected to $M_z(t)$ by the relation $\dot{M}_\mathrm{r} \propto M_z$ and has a phase difference of $\pi/2$, so that the magnetization vector performs an elliptical precessional path as shown in [9]. Thus, in this linear limit, it is sufficient to consider only the z-component of the magnetization in order to have the full knowledge of the modal structure. We have measured only the component that could be detected with the greatest precision, namely the one perpendicular to the disk (M_z).

Some information can be gained by analyzing the average perpendicular Kerr signal over the whole element as shown in Fig. 5a. We clearly observe oscillatory behavior as a function of time and can see at first glance that

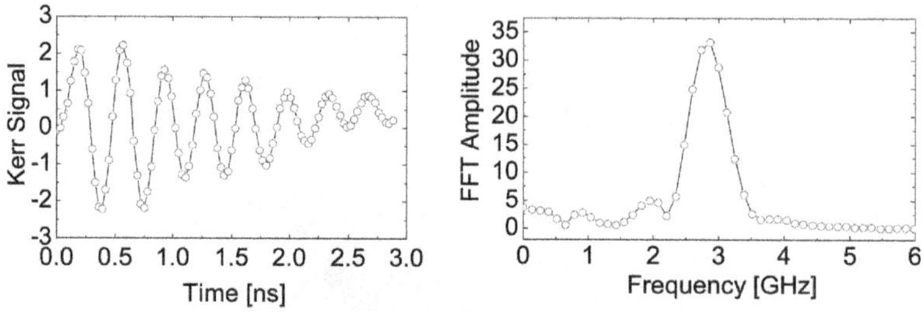

Fig. 5. (**a**) Average Kerr signal of the data from Fig. 4. (**b**) Fourier transformation of the data of (a)

this behavior cannot be described by a single damped harmonic oscillator. The characteristic period of the order of 350 ps of the signal hints at dipolar interactions as the main source for the observed modes. In Fig. 5b a Fourier spectrum of this data is shown that reveals distinct peaks at 2.8 GHz and 3.9 GHz together with a broad structure around 2 GHz. One of the difficulties in identifying the dynamical eigenmodes is that the excitation by, e.g., a short magnetic field pulse results in a complicated spin motion where several modes are superposed in an intricate way. This difficulty is circumvented by using a phase-sensitive Fourier transformation (FT) technique [14, 76]. Frequency-domain analysis of the time-sequenced images reveals resonances corresponding to the eigenmodes defined by the lithographically produced elements.

A detailed view of the various eigenmodes driving the spin motion in Fig. 4 is obtained by Fourier transforming the time-domain signal recorded at each location into the frequency domain. This is referred to as the local Fourier transform. Of each Fourier transform (FT), not only the amplitude – as in [14,17] – is retained but also the phase. An illustration of the procedure is shown in Fig. 6. The experimental spatiotemporal response of the magnetic system to the field pulse is available as an image sequence, i.e., spatial maps of one component of $\boldsymbol{M}$ as a function of time. Figure 6 illustrates the process of the local Fourier transform, which results in maps of the amplitude and phase as a function of frequency. The Fourier spectrum consists of a sequence of resonances along the frequency axis marking the characteristic eigenmodes of the circular sample. The location of the maximum of these resonances can be determined with an accuracy of 0.2 GHz (corresponding to the separation of data points in the FT). The typical width of the resonance curves is 0.5 GHz to 1.0 GHz. The values of the resonance frequencies have been accurately reproduced by an analytical calculation based on [30] and by a micromagnetic simulation [77]. Images of the amplitude (top) and the phase (bottom) at resonance of various eigenmodes are displayed in Fig. 7. One recognizes two types of modes: Some are organized into concentric rings with circular nodes. The spectral weight and phase distribution (Fig. 7a to Fig. 7c) obey the axial symmetry of the ground state. The modes in Fig. 7d and Fig. 7e instead,

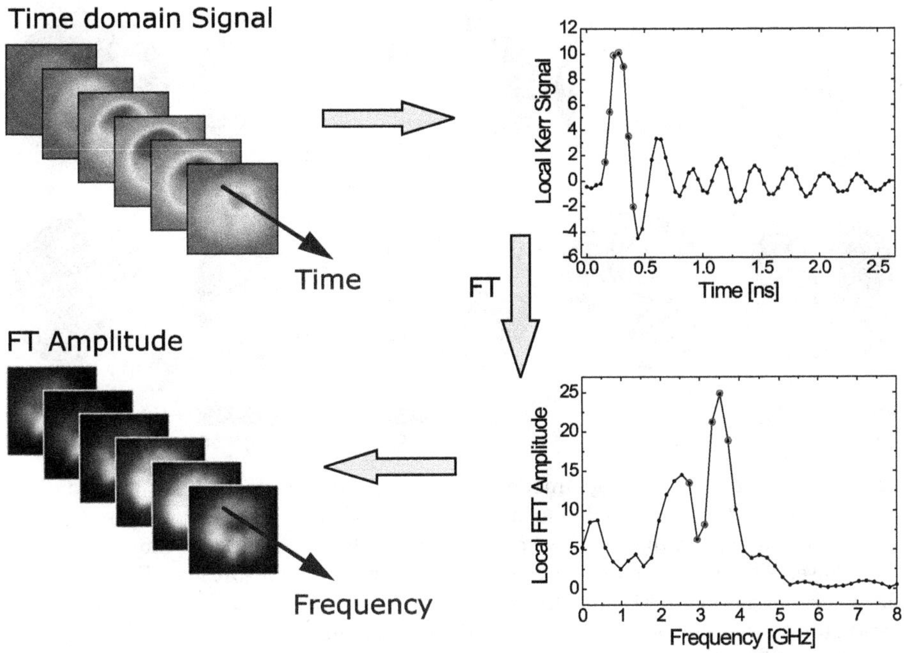

Fig. 6. Illustration of the local Fourier-transform process. *Top:* For each position in space, the time-domain signal is extracted from the magnetic image sequence. These time signals are Fourier transformed. *Bottom:* The local spectra are recombined in the frequency domain and form once again a sequence of images that yield the local distribution of the amplitude (and phase) at the different frequencies

have one, respectively two, diametric nodes and break the axial symmetry. The various nodal lines are easily identified because at their spatial location the spectral weight is small and the phase jumps by π.

Let us first consider the axially symmetric modes. Fourier-amplitude (left) and -phase (right) images of the three low-lying axially symmetric modes are displayed in Figs. 7a–c. The left-hand side of each image shows the result of micromagnetic simulations, the right-hand side is the amplitude (or phase) of the Fourier transform of the experimental time-domain images. The modes can be classified according to the number of nodes: the fundamental and highest-amplitude mode has a node only in the center of the disk and at its border ($n = 1$, Fig. 7a). The phase is uniform over the disk. The next mode has a node at approximately half-way between the core and the boundary ($n = 2$, Fig. 7b). Going across the node the temporal phase changes by π, just as in any standing wave. Finally, we observe a third mode with two nodes within the disk ($n = 3$, Fig. 7c) and with a π phase jump across each node. This modal structure is well described by the first-order Bessel function $J_1(k_n r)$, where k_n assumes discrete values determined by the boundary conditions and r is the radius. Such modes are natural basis functions in a

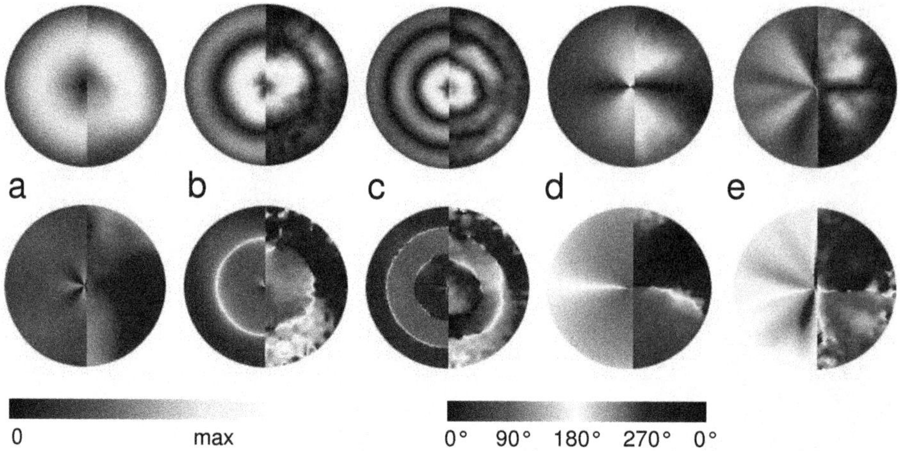

Fig. 7. Images of the eigenmodes for a disk with radius $R = 3\,\mu\text{m}$. FT of Fig. 4 at each location results in a spectrum with five resonances. The *top row* shows the absolute value of the Fourier amplitude at resonance, the *bottom row* the phase. The modal maps are composed from two half-images: the left from the micromagnetic simulation, the right from the experiment. For the micromagnetic simulation the sample was divided into $10.6\,\text{nm} \times 10.6\,\text{nm} \times 15.0\,\text{nm}$ pixels in a 375×375 array. The saturation magnetization, exchange stiffness, gyromagnetic frequency and damping constant are 860 emu/cc, 1.3 μerg/cm, 17.6 MHz/Oe and 0.008, respectively. The equations of motion were integrated in 0.4 ps steps from $t = 0\,\text{ns}$ to $t = 20\,\text{ns}$. The data were stored for the entire array and for the average value of each magnetization component. (**a**)–(**c**): Axially symmetric modes showing concentric nodes ($n = 1, 2, 3$, $m = 0$). Mode (**a**) at 2.80 GHz has the largest spectral weight, i.e., it is dominating the motion and is responsible for the overall periodicity apparent from Fig. 4. (**b**): 3.91 GHz and (**c**): 4.49 GHz are higher modes with regions exhibiting different phases. (**d**), (**e**): These modes have nodal lines going across the image. (**d**): 2.08 GHz, one azimuthal node $(1, 1)$. (**e**): 1.69 GHz, two azimuthal nodes $(1, 2)$. This mode consists of four regions, oscillating in pairs in phase. Notice that across the nodal lines the phase changes by 180°. A Hamming cutoff window was used to obtain the FFT data from the 3.33 ns long scans. Although the spacing in the frequency domain of the raw data is $df = 1/(T_{\text{end}} - T_{\text{start}}) = 1/3.33\,\text{ns} = 0.30\,\text{GHz}$, the exact location of the maximum of these resonances can be determined with a higher accuracy when using zero filling. The half-images from the time-domain micromagnetic simulation data were obtained using the same procedure

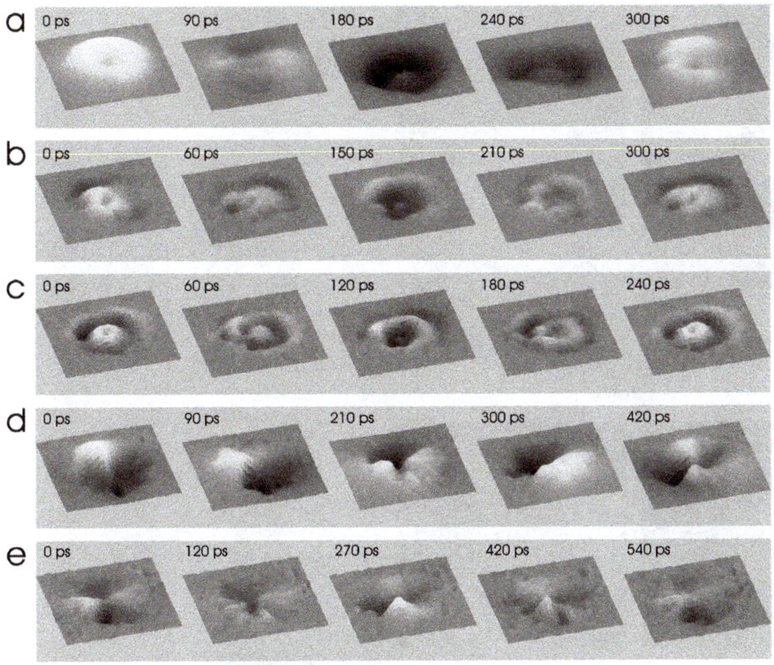

Fig. 8. Back transformation into the time domain of the modes at resonance is an alternative way of displaying the motion of the individual modes. Each row containing five movie frames (**a**)–(**e**) contains the same information as the experimental modal maps of Figs. 7a–e. The frames are labeled with the time difference related to the first image

circular membrane [78]. The frequencies can be calculated by a surprisingly simple analytical model described in detail in [30, 62].

In the Fourier-transform method used above, the signal $M_z(x, y, t)$ is Fourier transformed along the time axis to give the complex-valued spectral maps $S(x, y, f)$. An alternative way of interpreting the phase and amplitude images at a certain frequency f^* is to visualize it as a motion in the time domain again. This is done by setting all $S(x, y, f)$ to zero except for f^*. Usually f^* is the frequency at the peak maximum of the amplitude spectrum. In contrast to Fig. 4, the time sequence of Fig. 8 images only the specific mode under consideration. As the time-domain signal is Fourier transformed, retaining both the amplitude and the phase of the FT, this back-transformation of the complex FT at resonance allows mode-by-mode reconstruction. Figure 8 shows selected images from the back-transformation sequence of the experimental modes (a)–(e) displayed in Fig. 7.

Only axially symmetric modes are expected if the tipping pulse is uniform over the disk and all geometries are perfectly axially symmetric. Symmetry-breaking modes, instead, require a nonuniform tipping pulse or a deviation

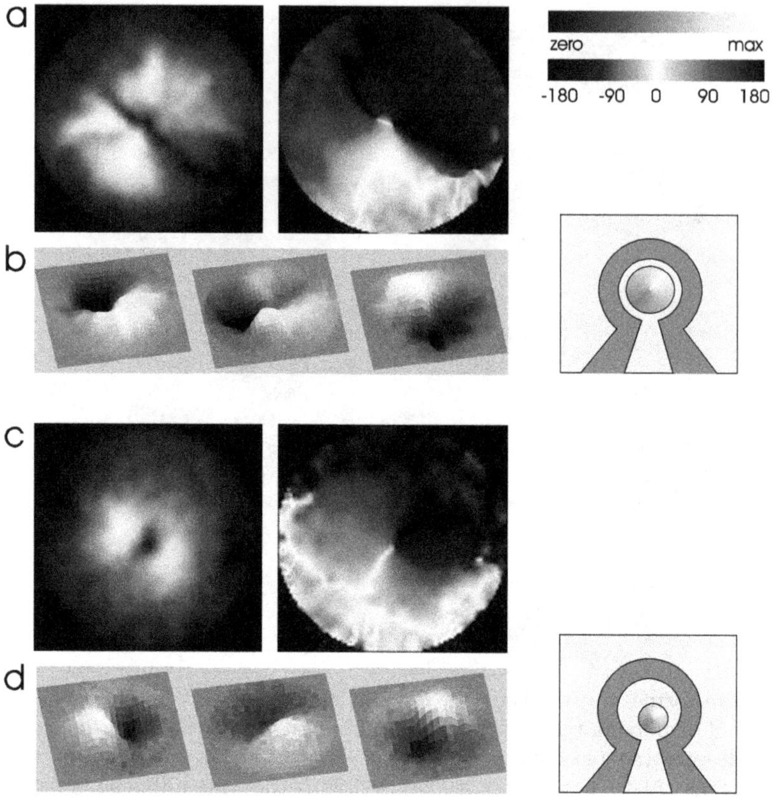

Fig. 9. Single-node symmetric mode ($n = 1, m = 1$). (**a**): On the left the amplitude, on the right the phase of the mode from the 6 μm diameter disk at 1.9 GHz is displayed. The mode consists of two regions divided by a node and oscillating with equal Fourier amplitude, but roughly opposite phase. Three images of the temporal evolution of this mode by back transformation are given in (**b**): Time interval between the middle (*right*) image and the left image: 120 ps and 270 ps. (**c**): The single-node nonaxially symmetric mode in a 3 μm disk occurs at 2.8 GHz. The node is barely visible (*left*) and the phase (*right*) changes gradually along a trajectory surrounding the core center. Correspondingly, the temporal sequence obtained by back transformation – given in (**d**) after a time interval of 125 ps and 250 ps consists of a bright-dark doublet rotating around the disk center

of the sample from a perfect cylindrical shape. Our computation of the field configuration arising from the single-turn coil reveals that the tipping pulse has a sizeable gradient in the plane of the vortex, owing to the coil opening towards the leads. The maximum difference in tipping field amplitude reaches 30 %. As revealed by our micromagnetic simulations [77], this asymmetry is capable of exciting nonaxially symmetric modes. Figure 7d shows Fourier amplitude (top) and Fourier phase (bottom) of a single-node, nonaxially symmetric mode at 2.08 GHz. The lowest lying nonaxially symmetric mode is imaged in Fig. 7e (top: amplitude, bottom: phase). This mode has two mutually orthogonal radial nodal lines (top). In going across each node the phase changes by π (bottom). Again, images of these axially nonsymmetric modes in the time domain can be reconstructed by back transformation. Snapshots of the sequence are shown in Fig. 8d and Fig. 8e.

We observe, for the single-node nonaxial mode, a systematic striking behavior, illustrated in Fig. 9 for a 6 µm (top), respectively 3 µm, diameter disk (bottom). The Fourier amplitude is on the left of Fig. 9a and Fig. 9c, the Fourier phase is on the right and the reconstructed time sequences in Fig. 9b and Fig. 9d. The phase changes in both cases, but in particular in Fig. 9c, almost continuously along a circular path surrounding the center, in contrast to Fig. 9a. Correspondingly, the time-domain sequences of Fig. 9b and Fig. 9d reveal that this mode is traveling along a circular trajectory surrounding the core. We explain this motion by noticing that the symmetry of this problem allows the existence of two mutually orthogonal degenerate single-node modes that can be simultaneously excited, their relative weight depending on the actual geometry of the tipping pulse. Superposition of such modes can produce the observed "circularly polarized" eigenmode – just as, by analogy, linearly polarized light can be superposed to yield circular polarization. In these specific samples, optical micrographs reveal that the 3 µm disk is displaced from the coil center while the 6 µm disk is not: This explains why the "circularly polarized" mode is more pronounced in Fig. 9c and Fig. 9d than in Fig. 9a and Fig. 9b.

6 Conclusions

There are two major points one may consider as conclusions: 1. with the available experimental techniques it is possible to not only determine the frequency spectrum of normal modes in confined magnetic structures, but at least for the size range around and above 1 µm it is now possible to obtain spatial maps of the modal structure of the dynamic magnetization. Further improvement of the spatial resolution is necessary to investigate a size range where exchange interaction starts to take a more prominent role. 2. The comparison to experiments and the evident excellent agreement with experiments gives micromagnetic simulations, when used in the proper way, a significant weight in the investigation of the magnetization dynamics of magnetic nanostructures. They can be used and trusted to guide the experimentalist to new and exciting phenomena in magnetization dynamics.

References

[1] A. Y. Elezzabi, M. R. Freeman, M. Johnson: Direct measurement of the conduction electron spin-lattice relaxation time t1 in gold, Phys. Rev. Lett. **77**, 3220 (1996)

[2] W. K. Hiebert, A. Stankiewicz, M. R. Freeman: Direct observation of magnetic relaxation in a small permalloy disk by time-resolved scanning Kerr microscopy, Phys. Rev. Lett. **79**, 1134 (1997)

[3] M. R. Freeman, W. K. Hiebert, A. Stankiewicz: Time-resolved scanning Kerr microscopy of ferromagnetic structures (invited), J. Appl. Phys. **83**, 6217 (1998)

[4] R. H. Koch, D. W. Abraham, P. L. Trouilloud, R. A. Altman, Y. Lu, W. J. Gallagher, R. E. Scheuerlin, K. P. Roche, S. S. P. Parkin: Magnetization reversal in micron-sized magnetic thin films, Phys. Rev. Lett. **81**, 4512 (1998)

[5] T. M. Crawford, T. J. Silva, C. W. Teplin, C. W. Rogers: Subnanosecond magnetization dynamics measured by the second-harmonic magneto-optic Kerr effect, Appl. Phys. Lett. **74**, 3386 (1999)

[6] G. Ju, A. V. Nurmikko, R. F. C. Farrow, R. F. Marks, M. J. Carey, B. A. Gurney: Ultrafast time resolved photoinduced magnetization rotation in a ferromagnetic/antiferromagnetic exchange coupled system, Phys. Rev. Lett. **82**, 3705 (1999)

[7] R. J. Hicken, J. Wu: Observation of ferromagnetic resonance in the time domain, J. Appl. Phys. **85**, 4580 (1999)

[8] G. E. Ballentine, W. K. Hiebert, A. Stankiewicz, M. R. Freeman: Ultrafast microscopy and numerical simulation study of magnetization reversal dynamics in permalloy, J. Appl. Phys. **87**, 6830 (2000)

[9] Y. Acremann, C. H. Back, M. Buess, O. Portmann, A. Vaterlaus, D. Pescia, H. Melchior: Imaging precessional motion of the magnetization vector, Science **290**, 492 (2000)

[10] T. M. Crawford, P. Kabos, T. J. Silva: Coherent control of precessional dynamics in thin film permalloy, Appl. Phys. Lett. **76**, 2113 (2000)

[11] M. Bauer, R. Lopusnik, J. Fassbender, B. Hillebrands: Suppression of magnetic-field pulse-induced magnetization precession by pulse tailoring, Appl. Phys. Lett. **76**, 2758 (2000)

[12] B. C. Choi, M. Belov, W. K. Hiebert, G. E. Ballentine, M. R. Freeman: Ultrafast magnetization reversal dynamics investigated by time domain imaging, Phys. Rev. Lett. **86**, 708 (2001)

[13] J. P. Park, P. Eames, D. M. Engebretson, J. Berezovsky, P. A. Crowell: Spatially resolved dynamics of localized spin-wave modes in ferromagnetic wires, Phys. Rev. Lett. **89**, 277201 (2002)

[14] J. P. Park, P. Eames, D. M. Engebretson, J. Berezovsky, P. A. Crowell: Imaging of spin dynamics in closure domain and vortex structures, Phys. Rev. B **67**, 020403(R) (2003)

[15] S. Tamaru, J. A. Bain, R. J. M. van de Veerdonk, T. M. Crawford, M. Covington, M. H. Kryder: Imaging of quantized magnetostatic modes using spatially resolved ferromagnetic resonance, J. Appl. Phys. **91**, 8034 (2002)

[16] J. Jorzick, S. O. Demokritov, B. Hillebrands, M. Bailleul, C. Fermon, K. Y. Guslienko, A. N. Slavin, D. V. Berkov, N. L. Gorn: Spin wave wells in nonellipsoidal micrometer size magnetic elements, Phys. Rev. Lett. **88**, 047204 (2002)

[17] R. J. Hicken, A. Barman, V. V. Kruglyak, S. Ladak: Optical ferromagnetic resonance studies of thin film magnetic structures, J. Phys. D: Appl. Phys. **36**, 2183 (2003)

[18] G. Gubbiotti, G. Carlotti, T. Okuno, T. Shinjo, F. Nizzoli, R. Zivieri: Brillouin light scattering investigation of dynamic spin modes confined in cylindrical permalloy dots, Phys. Rev. B **68**, 1844091 (2003)

[19] L. Giovannini, F. Montoncello, F. Nizzoli, G. Gubbiotti, G. Carlotti, T. Okuno, T. Shinjo, M. Grimsditch: Spin excitations of nanometric cylindrical dots in vortex and saturated magnetic states, Phys. Rev. B **70**, 172404 (2004)

[20] H. W. Schumacher, C. Chappert, P. Crozat, R. C. Sousa, P. P. Freitas, J. Miltat, J. Fassbender, B. Hillebrands: Phase coherent precessional magnetization reversal in microscopic spin valve elements, Phys. Rev. Lett. **90**, 017201 (2003)

[21] H. W. Schumacher, C. Chappert, R. C. Sousa, P. P. Freitas, J. Miltat: Quasi-ballistic magnetization reversal, Phys. Rev. Lett. **90**, 017204 (2003)

[22] T. Gerrits, H. A. M. van den Berg, J. Hohlfeld, L. Br, T. Rasing: Ultrafast precessional magnetization reversal by picosecond magnetic field pulse shaping, Nature **418**, 509 (2002)

[23] W. K. Hiebert, L. Lagae, J. D. Boeck: Spatially inhomogeneous ultrafast precessional magnetization reversal, Phys. Rev. B **68**, 020402(R) (2003)

[24] S. O. Demokritov: Dynamic eigen-modes in magnetic stripes and dots, J. Phys.: Condens. Matter **15**, 2575 (2003)

[25] S. O. Demokritov, B. Hillebrands: Spinwaves in laterally confined magnetic structures, in B. Hillebrands, K. Ounadjela (Eds.): *Spin Dynamics in Confined Magnetic Structures I*, vol. 83, Top. Appl. Phys. (Springer-Verlag, Berlin, Heidelberg 2002) p. 65

[26] U. Ebels, L. D. Buda, K. Ounadjela, P. E. Wigen: Small amplitude dynamics of nonhomogeneous magnetization distributions, in B. Hillebrands, K. Ounadjela (Eds.): *Spin Dynamics in Confined Magnetic Structures I*, vol. 83, Top. Appl. Phys. (Springer, Berlin, Heidelberg 2002) p. 167

[27] J. Jorzick, S. O. Demokritov, B. Hillebrands, B. Bartenlian, C. Chappert, D. Decanini, F. Rousseaux, E. Cambrill: Spin-wave quantization and dynamic coupling in micron-size circular magnetic dots, Appl. Phys. Lett. **75**, 3859 (1999)

[28] J. Wu, D. S. Schmool, N. D. Hughes, J. R. Moore, R. J. Hicken: Picosecond large angle reorientation of the magnetization in $Ni_{81}Fe_{19}$ circular thin-film elements, J. Appl. Phys. **91**, 278 (2002)

[29] V. Novosad, M. Grimsditch, K. Y. Guslienko, P. Vavassori, Y. Otani, S. D. Bader: Spin excitations of magnetic vortices in ferromagnetic nanodots, Phys. Rev. B **66**, 052407 (2002)

[30] M. Buess, Y. Acremann, A. Kashuba, C. H. Back, D. Pescia: Pulsed precessional motion on the "back of an envelope", J. Phys.: Condens. Matter **15**, R1093 (2003)

[31] B. Hillebrands, K. Ounadjela (Eds.): *Spin Dynamics in Confined Magnetic Structures I* (Springer, Berlin 2002)

[32] K. Y. Guslienko, S. O. Demokritov, B. Hillebrands, A. N. Slavin: Effective dipolar boundary conditions for dynamic magnetization in thin magnetic stripes, Phys. Rev. B **66**, 132402 (2002)
[33] C. Bayer, J. P. Park, H. Wang, M. Yan, C. E. Campbell, P. A. Crowell: Spin waves in an inhomogeneously magnetized stripe, Phys. Rev. B **69**, 134401 (2004)
[34] W. Dietrich, W. E. Proebster, P. Wolf: Nanosecond switching in thin films, IBM J. **189**, 189 (1960)
[35] S. E. Russek, R. D. McMichael, M. J. Donahue: High speed switching and rotational dynamics in small magnetic thin film devinces, in B. Hillebrands, K. Ounadjela (Eds.): *Spin Dynamics in Confined Magnetic Structures II*, vol. 87, Top. Appl. Phys. (Springer, Berlin, Heidelberg 2003) p. 93
[36] T. J. Silva, C. S. Lee, T. M. Crawford, C. T. Rogers: Inductive measurement of ultrafast magnetization dynamics in thin-film permalloy, J. Appl. Phys. **85**, 7849 (1999)
[37] M. Bailleul, D. Olligs, C. Fermon: Micromagnetic phase transitions and spin wave excitations in a ferromagnetic stripe, Phys. Rev. Lett. **91**, 137204 (2003)
[38] M. Bailleul, D. Olligs, C. Fermon: Propagating spin wave spectroscopy in a permalloy film: A quantitative analysis, Appl. Phys. Lett. **83**, 972 (2003)
[39] F. Giesen, J. Podbielski, T. Korn, M. Steiner, D. Grundlery: Hysteresis and control of ferromagnetic resonances in rings, Appl. Phys. Lett. **86**, 112510 (2005)
[40] B. E. Argyle, E. Terrenzio, J. C. Slonczewski: Magnetic vortex dynamics using the optical Cotton–Mouton effect, Phys. Rev. Lett. **53**, 190 (1984)
[41] A. Hubert, R. Schäfer: *Magnetic Domains – The Analysis of Magnetic Microstructures* (Springer, Berlin, Heidelberg 2000)
[42] F. B. Humphrey: Transverse flux change in soft ferromagnetics, J. Appl. Phys. **29**, 284 (1958)
[43] M. H. Kryder, F. B. Humphrey: A nanosecond Kerr magneto-optic camera, J. Appl. Phys. **38**, 829 (1969)
[44] M. R. Freeman, J. F. Smyth: Picosecond time-resolved magnetization dynamics of thin-film heads, J. Appl. Phys. **79**, 5898 (1996)
[45] M. R. Freeman, W. K. Hiebert: Stroboscopic microscopy of magnetic dynamics, in B. Hillebrands, K. Ounadjela (Eds.): *Spin Dynamics in Confined Magnetic Structures I*, vol. 83, Top. Appl. Phys. (Springer, Berlin, Heidelberg 2002) p. 93
[46] D. H. Auston: Impluse response of photoconductors in transmission lines, IEEE J. Quantum Electron. **19**, 639 (1983)
[47] A. Barman, V. V. Kruglyak, R. Hicken, A. Kundrotaite, M. Rahman: Anisotrophy, damping, and coherence of magnetization dynamics in a 10 μm square $Ni_{81}Fe_{19}$ element., Appl. Phys. Lett. **82**, 3065 (2003)
[48] D. Chumakov, J. McCord, R. Schäfer, H. Vinzelberg, R. Kaltofen, I. Münch, L. Schultz: Nanosecond time-scale switching of permalloy thin film elements studied by wide-field time-resolved Kerr microscopy, Phys. Rev. B **71**, 014410 (2005)
[49] S. Demokritov, B. Hillebrands, A. Slavin: Brillouin light scattering studies of confined spin waves: linear and nonlinear confinement, Phys. Rep. **348**, 441 (2001)

[50] B. Hillebrands: Brillouin light scattering from layered magnetic structures, in G. G. M. Cardona (Ed.): *Light Scattering in Solids VII*, vol. 75, Top. Appl. Phys. (Springer, Berlin, Heidelberg 2000)

[51] J. R. Sandercock: Trends in Brillouin light scattering: Studies of Opaque Materials, Supperted Films, and Central Modes, in M. Cardona, G. Güntherodt (Eds.): *Light Scattering in Solids III* (Springer, Berlin, Heidelberg 1982)

[52] V. E. Demidov, S. O. Demokritov, B. Hillebrands, M. Laufenberg: Radiation of spin waves by a single micrometer-sized magnetic element, Appl. Phys. Lett. **85**, 2866 (2004)

[53] K. Perzelmaier, V. Demidov, M. Buess: Spin-wave eigenmodes of permalloy squares with closure domain structure, Phys. Rev. Lett. **94**, 057202 (2005)

[54] G. Schütz, W. Wagner, W. Wilhelm, P. Kienle, R. Zeller, R. Frahm, G. Materlik: Absorption of circularly polarized X-rays in iron, Phys. Rev. Lett. **58**, 737 (1987)

[55] S. Pizzini, J. Vogel, M. Bonfim, A. Fontaine: Time-resolved X-ray magnetic circluar dichroism, in B. Hillebrands, K. Ounadjela (Eds.): *Spin Dynamics in Confined Magnetic Structures II*, vol. 87, Top. Appl. Phys. (Springer, Berlin, Heidelberg 2002) p. 155

[56] J. Stöhr, Y. Wu, B. D. Hermsmeier, M. G. Samant, G. R. Harp, S. Koranda, D. Dunham, B. P. Tonner: Element specific magnetic microscopy with circularly polarized X-rays, Science **259**, 658 (1993)

[57] M. Bonfim, G. Ghiringhelli, F. Montaigne, S. Pizzini, N. B. Brookes, F. Petroff, J. Vogel, J. Camarero, A. Fontaine: Element-selective nanosecond magnetization dynamics in magnetic heterostructures, Phys. Rev. Lett. **86**, 3646 (2001)

[58] A. Puzic, B. Van Wayenberge, Kang-Wie-Chow, P. Fischer, H. Stoll, G. Schütz, T. Tyliszczak, K. Rott, H. Brückl, G. Reiss, I. Neudecker, T. Haug, M. Buess, C. H. Back: J. Appl. Phys. **97**, 10E704 (2005)

[59] S.-B. Choe, Y. Acremann, A. Scholl, A. Bauer, A. Doran, J. Stöhr, H. A. Padmore: Vortex coredriven magnetization dynamics, Science **304**, 420 (2004)

[60] J. Raabe, C. Quitmann, C. H. Back, F. Nolting, S. Johnson, C. Buehler: Phys. Rev. Lett. **94**, 217204 (2005)

[61] M. Buess, R. Höllinger, T. Haug, K. Perzlmaier, U. Krey, D. Pescia, M. R. Scheinfein, D. Weiss, C. H. Back: Fourier transform imaging of spin vortex eigenmodes, Phys. Rev. Lett. **93**, 077207 (2004)

[62] M. Buess, T. P. J. Knowles, R. Höllinger, T. Haug, U. Krey, D. Weiss, D. Pescia, M. R. Scheinfein, C. H. Back: Excitations with negative dispersion in a spin vortex, Phys. Rev. B **71**, 104415 (2005)

[63] A. Y. Elezzabi, M. R. Freeman: Ultrafast magneto-optic sampling of picosecond current pulses, Appl. Phys. Lett. **68**, 3546 (1996)

[64] T. Gerrits, H. A. M. van den Berg, J. Hohlfeld, O. Gielkens, K. J. Veenstra, K. Bal, T. Rasing: Precession dynamics in N, Te thin films, induced by short magnetic in-plane field pulses generated by a photoconductive switch, J. Mag. Soc. Jap. **25**, 192 (2001)

[65] T. Shinjo, T. Okuno, R. Hassdorf, K. Shigeto, T. Ono: Magnetic vortex core observation in circular dots of Permalloy, Science **289**, 930 (2000)

[66] J. Raabe, R. Pulwey, R. Sattler, T. Schweinböck, J. Zweck, D. Weiss: Magnetization pattern of ferromagnetic nanodisks, J. Appl. Phys. **88**, 4437 (2000)

[67] A. Wachowiak, J. Wiebe, M. Bode, O. Pietzsch, M. Morgenstern, R. Wiesendanger: Science **298**, 577 (2002)

[68] A. Yamasaki, W. Wulfhekel, R. Hertel, S. Suga, J. Kirschner: Direct observation of the single-domain limit of Fe nanomagnets by spin-polarized scanning tunneling spectroscopy, Phys. Rev. Lett. **91**, 127201 (2003)
[69] K. Y. Guslienko, A. N. Slavin: Spin-waves in cylindrical magnetic dot arrays with in-plane magnetization, J. Appl. Phys. **87**, 6337 (2000)
[70] M. Rahm, M. Schneider, J. Biberger, R. Pulwey, J. Zweck, D. Weiss, V. Umansky: Vortex nucleation in submicrometer ferromagnetic disks, Appl. Phys. Lett. **82**, 4110 (2003)
[71] J. F. Dillon, Jr.: Magnetostatic modes in disks and rods, J. Appl. Phys. **31**, 1605 (1960)
[72] A. A. Thiele: Steady-state motion of magnetic domains, Phys. Rev. Lett. **30**, 230 (1973)
[73] D. L. Huber: Equation of motion of a spin vortex in a two-dimensional planar magnet, J. Appl. Phys. **53**, 1899 (1982)
[74] B. A. Ivanov, C. E. Zaspel: Magnon modes for thin circular vortex-state magnetic dots, Appl. Phys. Lett. **81**, 1261 (2002)
[75] K. Y. Guslienko, B. A. Ivanov, V. Novosad, Y. Otani, H. Shima, K. Fukamichi: Eigenfrequencies of vortex state excitations in magnetic submicron-size disks, J. Appl. Phys. **91**, 8037 (2002)
[76] M. Buess, T. P. J. Knowles, U. Ramsperger, D. Pescia, C. H. Back: Phase-resolved pulsed precessional motion at a schottky barrier, Phys. Rev. B **69**, 174422 (2004)
[77] LLG: The LLG code can be found at URL `http://llgmicro.home.mindspring.com/`
[78] M. Sparks: Magnetostatic modes in an infinite circular disk, Solid State Commun. **8**, 731 (1970)

Index

Domain-Wall Dynamics in Nanowires and Nanostrips

André Thiaville[1] and Yoshinobu Nakatani[2]

[1] Laboratoire de physique des solides, CNRS – Université Paris-sud, 91405 Orsay Cedex, France
andre.thiaville@u-psud.fr

[2] Department of Computer Science,University of Electro-communications, Chofu, 182-8585 Tokyo, Japan
nakatani@cs.uec.ac.jp

Abstract. In nanowires or nanostrips, the magnetic domain walls have structures that depend sensitively on the transverse sizes of these samples. These domain-wall structures govern the dynamics of the walls under an applied field, so that rather different wall velocities can result from small changes in the sample transverse dimensions. These phenomena are first presented and discussed using results of micromagnetic simulations. In order to understand them, a model – valid in the limit of vanishing transverse dimensions – is introduced. It is shown that, however, this model applies also to the more complex domain-wall structures that exist at larger transverse sizes, qualitatively but even quantitatively in some cases. A procedure for determining the effective parameters of this model is given. Comparison to available experimental results is performed whenever possible.

1 Introduction

The dynamics of magnetic domain walls had started to be studied even before the domains and walls were actually observed. In the famous Sixtus–Tonks experiments [1, 2], the sample was a $Fe_{86}Ni_{18}$ alloy wire of diameter 380 µm submitted to a tensile stress that created a strong uniaxial anisotropy along the wire axis, and the domain wall was detected by pickup coils wound around the wire. Some basic phenomena of the domain-wall dynamics were established by these studies: the existence of a domain-wall nucleation and propagation field, and the definition of a domain-wall mobility (ratio of velocity to applied field, in a quasilinear regime).

Fast magnetization dynamics started to be studied in the 1950s and, at the end of the decade, *Walker* invented (but did not publish [3,4]) his solution for the dynamics of a 1D Bloch wall, the dynamic counterpart of the famous static calculation by *Bloch* [5]. This analytic solution separated two regimes, one with constant and high mobility at low field, and one with periodic Bloch-wall magnetization reversal together with a low mobility, at high fields [6]. This solution had been preceded by the first calculation of the Bloch-wall low-field motion by *Landau* and *Lifshitz* [7], and by the calculation of the

B. Hillebrands, A. Thiaville (Eds.): Spin Dynamics in Confined Magnetic Structures III, Topics Appl. Physics **101**, 161–206 (2006)

steady-state domain-wall structure at low velocity by *Döring* [8], introducing the concept of the domain-wall mass.

The research related to the development of magnetic bubble memories, in which cylindrical magnetic domains have to be displaced in order to access the information, led to a deep understanding of the domain-wall dynamics in "bubble" thin films. These films, with a perpendicular uniaxial anisotropy, support Bloch walls with Néel regions at the film surfaces due to the domains' stray field. The theory of domain-wall dynamics in these samples, mainly developped by *Malozemoff* and *Slonczewski* [9], could account very well for the physics of the dynamics of such domain walls. In particular, it showed that the Walker limit was very hard to reach, as modifications of the internal structure of the domain walls often occurred at lower fields. This knowledge is exposed in a book [9] that is still a reference work on the subject of magnetic domain-wall dynamics (other reviews, of smaller scope, however, exist [4, 10–12], as well as one specialized in the weak ferromagnets [13]).

The recent progress in the fabrication and direct synthesis of laterally confined structures has allowed new artificial objects to be created, known generally as nanowires when confinement occurs along two dimensions whereas the axial size is much larger. These structures allow very fundamental problems to be addressed, such as the nature of a 1D magnetism [14], and its excitation spectrum [14, 15] (see also, about that last point, the Chapter in this volume by *Bayer* et al.). The domain-wall dynamics can be viewed as a long-wavelength probe of these fundamental questions, that are part of statistical physics. However, from all the past developments in the physics of magnetic domain walls, it has also an interest in its own. Briefly stated, one can expect the lateral confinement to, partly or totally, suppress the instabilities within the domain wall itself that were leading to a complex dynamics at rather low fields. The knowledge of the domain-wall displacement regimes in nanowires can also be of direct interest for possible devices, either industrial or in other branches of physics like the so-called mesoscopic physics [16].

These general ideas, together with the first experiments performed using such structures [17–20] have led us into this research subject [21, 22]. There are now many experimental and theoretical papers published on the subject, and it is still advancing, so that a self-contained presentation of the physics involved has been felt to be timely. The purpose of this Chapter within this book series is to provide an understanding, as complete and accurate as possible, of the sometimes counterintuitive behavior of domain walls under a field.

We made the deliberate choice not to consider spin-transfer effects. The Chapter by *Stiles* and *Miltat* in this book provides an indepth review of the phenomena expected, and seen, in a geometry with two magnetic layers separated by a nonmagnetic spacer. The situation appears now to be fairly well understood, although not fully yet. For example, there is no LLG equation incorporating the spin-transfer torque that is as firmly established as the LLG equation itself. Even if this equation has been straightforwardly extended from two layers to a wire containing a domain wall, both theoretically and as a support to experiments, we felt that the subject was not mature enough for inclusion in the present review.

The Chapter will start (Sect. 2) by a presentation of the simplest domain-wall structures, as they depend on the sample geometry. A very simple effective model will be introduced, valid in the 1D limit, and systematically compared to the more complex reality. Then, for each domain-wall structure, the dynamics will be described as it is predicted by numerical calculations, this being compared to the outcome of the 1D model and, when possible, to experiment. This case-to-case treatment is contained in Sect. 3–6. The general idea we would like to convey is that the 1D effective model is very good qualitatively and also (for small sizes) quantitatively. However, the nature of the associated Walker breakdown depends a lot on the sample shape, which is why the cases need to be treated in sequence. Section 7 closes the chapter by the derivation of several general relations valid for domain-wall motion at constant velocity, that help understanding some features seen earlier and that are beyond the effective 1D model. Finally, Sect. 8 summarizes the chapter and proposes a perspective for future research in this field.

2 Types of Samples and Domain Walls

We shall make a distinction between nanowires, whose cross sections have an aspect ratio of order unity (a circle, a square, essentially) and nanostrips, that have a cross section with a large aspect ratio (the strip thickness being much smaller than the strip width). This corresponds to different fabrication routes, e.g., growth in nanopores [23] or self-organized chemical growth [24] for nanowires, and thin-film patterning for nanostrips [25]. In the latter case the thickness is a few nanometers but the strip width is very large, of the order of a hundred nanometers, typically. This large dimension, in terms of the so-called micromagnetic exchange length Λ [defined later in (2)], gives to the domain wall (hereafter abbreviated as DW) additional degrees of freedom, namely the opportunity to create substructures across the strip width, that have important consequences on the DW dynamics. Thus, the structures and dynamics in nanowires are expected to be simpler.

The typical material considered in this chapter is magnetically soft, which means that 1. the coercive force is low, so that domain walls can propagate easily; 2. the structures are largely defined by magnetostatics. Most experiments on DW dynamics in such one-dimensional structures have indeed relied on soft materials, the archetype of which is $Ni_{81}Fe_{19}$ (Permalloy hereafter).

The main effect of the nanowire or nanostrip geometry is to confine the magnetization along the wire axis, so as to minimize the demagnetizing energy. Walls consequently bear a magnetostatic charge (positive for a head-to-head and negative for a tail-to-tail wall). In thick soft magnetic films, charged walls are often avoided by a zig-zag deformation of the wall plane, in order to decrease the magnetostatic cost associated with this charge [11]. In the nanowire/nanostrip geometry this deformation is (nearly) suppressed, because of the small thickness of the sample. Thus, walls in nanowires and nanostrips represent objects that have been little investigated before. This chapter will, however, stress that concepts developed in the past for non-charged walls can be applied with much physical insight.

2.1 Wall Types in Nanowires

We start the description of the simple wall types at equilibrium with the nanowire geometry, which is closest to the one-dimensional limit.

Numerically calculated structures will be shown first. They are the outcome of a micromagnetic code developed by *Nakatani* and *Hayashi* [21, 26–29]. The adaptation to the wire geometry consisted in removing the end effects so as to simulate an infinite wire (suppression of the surface charges at wire ends and assumption of a uniform axial magnetization outside the calculated wire section). Long-time dynamics is enabled by displacing the calculation region so as to keep the DW always in its center. The mesh parameter chosen was equal to the exchange length of Permalloy [$\Lambda = 5\,\mathrm{nm}$, see definition in (2)] or below, as this length dictates the scale over which the magnetization may deviate.

These structures are then compared to a 1D DW model. For each case, the only parameter of this model (the wall-width parameter Δ) will be determined by fitting, and the quality of this fit will be discussed. The parameter Δ is indeed of paramount importance for the DW dynamics, as it controls the DW linear mobility (ratio of velocity to field at vanishing applied fields).

In a second part, this model is shown to result from an approximation valid at small wire diameters.

2.1.1 Numerical Results

Consider a nanowire of square cross section, with edge size a. This sample is (not yet!) realistic, yet very convenient for calculation and didactical purposes.

Transverse Wall

The simplest DW structure, shown in Fig. 1, is called the transverse wall (TW), as the average DW magnetization is nonzero and oriented perpendicular to the wire axis. In the case of zero magnetic anisotropy, the DW structure depends solely on the ratio a/Λ, where Λ is a characteristic length of micromagnetics whose definition is recalled later (2). As the edge size increases, the magnetization distribution within the DW becomes less uniform, in response to the nonuniform demagnetizing field associated to a uniformly magnetized DW. The cross sections perpendicular to the wire axis (see the

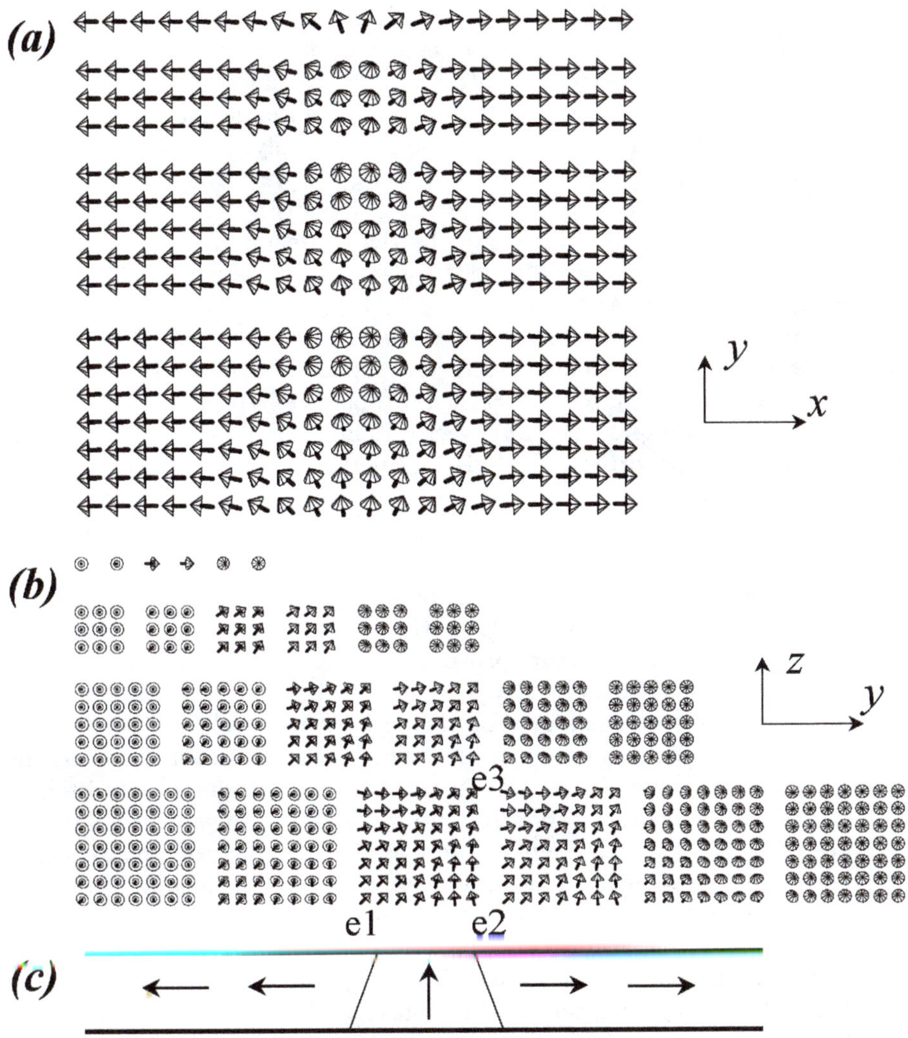

Fig. 1. Computed domain-wall structures in Permalloy nanowires of square cross section with edges $a = 5\,\mathrm{nm}, 15\,\mathrm{nm}, 25\,\mathrm{nm}$ and $35\,\mathrm{nm}$, featuring a tail-to-tail transverse wall. The mesh size is 5 nm, which is the value of the exchange length Λ in Permalloy. Cross sections along the wire axis, passing through the center of the wire, are shown in (**a**), with one arrow per mesh cell. Cross sections perpendicular to the wire axis, for several consecutive cell planes, are displayed in (**b**). Positions e1–e3 are defined for later use (Fig. 4). A schematic of the structures in (a) is presented in (**c**)

central drawings Fig. 1b, especially at large size) reveal that this distribution takes the form of an arrow. It may be tempting to compare this distribution with those observed in elements patterend out of thin films [30]. There is, however, no direct correspondence to be found, as the deviations of magnetization are mainly caused by the DW charge, and only secondarily modified by the sample shape. The cross sections along the wire axis also show that the DW region tends to assume a triangular shape (Fig. 1a). The schematic structure (Fig. 1c) explains this shape by the tendency to close the magnetization flux, which is the main driving force for the existence of domains in sufficiently large samples.

Figure 1 depicts the case of a tail-to-tail wall only. There is, in fact, no real difference from the head-to-head wall, contrary to appearance. Indeed, as no field is applied, if $\boldsymbol{m}(\boldsymbol{r})$ is a solution then its opposite $-\boldsymbol{m}(\boldsymbol{r})$ will also be one (throughout this Chapter, the unit vector defining the local magnetization orientation will be written $\boldsymbol{m}$). This symmetry exchanges tail-to-tail and head-to-head walls.

Bloch-Point Wall

For wider wires, a second structure is found to be more stable (Fig. 2). Indeed, the TW structure poorly closes the magnetization flux at the wire edges, whereas the new structure proves more efficient via a vortex-like magnetization rotation. The associated cost in exchange energy is balanced by a gain in magnetostatic energy. For this reason, this wall has been called a vortex wall by some authors [31, 32]. There are two problems with this denomination, however: 1. the DW structure does not simply consist in a vortex; 2. there is another DW structure in nanostrips also called a vortex wall (see Sect. 2.3). Looking more closely at the structure, one realizes that the core of the vortex is not that of a usual vortex, as the magnetization reverses between both wall sides. In fact, a structure called a Bloch point (BP) sits in the center of this DW, so that we call this wall type a Bloch-point wall (BPW). The Bloch point is the elementary singularity of micromagnetics [33], and the magnetization modulus has to fall to zero at this point (so that the BP will never sit on a mesh point in numerical calculations). The applicability of numerical micromagnetics to Bloch points, surprising at first sight, has been discussed in depth for the case of vortex-core reversal [34]. The cross sections perpendicular to the wire axis (Fig. 2b) reveal in addition that the magnetization does not rotate in circles but rather in spirals around the BP: a small radial component is present. The reason for this is apparent on the schematic drawing of the DW (Fig. 2c). The BP bears a (monopolar) magnetic charge (a concentration of the DW charge), and the associated field twists the magnetization vectors in its vicinity. The dynamic effect of this deformation will be discussed below. The wire cross sections along the wire axis (Fig. 2a) also show that the DW width is smaller in the center than at the wire edges (indeed, passing through the Bloch point would give a DW width of atomic size).

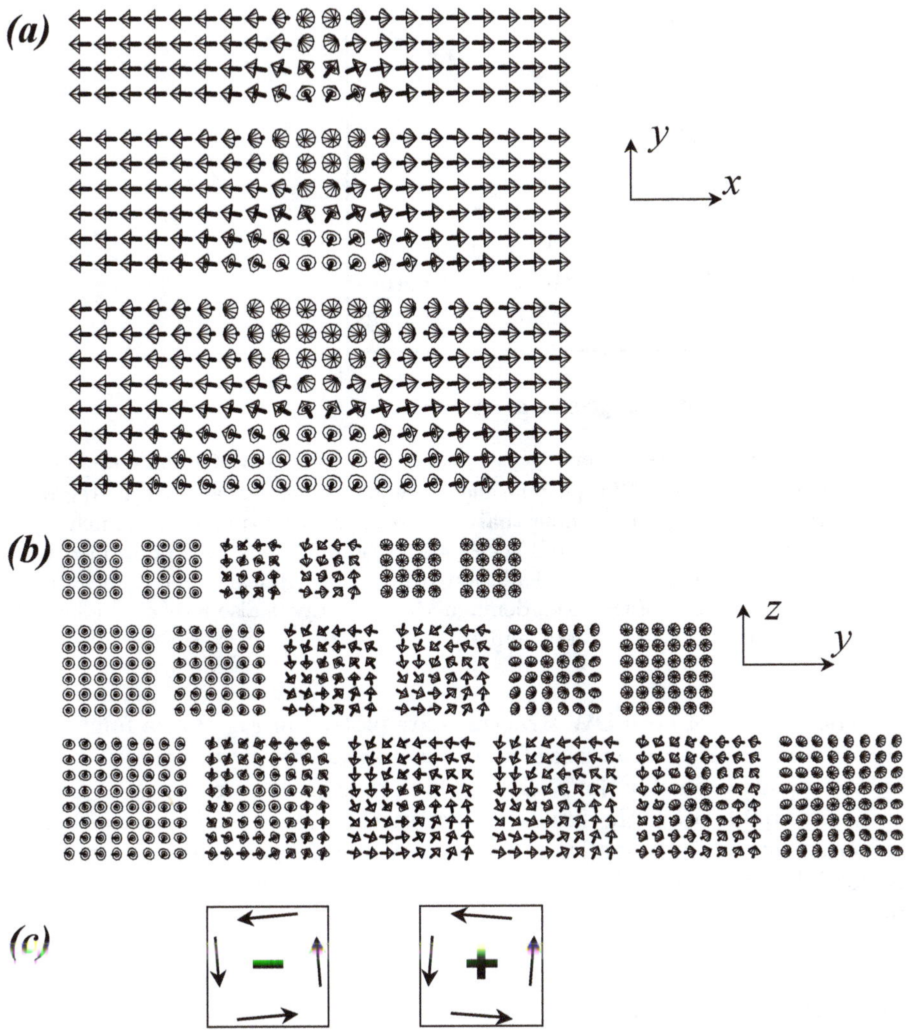

Fig. 2. Computed domain-wall structures in Permalloy nanowires of square cross section with edges $a = 20\,\mathrm{nm}$, $30\,\mathrm{nm}$ and $40\,\mathrm{nm}$, featuring a Bloch-point wall. The mesh size is $5\,\mathrm{nm}$ (an even number of cells is mandatory for the Bloch point to sit at the wire center, in between the mesh points). **(a)** cross sections along the wire axis, passing at a half-mesh distance from the wire axis (the Bloch point sits one half-mesh distance above the center of each figure); **(b)** cross sections perpendicular to the wire axis, for several consecutive cell planes (the Bloch point sits at the center and in between the two central cross sections); **(c)** *left*: schematic for (b); *right*: same for a head-to-head wall

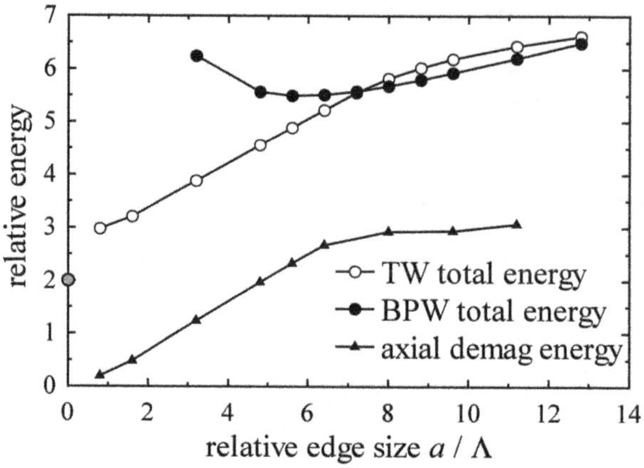

Fig. 3. Evolution of the DW energy in a square nanowire with edge size a and no anisotropy, depending on DW type. The energy density is defined as $\mathcal{E}/\left(a^2\sqrt{A\mu_0 M_s^2}\right)$, with $\mathcal{E}$ the total energy (remember that, due to the cancellation of the end charges, in the absence of a DW one would get $\mathcal{E}=0$). Computations were performed with a 2 nm cubic cell, keeping the calculation region eight times longer than its edge size. The contribution of the axial demagnetizing energy is also indicated (the TW case, but values for the BPW are similar)

The energies of both DW structures are plotted in Fig. 3 as a function of the wire edge size a, in the case of zero anisotropy. They prove equal at $a \approx 7\Lambda$ for this square geometry (for a circular wire, the equality was found at a lower value $a \approx 6.2\Lambda$, using the finite element method [31]). Note that both DW structures may exist beyond the equal energy point, a proof that metastability is possible when the DW structure is not of lowest energy. The graph also indicates the simplest analytic limit for the TW energy at zero wire width, namely 2 in these reduced units (see Sect. 2.2). The inclusion of a uniaxial anisotropy (longitudinal easy axis with anistropy constant K_u) displaces the isoenergy size to larger values. For example, at $K_u = \mu_0 M_s^2/8$ both DWs have the same energy at $a/\Lambda = 7.6$. The linear initial slope in the case of the TW is not related to the DW energy itself (it should decrease linearly, see Sect. 2.2), but to the presence of an axial demagnetizing field H_d^x due to the wall intrinsic charge. As a calculation box with constant aspect ratio was taken, a linear increase with a of the reduced axial demagnetizing energy is expected (for scale-invariant DW structures at least), and observed. On the other hand, the BPW relative energy has a $1/a$ variation at small sizes a, as readily found by an exchange-energy calculation for a schematic structure of the BPW. If the wire size is increased further, more complex structures appear, either statically or under a field for a moving wall [35, 36]. We will not consider them any further, confining the discussion to these simplest structures, that are relevant for the smallest nanowires.

2.1.2 Comparison to 1D Profiles

We now compare these magnetization distributions to a profile obtained in the one-dimensional (1D) limit, namely the Bloch-wall profile. This profile was originally derived for a divergence-free magnetization rotation in the presence of a uniaxial anisotropy [5], which is not at all the case addressed here. We will show, however, that this profile is more general, so that it applies also to the present case. According to this profile, the longitudinal (m_x) and transverse (m_t) components of the unit magnetization vector are given, for a tail-to-tail wall, by

$$m_x = \tanh(x/\Delta)\,; \quad m_\mathrm{t} = 1/\cosh(x/\Delta)\,. \tag{1}$$

The profile is described by the parameter Δ, called the wall-width parameter (the full wall width computed on the magnetization angle from its tangent at the wall center being $\pi\Delta$).

Figure 4 shows that for a narrow nanowire (5 nm edge, which is the exchange length Λ of the material considered) and a transverse wall, the agreement is nearly perfect. It becomes slightly worse at a larger size (35 nm), because of some nonuniformity along y and z of the magnetization profiles (see the profiles at three different positions in Figs. 4c–d). The main difference lies in the negative wings of m_t, that occur in response to the stray field of the transverse magnetization. These are forbidden in the Bloch-wall model (1), so that the fit will result into a smaller Δ when compared to the width derived from the longitudinal component.

Thus, the applicability of the 1D profile can be monitored by the difference of the values of Δ deduced from $< m_x >$ and $< m_\mathrm{t} >$. The value of $< m_\mathrm{t} >$ at the wall center is another confidence index (a value lower than 1 directly shows the nonuniformity of the transverse component at the wall center).

Some care is necessary in order to apply the same analysis to a BPW. Instead of the transverse y and z components, one has to consider the orthoradial and radial projections of the magnetization [defined as $m_\mathrm{o} = (ym_z - zm_y)\,/\sqrt{(y^2+z^2)}$ and $m_\mathrm{r} = (ym_y + zm_z)\,/\sqrt{(y^2+z^2)}$, respectively]. Such profiles are plotted in Fig. 5. The fit of the averaged values (Figs. 5a and b) is rather satisfactory, considering the criteria discussed above. Note the small radial component mentioned above as a result of the BP charge. The profiles at selected positions (Figs. 5c and d) reveal, however, that this global result is the average of very different behaviors at the center and periphery.

2.2 One-Dimensional Effective Model

In this section, the 1D profile (1) will be shown to result from an approximate solution of the full micromagnetic problem.

The first approximation is to assume that the magnetization $\boldsymbol{m}$ depends only on the axial coordinate x. We have seen above that the driving force

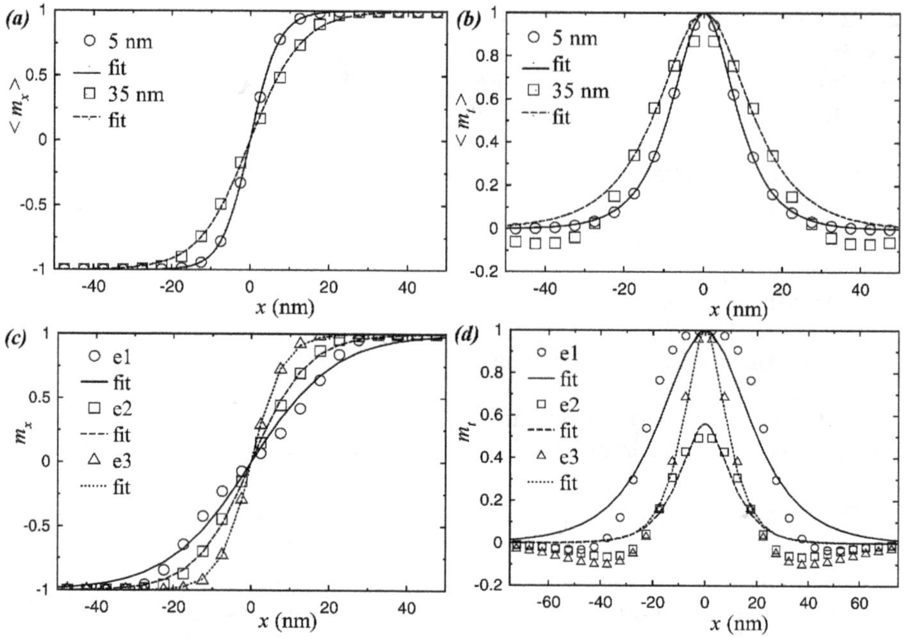

Fig. 4. Transverse wall in a square Permalloy nanowire with edge sizes 5 nm and 35 nm (mesh size: 5 nm). Comparison of the profiles for the average (over y and z) axial [m_x, **(a)**] and transverse components [m_t, here oriented along the square diagonal, **(b)**] of the magnetization, to the Bloch-wall-type profile. The fitted wall-width parameters Δ are 7.202 nm (a) and 7.089 nm (b) for the 5 nm sample, and 13.03 nm (a) and 10.07 nm (b) for the 35 nm nanowire. Nonaveraged magnetization profiles are then shown [axial **(c)** and transverse **(d)** components], at three selected edge positions e1, e2 and e3 that are displayed in Fig. 1b. Position e3, at the top of the schematic triangle shown in Fig. 1c shows the smallest width parameter. Position e1, at the base of this triangle, has conversely the largest width parameter. Position e2, located half-way, has an intermediate width but also a reduced amplitude of the transverse magnetization component as an effect of the nonuniform DW magnetization structure (Fig. 1b). The fitted wall-width parameters are 21.36 nm, 14.20 nm and 8.07 nm in **(c)** and 15.19 nm, 8.45 nm (with an amplitude of 0.56) and 7.17 nm in (d) for the positions e1, e2 and e3, respectively. Note the change in x scale for (d)

for variations of $\boldsymbol{m}$ along the transverse directions is flux closure. The competition of this tendency with the decrease of the exchange energy leads, in micromagnetics [11], to the definition of the so-called exchange length

$$\Lambda = \sqrt{\frac{2\,\mathrm{A}}{\mu_0 M_{\mathrm{s}}^2}}\,. \tag{2}$$

Therefore, for wire edge sizes not much larger than Λ, the $\boldsymbol{m} = \boldsymbol{m}(x)$ approximation is justified.

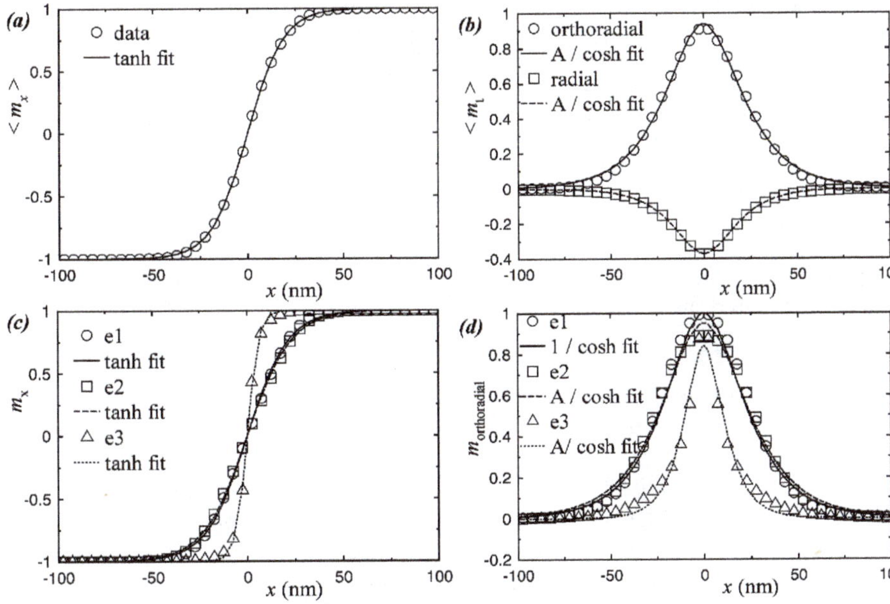

Fig. 5. Bloch-point wall in a square nanowire (edge size 40 nm, mesh size 5 nm). Comparison of the profiles for the average axial [m_x, (**a**)] and transverse components [m_t, separated into the orthoradial and radial components, (**b**)] of the magnetization, to the Bloch-wall-type profile. The fitted wall-width parameters Δ are 18.95 nm for the axial, 17.86 nm (amplitude 0.94) for the orthoradial and 14.51 nm (amplitude −0.37) for the radial components. The Magnetization profiles [axial (**c**) and orthoradial (**d**) components] are then shown at three selected positions e1, e2 and e3. Position e1 is at the wire edge, e2 half-way inside and e3 at a cell closest to the center. The width parameter fitted on the axial magnetization shows a marked decrease at the center, due to the Bloch-point structure. The orthoradial component is seen to decrease from the edge towards the center, as it is progressively replaced by a radial component that develops as a response to the wall charge (see Fig. 2c). The fitted wall-width parameters are 21.44 nm, 22.98 nm and 6.11 nm in (c) and 18.29 nm, 19.72 nm (amplitude 0.95) and 9.17 nm (amplitude 0.84) in (d) for the positions e1, e2 and e3, respectively

Given a magnetic configuration of a DW in a nanowire, the calculation of its magnetostatic energy is a heavy task, in most cases only amenable to numerical calculations. However, when the DW width is much larger than the nanowire edge length, the approximation of a local demagnetizing field becomes valid. For a square or circular cylinder, as the transverse demagnetizing factor is 1/2, one then writes

$$\boldsymbol{H}_{\mathrm{d}}(x) = -\frac{M_{\mathrm{s}}}{2}\boldsymbol{m}_{\mathrm{t}}(x)\,, \tag{3}$$

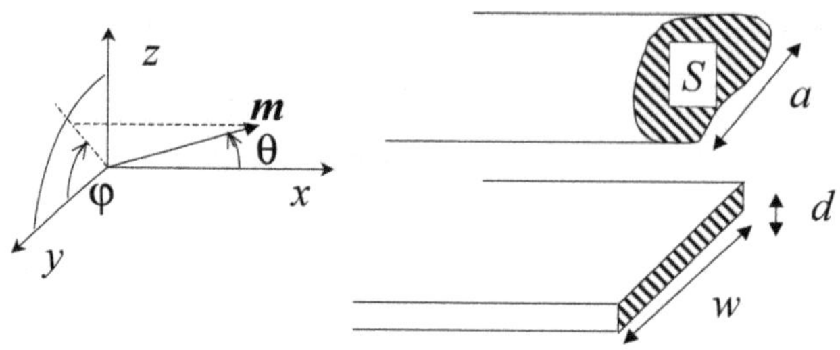

Fig. 6. Definition of axes, nanowire or nanostrip dimensions and magnetization angles. S denotes the cross-section at area

where $\boldsymbol{m}_{\mathrm{t}}$ is the transverse magnetization vector $\boldsymbol{m}_{\mathrm{t}} = \boldsymbol{m} - \hat{x}\,(\boldsymbol{m}\cdot\hat{x})$. This relation leaves aside the axial field that was discussed earlier (Fig. 3). This field is zero at DW center and builds up progressively when moving out of the DW center plane. Thus, in a linear approximation around the wall center ($\boldsymbol{H}_{\mathrm{d}}\hat{x} = -\nu M_{\mathrm{s}} m_x$), the axial field results in an energy term similar to that of the transverse demagnetizing field, so that it can be absorbed into it. Far from the wall, however, the field decreases to zero. The far-field contribution gives rise to a part of the total energy that does not belong to the DW energy (it also grows logarithmically with the wire length), called C in (42). The classical Bloch-wall width calculation [11] augmented with the energy term from (3), yields the profile (1) with a width parameter $\Delta = \sqrt{(4\,\mathrm{A})/(\mu_0 M_{\mathrm{s}}^2)} = \sqrt{2}\Lambda$. The approximation (3) is therefore valid for wires with, again, transverse sizes up to the exchange length Λ. For a wire with rectangular cross section, the demagnetizing factor N differs from 1/2 [37,38]. Moreover, as it seems at present unlikely to have wire widths much smaller than the DW width, corrections for a finite length of the cylinder should be included [39]. Numerically computed DW widths have been satisfactorily compared to the relation $\Delta = \Lambda/\sqrt{N}$ in the case of Permalloy rectangular nanowires [39]. From (3) and its generalization to a rectangle it becomes clear that the magnetostatic term is equivalent in the 1D limit to a transverse anisotropy term. A true transverse anisotropy may also really exist, for example in cobalt nanowires in the case where the crystalline c-axis is oriented transversally [40].

Altogether, the simplest 1D model of a nanowire incorporates two effective anisotropy terms, whose origin is magnetocrystalline, magnetostatic as well as exchange energy. The latter contribution arises from the nonuniformity of the magnetization in the cross section (this is the configurational anisotropy known in the thin-film case [30]). Denoting by θ the magnetization polar angle (the polar axis being the wire axis), and by φ the azimuthal angle describing the orientation of the magnetization projection on the cross-sectional plane (Fig. 6), the general effective anisotropy density reads

$$E_{\text{an, eff}} = K_0 \sin^2\theta + \sum_{n=1}^{\infty} K_n \sin^{2n}\theta \sin^2(2n\varphi + \varphi_n) \,. \tag{4}$$

For most cases, the sum is limited to $n = 1$ and then K_1 is denoted simply by K, with $\varphi_1 = 0$.

The exchange energy for magnetization variations along x only (1D model) is

$$E_{\text{ex}} = A\left(\frac{\mathrm{d}\boldsymbol{m}}{\mathrm{d}x}\right)^2 = A\left[\left(\frac{\mathrm{d}\theta}{\mathrm{d}x}\right)^2 + \sin^2\theta\left(\frac{\mathrm{d}\varphi}{\mathrm{d}x}\right)^2\right] . \tag{5}$$

Summing both terms (4) and (5) we see that there is no driving force for a variation of φ with x (i.e., a helical configuration), for an anisotropy function limited to one term in the sum of (4). Therefore, we posit that φ is constant $\varphi(x) = \phi$. Then, the application of the calculation of the Bloch wall to the total energy is straightforward. The magnetization profile is that given by (1), with $m_x = \cos\theta$ and $m_{\mathrm{t}} = \sin\theta$, the transverse-component orientation having an azimuth angle ϕ. The optimal wall-width parameter depends on ϕ and reads

$$\Delta = \sqrt{\frac{A}{K_0 + K\sin^2\phi}} \,. \tag{6}$$

Some useful properties of this Bloch-wall-type solution are: the total exchange energy is $2\,\mathrm{A}/\Delta$, the total anisotropy energy is $2\Delta\left(K_0 + K\sin^2\phi\right)$, the total energy is $4\sqrt{A\left(K_0 + K\sin^2\phi\right)}$, the integral of m_{t} is $\pi\Delta$ and that of m_{t}^2 is 2Δ.

When the anisotropy function incorporates several terms in the sum over n in (4), the profile is not so easily found. The problem was investigated for Bloch walls in cubic materials, see for example [11].

2.3 Wall Types in Nanostrips

We now consider a strip of geometry as defined in Fig. 6. Similarly to the previous section, numerical calculation results will be shown first and then compared to the 1D model. From the conclusions of the previous section, we expect that this 1D model will prove too simple, as the strip width is much larger than the exchange length Λ.

Transverse Wall

The simplest wall structure is again the transverse wall, with a wall magnetization oriented along the y-axis, as opposed to the z-direction, for magnetostatic reasons. However, the strip geometry offers room for a wall distortion in the y-direction. The structure calculated for a 120 nm × 5 nm Permalloy strip is shown in Fig. 7. Similarly to a TW in a nanowire, the TW displays a triangular shape. The fitting by the 1D profile gives results similar to those for a TW in a nanowire with similar cross-sectional area S, even a little more satisfactory. As mentioned before, this arises from the very small strip thickness that reduces the magnetostatic interaction. Global results (Fig. 7b) are rather encouraging, with a 16 % decrease of Δ between the $< m_x >$ and $< m_y >$ profiles. But fits at different y positions (Fig. 7c,d) reveal large variations, showing that the triangular wall shape is well developed.

Vortex Wall

Increasing the strip width or thickness introduces another structure, called a vortex wall (VW) (Fig. 8). As for the nanowire, the increased strip cross section allows for a flux-closure structure to develop. There is no singularity involved here, however, the moments at the vortex core pointing perpendicularly to the strip plane, either upwards or downwards. For the same directions of the domain magnetizations four possible VW exist, as there are two core magnetization orientations and two winding directions for the vortex (Fig. 8b). A closer examination of the magnetization structure reveals that the vortex is flanked by two partial antivortices, one at each strip edge (see also the schematic structure Fig. 8b). On the $m_x(x)$ plots (y-averaged or not), the difference with the 1D Bloch-wall profile is clearly seen, showing that the VW is a truly different structure.

Phase Diagram

The phase diagram of the DW structures in a nanostrip, first investigated by *McMichael* and *Donahue* [41], shows a TW-VW equal-energy line obeying $wt = (60 \pm 2)\Lambda^2$. A refined calculation has shown that, in addition, a tilted TW phase is present close to this line [42] (see Fig. 9).

This structure, called the asymmetric transverse wall (ATW) can be seen as a first step towards the zig-zag structure of charged walls in thin films. The transition ATW-TW is second order [42] so that under a field, for example, both structures cannot be separated. On the other hand, the (A)TW-VW transition is first order. As a result, DW structures can be metastable, i.e., they exist outside their region of stability shown in Fig. 9. This was demonstrated experimentally and also by direct calculation recently [43].

3 Dynamics of a Transverse Wall in a Nanowire

We turn now to the examination of how the DW structures described above move under an applied field. Having seen that the transverse wall in a nanowire is the simplest domain-wall structure, and closest to the 1D analytical profile, we start by investigating its dynamics.

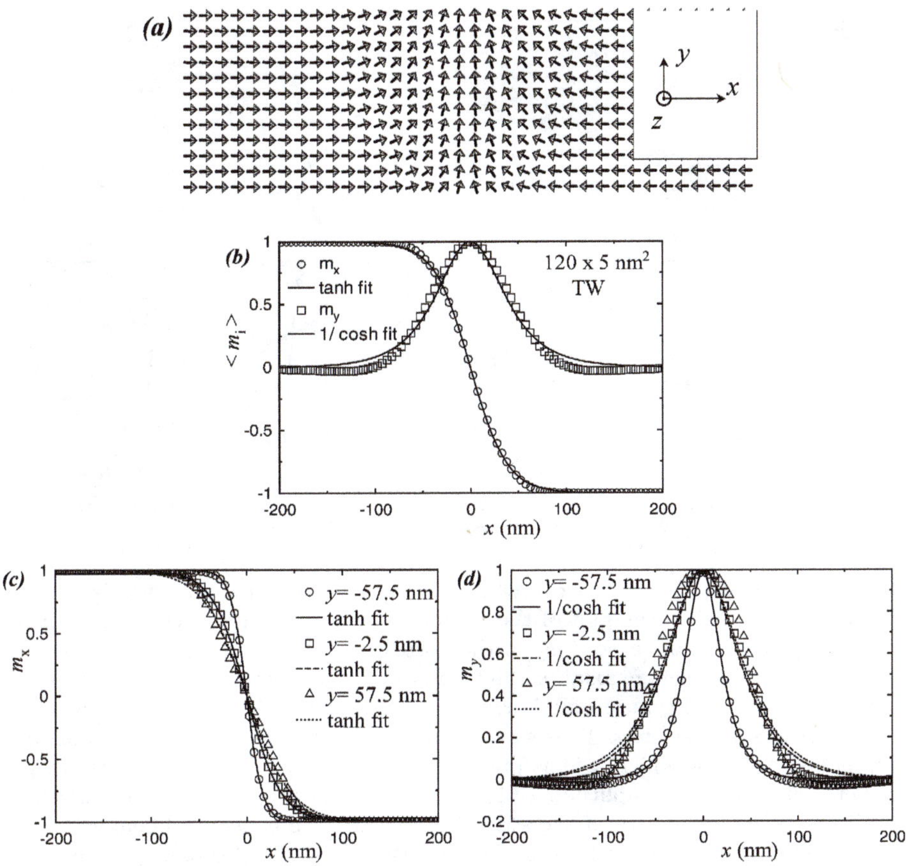

Fig. 7. Transverse wall in a nanostrip (width 120 nm, thickness 5 nm, mesh size 5 nm). **(a)** arrow plot of the magnetization distribution. **(b)** comparison of the profiles for the average axial (m_x) and transverse components (m_y) of the magnetization, to the Bloch-wall-type profile. The fitted wall-width parameters Δ are 37.67 nm for the axial component and 31.31 nm for the transverse. Nonaveraged profiles are shown additionally in **(c)** and **(d)** at three positions across the strip width (bottom edge, center, top edge). For the axial magnetization (c), the profile width increases steadily with distance to the wall apex, the fitted values for Δ being 15.97 nm, 35.42 nm and 45.34 nm at $y \approx -60$ nm, 0 nm and 60 nm, respectively. For the transverse magnetization, the effect is nearly the same (the fitted width parameters being 16.53 nm, 33.08 nm and 35.14 nm). The m_z component is virtually zero everywhere (below 10^{-6})

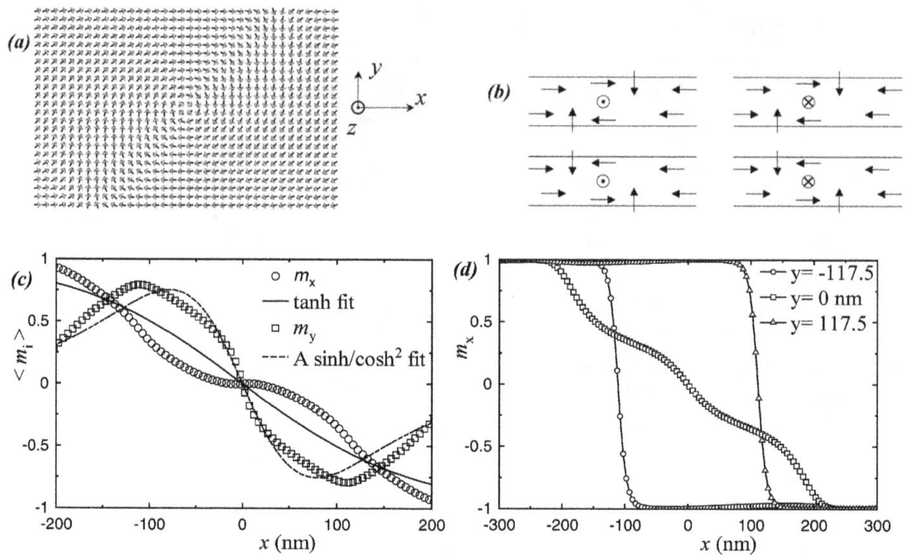

Fig. 8. Vortex wall in a nanostrip (width 240 nm, thickness 10 nm, mesh size 5 nm × 5 nm × 10 nm). **(a)** arrow plot of the magnetization distribution. **(b)** schematic of the four possible walls according to the vortex-core direction and rotation sense. **(c)** profiles of the average axial m_x and transverse m_y magnetization. The fit of m_x with the Bloch-wall-type profile is realized with a value $\Delta = 180.4$ nm, but obviously misses some features. This is explained by the consideration of the transverse component m_y (for which a fit is proposed as a guide to the eye, with width parameter 89.4 nm and amplitude 1.51). **(d)** profiles of the magnetization at three positions across the width of the strip (bottom and top edges, and center), showing in detail the difference from a one-dimensional profile (for example, the edge profiles show a very sharp transition when crossing the half-antivortex). A larger horizontal scale was chosen for **(c)**

3.1 Numerical Results

Consider first a 5 nm × 5 nm Permalloy nanowire, with a TW located at position $x = 0$. At time $t = 0$, a field H_a is applied along the wire axis. After some transient time, a stationary DW velocity is reached that increases with field (see Figs. 10a and b). Although the wall is advancing regularly, the plot of the transverse magnetization components reveals that the DW magnetization is rotating continuously around the wire axis. Under a larger field this precession is faster, and the wall also moves faster, but still with extremely low velocities compared to other walls. A larger Gilbert damping constant α gives rise to a larger wall velocity (in fact, no wall motion occurs for zero damping!). But this case is an extreme one, in the sense that no transverse anisotropy interferes with this precession. This behavior, found in several studies using different methods [31, 32, 44] has been dubbed once "corkscrew" motion [35].

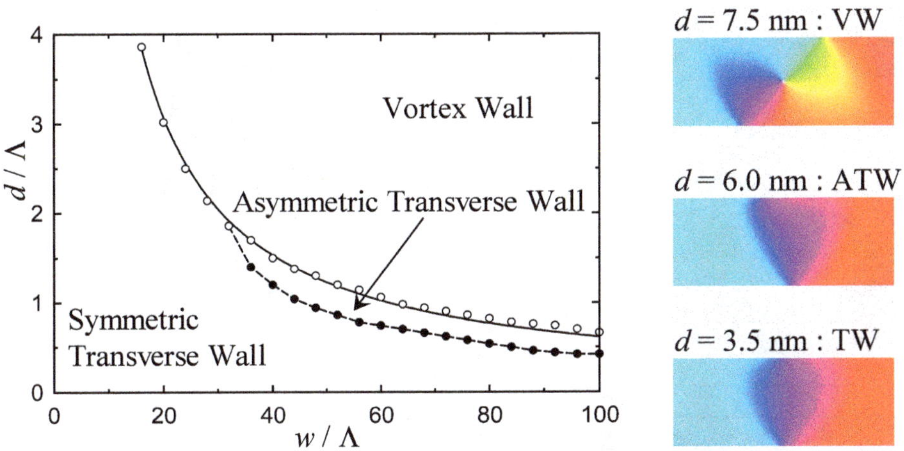

Fig. 9. Phase diagram of the stable domain-wall structures in a soft nanostrip (no anisotropy) of width w and thickness t, the exchange length being Λ. The main structures are the transverse wall, stable at low thickness and width, and the vortex wall, stable at large sizes. The phase boundary is close to a hyperbola $wd = 61.37\Lambda^2$, shown as a full line. An asymmetric transverse wall is stable in a small region of the parameter space [42]. Representative images of the three structures are shown on the right for a 240 nm wide strip, with color coding as in Fig. 14

Let us therefore consider a structure with a shape-induced transverse anisotropy, namely a 10 nm × 5 nm Permalloy nanowire (Fig. 11). For the same low field ($\mu_0 H_a = 1\,\mathrm{mT}$, Fig. 11a) the DW velocity is much larger (by a factor of nearly 10^4 for $\alpha = 0.01$), and no precession of the integrated wall magnetization occurs. The velocity increases linearly with field at low fields, and is now inversely proportional to the damping constant α. However, there is a maximum field above which the behavior seen in Fig. 10 sets in, together with a big decrease of velocity (Fig. 11b).

This behavior is in fact what was predicted for the dynamics of a 1D Bloch wall by *Walker* [3,6] (this calculation will be developped in the next section). It has, however, never been observed for bulk samples as other instabilities were taking place at fields below this so-called Walker threshold [9]. In ultrathin films with perpendicular anisotropy, however, (Au/Co/Au [45,46]) a velocity plateau was observed and attributed to the Walker maximum velocity. In nanowires, one now can hope to observe all features of the Walker DW dynamics, as presented below.

3.2 Analytical Model

We now describe how the 1D analytical model can account for the numerical results shown above.

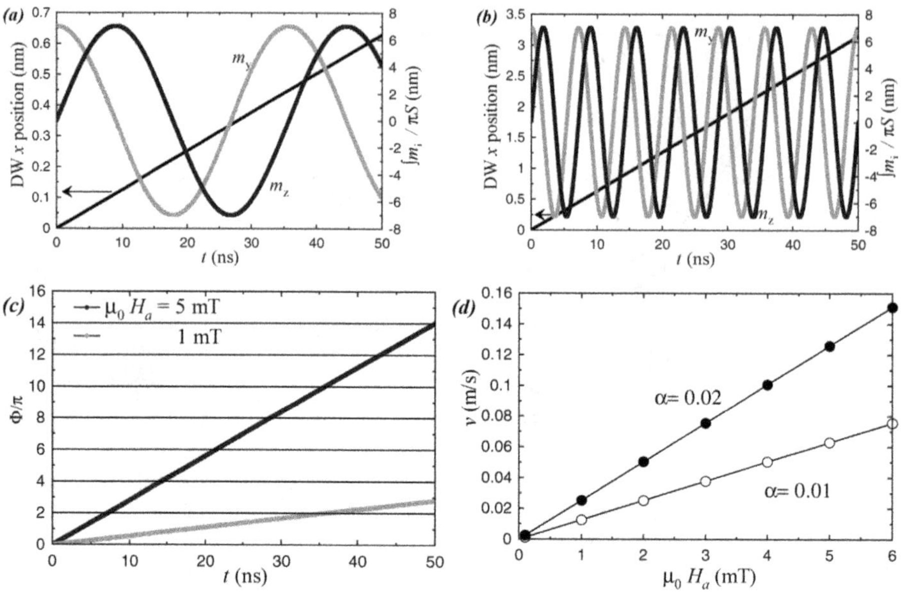

Fig. 10. Transverse-wall dynamics in a square nanowire (edge size 5 nm, mesh size 5 nm). The field H_a applied along the wire axis is **(a)** 10 Oe ($\mu_0 H_a = 1\,\text{mT}$) and **(b)** 50 Oe ($\mu_0 H_a = 5\,\text{mT}$), the damping constant is $\alpha = 0.01$. The DW position is plotted as well as the integrated transverse magnetization components (normalized so as to be equal to Δ in the 1D model). The DW velocity is very low, whereas the integrated DW magnetization is precessing continuously. This precession is displayed in **(c)**. It is perfectly reproduced by the analytical relation (29). In **(d)**, the DW velocity is plotted versus applied field H_a, with the damping coefficient α as a parameter. The lines are drawn according to the analytical relation (30) with a constant wall-width parameter $\Delta_0 = 7.2\,\text{nm}$ (Table 1)

Slonczewski-like Equations

The basic equation of magnetization dynamics is that of *Landau* and *Lifshitz* [7] with the damping formulated by *Gilbert* [47] called the LLG equation [48] that reads

$$\frac{\partial \boldsymbol{m}}{\partial t} = \gamma_0 \boldsymbol{H}_{\text{eff}} \times \boldsymbol{m} + \alpha \boldsymbol{m} \times \frac{\partial \boldsymbol{m}}{\partial t}, \tag{7}$$

where $\gamma_0 = \mu_0 |\gamma|$ with γ the gyromagnetic ratio, and $\boldsymbol{H}_{\text{eff}}$ is the effective field of micromagnetics defined generally as

$$\boldsymbol{H}_{\text{eff}} = -\frac{1}{\mu_0 M_s} \frac{\delta E}{\delta \boldsymbol{m}}. \tag{8}$$

The local magnetization $\boldsymbol{M}$ is written $M_s \boldsymbol{m}$ in this continuous model, the modulus M_s being constant whereas the orientation can change, as described

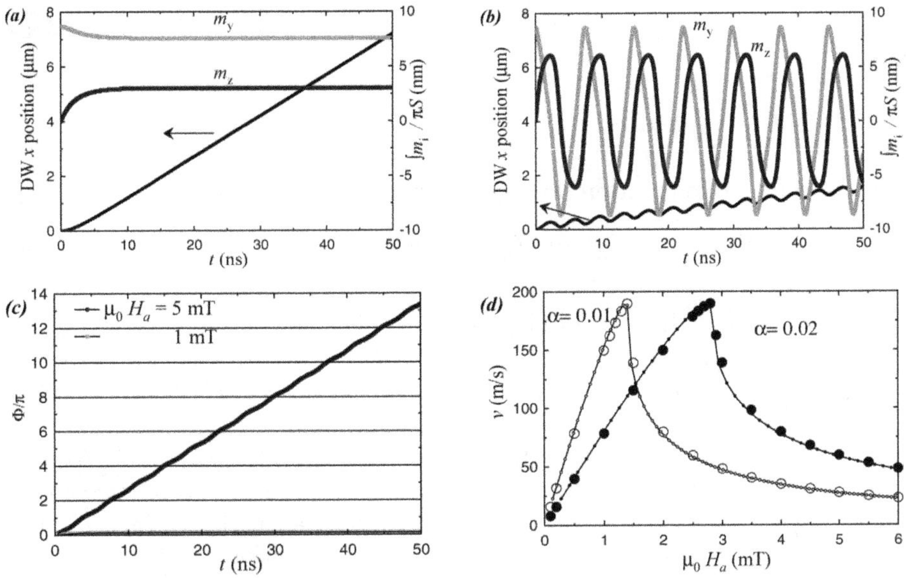

Fig. 11. Transverse-wall dynamics in a rectangular nanowire (edge sizes 10 nm and 5 nm, mesh size 5 nm). The field H_a applied along the wire axis is (**a**) 10 Oe ($\mu_0 H_a$ = 1 mT) and (**b**) 50 Oe ($\mu_0 H_a$ = 5 mT), the damping constant is $\alpha = 0.01$. The DW position is plotted as well as the integrated transverse magnetization components. The DW velocity is high at the low field (**a**), with no steady DW magnetization precession. At the higher field (**b**) this precession appears and the DW velocity drops and displays oscillations. The angle Φ of the DW magnetization orientation is displayed in (**c**) and the DW velocity in (**d**), versus applied field H_a, with the damping coefficient α as a parameter. The analytical results (full thin lines and small dots) are drawn in (**c**) and (**d**), using the parameters of Table 1

by the unit vector $\boldsymbol{m}$ [48]. The micromagnetic energy density is denoted by E, and the derivative indicated by δ is a functional derivative that results from variations of the total energy density with respect to $\boldsymbol{m}$. In terms of the angles θ and φ defined in Fig. 6, the LLG equation becomes

$$\begin{aligned} \dot{\theta} + \alpha \sin\theta \dot{\varphi} &= \gamma_0 H_\varphi \,, \\ \alpha\dot{\theta} - \sin\theta \dot{\varphi} &= \gamma_0 H_\theta \,, \end{aligned} \tag{9}$$

where from now on we adopt the overdot to denote a time derivative. The two components of the effective field that appear in (9) are

$$H_\theta = -\frac{1}{\mu_0 M_s}\frac{\delta E}{\delta\theta}\,; \quad H_\varphi = -\frac{1}{\mu_0 M_s \sin\theta}\frac{\delta E}{\delta\varphi}\,. \tag{10}$$

Consider now the following function L that is called the Lagrangian density [8]

$$L = E + \frac{\mu_0 M_\mathrm{s}}{\gamma_0} \dot{\varphi} \cos\theta \,, \tag{11}$$

and the dissipative function density F given by [47, 49–51]

$$F = \frac{\alpha \mu_0 M_\mathrm{s}}{2\gamma_0} (\dot{\boldsymbol{m}})^2 = \frac{\alpha \mu_0 M_\mathrm{s}}{2\gamma_0} \left(\dot{\theta}^2 + \sin^2\theta \dot{\varphi}^2 \right) . \tag{12}$$

It is easy to check that the LLG equation (9) is identical to

$$\frac{\delta L}{\delta X} - \frac{\mathrm{d}}{\mathrm{d}t} \left(\frac{\delta L}{\delta \dot{X}} \right) + \frac{\partial F}{\partial \dot{X}} = 0 \,, \tag{13}$$

where $X = \theta$ or φ and the functional derivative is $\delta/\delta X = \partial/\partial X - \boldsymbol{\nabla} (\partial / \partial \boldsymbol{\nabla} X)$ (the Lagrangian formulation was first introduced by *Döring* [8] for stationary motions without damping and generalized by *Gilbert* [47, 49–51]). It means that, in the absence of damping, the action

$$\mathcal{L} = \int L\left(\boldsymbol{m}, t\right) \mathrm{d}^3\boldsymbol{r}\, \mathrm{d}t \tag{14}$$

is stationary with respect to any variation of the magnetization structure. When damping is nonzero, the action is no longer stationary and its change is computed using the dissipative function defined similarly to (14).

This general framework is very useful for constructing the LLG equation pertaining to a given class of magnetic structures. In the previous sections, we have seen that the DW structure could be approximated by a 1D profile with just a few parameters. If we restrict the configuration space of the magnetic structures to that class of profiles, the allowed variations of the magnetization are only those that result from a variation of the class parameters. In other words, the LLG equations within the class are directly obtained by setting X equal to each of these parameters, the integrated action $\mathcal{L}$ and dissipation function $\mathcal{F}$ having transformed to functions of these parameters.

Specifically, the 1D profile is described by 3 parameters, namely q the DW center position along the x-axis, Δ the DW width parameter and ϕ the angle of DW magnetization. The magnetization profile is given everywhere as

$$\theta(x, t) = 2 \tan^{-1} \exp \left(\frac{x - q(t)}{\Delta(t)} \right) , \tag{15}$$

$$\varphi(x, t) = \phi(t). \tag{16}$$

Note that (15) describes head-to-head walls (π has to be added to θ in order to get tail-to-tail walls), and that it has the useful property

$$\partial\theta/\partial x = \sin\theta/\Delta \,. \tag{17}$$

The space-integrated Lagrangian $\mathcal{L}$ has the following contributions [we integrate only along x, because of the 1D approximation; all integrals are very easy using (17)]

$$\text{exchange}: \int A\left(\frac{\partial\vartheta}{\partial x}\right)^2 \mathrm{d}x = \frac{2A}{\Delta}\,; \tag{18}$$

$$\text{effective anisotropy}: \int \sin^2\theta\left(K_0 + K\sin^2\varphi\right)\mathrm{d}x = 2\Delta\left(K_0 + K\sin^2\phi\right)\,; \tag{19}$$

$$\text{applied field}: -\mu_0 M_\mathrm{s}\int H_\mathrm{a}\cos\theta\,\mathrm{d}x = -2\mu_0 M_\mathrm{s} H_\mathrm{a} q\,; \tag{20}$$

$$\text{dynamic term}: -\frac{\mu_0 M_\mathrm{s}}{\gamma_0}\int \dot\varphi\cos\theta\,\mathrm{d}x = \frac{2\mu_0 M_\mathrm{s}}{\gamma_0}\dot\phi q\,. \tag{21}$$

It will be convenient for the following to define the ratio of the anisotropies

$$\kappa = K/K_0\,. \tag{22}$$

Similarly, the space-integrated dissipation function $\mathcal{F}$ consists in

$$\int F\,\mathrm{d}x = \frac{\alpha\mu_0 M_\mathrm{s}}{\gamma_0}\left(\Delta\dot\phi^2 + \frac{\dot q^2}{\Delta} + a\frac{\dot\Delta^2}{\Delta}\right), \tag{23}$$

where the pure number a is

$$a = \frac{1}{2\Delta}\int \sin^2\theta\left(\frac{x-q}{\Delta}\right)^2 \mathrm{d}x = \int_0^\infty \frac{x^2}{\cosh^2 x} = \frac{\pi^2}{12}\,. \tag{24}$$

The dynamic equations for the three parameters, obtained from (13) applied to (18)–(21) and (23), constitute the Slonczewski equations [9] for the 1D DW motion, generalized by the inclusion of a variable Δ. They read

$$\begin{aligned} \alpha\frac{\dot q}{\Delta} + \dot\phi &= \gamma_0 H_\mathrm{a}\,,\\ \frac{\dot q}{\Delta} - \alpha\dot\phi &= \gamma_0 H_\mathrm{K}\frac{\sin 2\phi}{2}\,,\\ \dot\Delta &= \frac{\gamma_0}{\alpha\mu_0 M_\mathrm{s} a}\left[\frac{A}{\Delta} - \left(K_0 + K\sin^2\phi\right)\Delta\right]\,. \end{aligned} \tag{25}$$

The first two equations are similar to the LLG equation (9). The third equation shows that Δ relaxes (without any precession effect!) towards its equilibrium value

$$\Delta^*(\phi) = \sqrt{\frac{A}{K_0 + K\sin^2\phi}} = \frac{\Delta_0}{\sqrt{1+\kappa\sin^2\phi}}\,. \tag{26}$$

The characteristic time τ of this relaxation is, by linearization close to equilibrium, given by

$$\tau = \frac{\alpha a}{\gamma_0 H_{\mathrm{K}}} \,. \tag{27}$$

As damping constants usually satisfy $\alpha \ll 1$, this time is very short (for $H_{\mathrm{K}} = M_{\mathrm{s}}/10$ in Permalloy with $\alpha = 0.01$ one gets $\tau = 0.5\,\mathrm{ps}$). Thus, in most practical cases a very good approximation will be realized with a "slave" Δ, i.e., $\Delta(t) = \Delta^* \left[\phi(t)\right]$.

Eliminating q in the two first equations of (25) provides a relation for ϕ only

$$\dot{\phi} = \frac{\gamma_0}{1+\alpha^2}\left(H_{\mathrm{a}} - \frac{\alpha}{2} H_{\mathrm{K}} \sin 2\phi\right) \,. \tag{28}$$

Discussion

The examination of the solutions of this simple equation forms the basis of the understanding of the dynamics of 1D domain walls.

Consider first the situation with no transverse anisotropy ($H_{\mathrm{K}} = 0$). Then from (28) one obtains that ϕ precesses at a uniform rate under a constant applied field

$$\phi = \frac{\gamma_0 H_{\mathrm{a}} t}{1+\alpha^2} \,. \tag{29}$$

The resulting wall velocity reads, from (25),

$$\dot{q} = \gamma_0 \Delta_0 H_{\mathrm{a}} \frac{\alpha}{1+\alpha^2} \,. \tag{30}$$

The velocity increases linearly with field (constant mobility), but it is low by a factor of roughly α^2 compared to the anistropic case discussed below.

When some transverse anisotropy term is present, the dynamics changes radically. The solution of (28) depends on the value of H_{a} with respect to a field called the Walker field that was introduced by *Walker* in its solution of the 1D Bloch-wall dynamics

$$H_{\mathrm{W}} = \alpha H_{\mathrm{K}}/2 \,. \tag{31}$$

If $|H_{\mathrm{a}}| < H_{\mathrm{W}}$ an equilibrium value ϕ^* exists, given by

$$\sin\left(2\phi^*\right) = H_{\mathrm{a}}/H_{\mathrm{W}} \,. \tag{32}$$

In this regime we have a steady-state DW motion, with a nonzero wall magnetization angle ϕ^*, and a constant wall velocity reading, from (25),

$$\dot{q} = \gamma_0 \Delta^*\left(\phi^*\right) H_{\mathrm{a}}/\alpha \,. \tag{33}$$

As the small damping constant α is in the denominator, the wall velocity is much higher than in the no-transverse anisotropy case (30). Note that the velocity-field relation is nonlinear now, because of the variation of the wall width. Eliminating ϕ^* for H_a gives indeed

$$\dot{q} = \frac{\gamma_0 \Delta_0}{\alpha} H_W h \Bigg/ \sqrt{1 + \kappa \left(\frac{1 - \sqrt{1 - h^2}}{2} \right)}, \tag{34}$$

where the anisotropy ratio κ was defined in (22) and h is the normalized applied field H_a/H_W. As might be expected, the nonlinearity effect becomes larger when the relative importance of the transverse anisotropy increases. Fig. 12a plots this relation. From (33) one defines the Walker velocity

$$v_W = \frac{\gamma_0 \Delta_0}{\alpha} H_W = \frac{\gamma_0 \Delta_0}{2} H_K \tag{35}$$

that, remarkably, is independent of α. When κ is not small, the maximum velocity is reached below the Walker field (Fig. 12a), at $h = 2(1+\kappa)^{\frac{1}{4}} \left(\sqrt{1+\kappa} - 1 \right) / \kappa$, and reads

$$v_{max} = 2 v_W \frac{\sqrt{1+\kappa} - 1}{\kappa} . \tag{36}$$

The effect of a nonzero κ is readily apparent in Fig. 12.

The description of the approach to steady-state conditions requires the solution of the differential system (25). In the case of small ϕ (i.e., small applied fields), an expansion of the trigonometric functions to first order gives

$$H_a - \frac{\alpha}{\gamma_0 \Delta_0} \dot{q} = \frac{1 + \alpha^2}{\gamma_0^2 \Delta_0 H_K} \ddot{q} . \tag{37}$$

This shows that the DW has a certain inertial mass, expressed by

$$m_D = \frac{\left(1 + \alpha^2\right) 2 \mu_0 M_s}{\gamma_0^2 \Delta_0 H_K} . \tag{38}$$

This concept was introduced by *Döring* [8] when analyzing the structure of a DW moving at constant velocity (see Sect. 7). Indeed, he found that the energy of such a DW increases in proportion to the square of velocity. Thus, this DW mass m_D is called the Döring mass (even if, in fact, it was first derived in the case of zero damping).

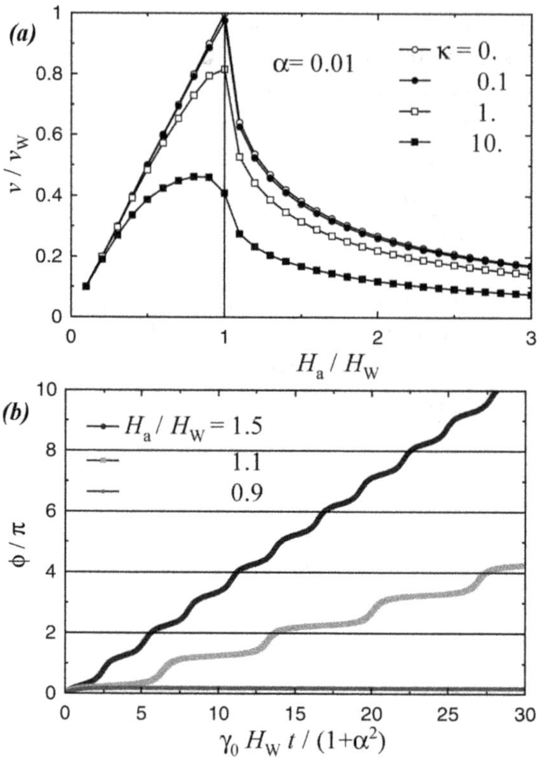

Fig. 12. One-dimensional model of the dynamics of a transverse wall in a nanowire, with a uniaxial transverse anisotropy given by the anisotropy ratio κ. **(a)** dependence of the DW velocity v on the applied field H_a. The velocity is normalized to the Walker velocity v_W, and the field to the Walker field H_W. Above the Walker field, the time-averaged velocity is plotted. All curves are drawn for a damping parameter $\alpha = 0.01$ (but they hardly change if $\alpha = 0.1$, because of the normalization), with κ as a parameter. A larger transverse anisotropy leads to nonlinear $v - H$ characteristics at low fields, a reduction of the maximum velocity below v_W and a downward shift of the maximum velocity field below H_W. **(b)** Time evolution of the wall-magnetization angle ϕ with time normalized according to (39). The normalization absorbs all dependence on α and κ

For applied fields larger than the Walker value, an equilibrium for ϕ is no longer possible. The angle ϕ precesses with time, albeit not linearly because of the transverse anisotropy term. The analytic solution for (28) reads

$$\frac{\mathrm{d}\phi}{h - \sin 2\phi} = \frac{\gamma_0 H_\mathrm{W}}{1+\alpha^2}\,\mathrm{d}t\,,$$

$$\frac{1}{\sqrt{h^2-1}} \tan^{-1}\left(\frac{h \tan\phi - 1}{\sqrt{h^2-1}}\right) = \frac{\gamma_0 H_\mathrm{W}}{1+\alpha^2} t + C^{\mathrm{st.}}\,, \tag{39}$$

where the constant is determined by the initial value of ϕ. Figure 12b plots this time evolution. From (39) one gets easily the period T of the precession

$$T = \frac{2\pi\left(1+\alpha^2\right)}{\gamma_0 H_{\mathrm{W}}\sqrt{h^2-1}}. \tag{40}$$

A numerical calculation is required for the wall velocity or position. Taking the average over one period of the first equation of (25), one can only get a relation that still contains the variation of the wall-width parameter

$$\left\langle \frac{\dot{q}}{\Delta} \right\rangle = \frac{\gamma_0 H_{\mathrm{a}} - \left\langle \dot{\phi} \right\rangle}{\alpha} = \frac{\gamma_0 H_{\mathrm{W}}}{1+\alpha^2}\left[h\left(1+\alpha^2\right) - \sqrt{h^2-1}\right]. \tag{41}$$

A sharp velocity drop occurs therefore above H_{W}, as shown in Fig. 12a (that figure contains the numerical evaluation from (25) and (26) of $< \dot{q} >$ above the Walker field).

Comparison to Numerical Results

We now compare this model to the numerical results presented before. As we introduced the 1D model with effective parameters (A, K_0 and K in the uniaxial case), we need to estimate them. For this, we make use of the relations obtained by minimization of the total energy $\mathcal{E}$ for a fixed ϕ:

$$\begin{aligned} \mathcal{E}_{\mathrm{wall}} &= 4\sqrt{A\left(K_0 + K\sin^2\phi\right)} + C\,, \\ \Delta &= \sqrt{A/\left(K_0 + K\sin^2\phi\right)}\,. \end{aligned} \tag{42}$$

The wall-width parameter Δ can be estimated from the computed magnetization structures in zero field, either from the axial or the transverse component, but rather from the axial values for the reasons exposed earlier (see Fig. 4). The total wall energy is also known from the calculations. The constant C is introduced to take into account an energy term, neglected in the 1D model, that is associated with the axial demagnetizing field existing because the wall is charged (see Sect. 2.2). The four quantities $E_{\min}\,(\phi=0)$, $\Delta_{\min}$, $E_{\max}\,(\phi=\pi/2)$ and $\Delta_{\max}$, determined from the minimum and maximum energy states, allow us to estimate the four parameters A, K_0, K and C. For symmetric wire shapes (such as a rectangular nanowire), the maximum energy state can be obtained easily by starting from an initial configuration with the same symmetry. In the general case this is not possible and the path technique [52] has to be used. Table 1 summarizes the parameters obtained in the cases considered before.

The calculated time evolutions have been superposed to the numerical results in Figs. 10c and d and Figs. 11c and d. The agreement is very good, showing the applicability of the 1D model to these cases. Additional calculations for larger transverse sizes show the progressive increase of the discrepancies, as well as a departure of the effective exchange constant A from the value input to the numerical calculations. For example, at a transverse size of 20 nm × 10 nm, the field for maximum velocity and that velocity are found numerically as $1.3\,\mathrm{mT} < \mu_0 H_\mathrm{a} < 1.4\,\mathrm{mT}$ and $v_\mathrm{max} = 200\,\mathrm{m/s}$, whereas from Table 1 the maximum values (with κ correction) are 1.42 mT and 228 m/s, respectively. At a transverse size of 40 nm × 20 nm, these values become $1.0\,\mathrm{mT} < \mu_0 H_\mathrm{a} < 1.1\,\mathrm{mT}$ and $v_\mathrm{max} = 180\,\mathrm{m/s}$, and 1.29 mT and 280 m/s, respectively.

Table 1. Static parameters (DW width and energy) obtained from numerical simulations of various DW structures in nanowires and nanostrips of different dimensions, parameters (A, K_0, K) of the 1D model extracted from them using (42), and computed dynamical parameters using (31) and (35) with $\alpha = 0.01$. For each dimension and DW structure, the parameters of the lowest and highest energy structures are given, on two successive lines, so as to determine the effective anisotropy K. The DW width Δ_T according to Thiele is defined later (58). Two dynamical parameters are given for the BPW (see text in Sect. 5). The star indicates a value that was assumed, not estimated

Cross sect. (nm^2)	Δ_x (nm)	Δ_t (nm)	Δ_T (nm)	$\mathcal{E}$ (10^{-18} J)	A (10^{-11} J/m)	K_0 ($10^5\,\mathrm{J/m^3}$)	K or K_1 ($10^5\,\mathrm{J/m^3}$)	$\mu_0 H_\mathrm{W}$ (mT)	v_W (m/s)
5 × 5	7.20	7.09	8.25	0.154	1(*)	1.93	0	0	0
TW									
10 × 5	9.28	8.89	9.90	0.325	1.02	1.18	1.12	1.4	229
TW	6.68	6.58	7.81	0.411					
20 × 10	10.99	9.89	10.9	1.81	1.13	0.94	1.19	1.49	286
TW	7.63	7.07	8.24	2.23					
40 × 20	15.36	12.19	12.53	11.01	1.52	0.70	1.14	1.42	369
TW	10.06	8.36	9.45	13.06					
25 × 25	11.05	9.11	10.10	6.023	1.04	0.85	0.075	0.125	24
TW	10.74	8.85	10.07	6.091					
40 × 40	18.95	17.86	17.26	17.9	5.5	1	1.5	1.9	635
BPW				46.2	±1		6	7.5	2 500
120 × 5	37.67	31.31	30.7	3.65	3.32	0.23	2.74	3.43	2 273
TW	10.57	11.49	11.64	9.09					
240 × 10	180	—	20.5	12.87	0.77	0.24	0.61	0.77	2 435
VW	95.1	—	12.3	49.57					

All the preceeding calculations were performed in the case of a uniaxial transverse anisotropy. However, the general anisotropy expression (4) when integrated along x, results in the more general expression

$$\mathcal{E}_\mathrm{an,eff} = 2\Delta \left[K_0 + K G(\phi)\right] . \tag{43}$$

The calculations that follow contain no difficulty. The generalized Walker field is now

$$H_\mathrm{W} = \frac{\alpha H_\mathrm{K}}{2} \max \left|G'(\phi)\right| , \tag{44}$$

but the calculation of the maximum velocity for a non-negligible transverse anisotropy has to be made numerically. One important special case is that of a square nanowire, where

$$E_{\mathrm{an,eff}} = K_0 \sin^2\theta + K_1 \sin^4\theta \sin^2(2\varphi) \ , \tag{45}$$

so that by integration one gets

$$\mathcal{E}_{\mathrm{an,eff}} = 2K_0\Delta + \frac{4K_1}{3}\Delta \sin^2(2\phi) \ , \tag{46}$$

from which it is convenient to define an associated anisotropy field as

$$H_1 = \frac{4K_1}{3\mu_0 M_\mathrm{s}} \ . \tag{47}$$

The evolution equation of the wall-magnetization angle transforms to

$$\dot{\phi} = \frac{\gamma_0}{1+\alpha^2}\left[H_\mathrm{a} - \alpha H_1 \sin(4\phi)\right] \ , \tag{48}$$

so that the Walker field now reads $H_\mathrm{W} = \alpha H_1$. With the substitution $\phi \to 2\phi$, this equation is the same as that for the uniaxial term (28). The effective values determined in the case of a TW in a 25 nm $\times$ 25 nm nanowire are given in Table 1, predicting a Walker field $\mu_0 H_\mathrm{W} = 0.125$ mT and a Walker velocity $v_\mathrm{W} = 24$ m/s. The numerical results are very close to these values: $0.135\,\mathrm{mT} < \mu_0 H_\mathrm{W} < 0.140\,\mathrm{mT}$ and 22.5 m/s, respectively.

We note that, at present, no experimental results exist to be confronted with these theoretical expectations, due to the lack of samples (see however [24]) and of experimental means to detect DW dynamics there.

4 Dynamics of a Transverse Wall in a Nanostrip

We turn now to the dynamics of the simplest DW structure in a nanostrip. Consider a 120 nm $\times$ 5 nm permalloy nanostrip, whose static DW structure has been presented in Sect. 2.

Numerical Results

Under low applied fields, the DW advances in steady-state conditions (Figure 13a), as seen previously for nanowires. At higher fields, however, the wall moves with some oscillations and the wall-magnetization precesses (Fig. 13b). The $v(H)$ curve (Fig. 13c) is also rather similar to the Walker behavior presented in Sect. 3.

However, the snapshots of Fig. 14 reveal that the precession of the DW magnetization is spatially nonuniform. In fact, the precession from $+y$ to $-y$ DW magnetization occurs through the displacement, across the strip width

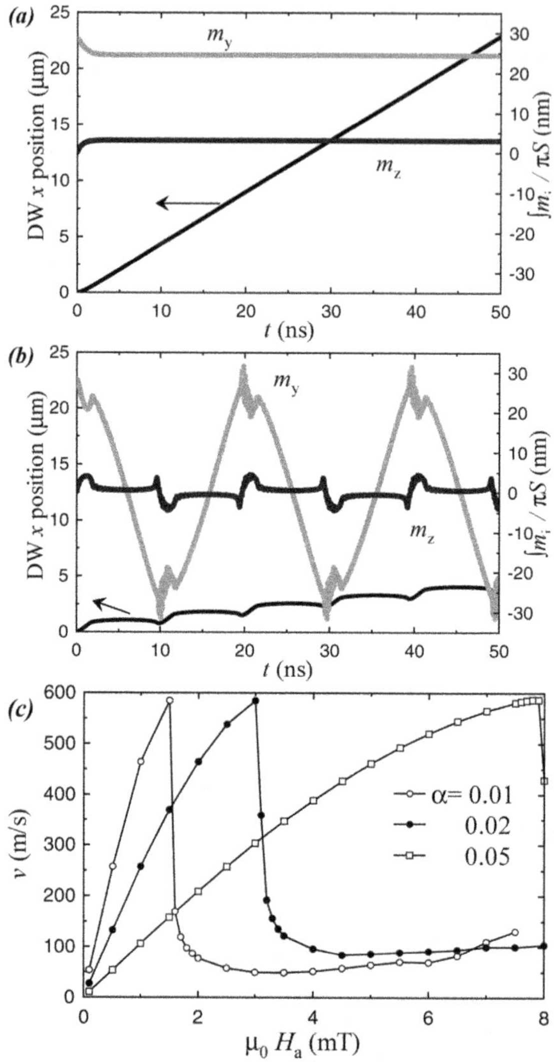

Fig. 13. Transverse-wall dynamics in a nanostrip (width 120 nm, thickness 5 nm, mesh size 5 nm). The field H_a applied along the wire axis is (**a**) 10 Oe ($\mu_0 H_a = 1$ mT) and (**b**) 20 Oe ($\mu_0 H_a = 2$ mT), the damping constant is $\alpha = 0.01$. The DW position is plotted as well as the integrated transverse magnetization components, using the same scales for these two figures. At low field (**a**) DW velocity is high with no evolution of the DW structure, after a transient period. At higher field (**b**) the DW velocity drops markedly and a periodic but complex wall transformation takes place. In contrast to Figs. 10 and 11, the DW magnetization does not keep its magnitude during precession. In (**c**), the DW velocity is plotted versus applied field H_a, with the damping coefficient α as a parameter. Note the remarkable independence of the maximum velocity on α, as predicted in the 1D model (35), (36) and further discussed in Sect. 7.2

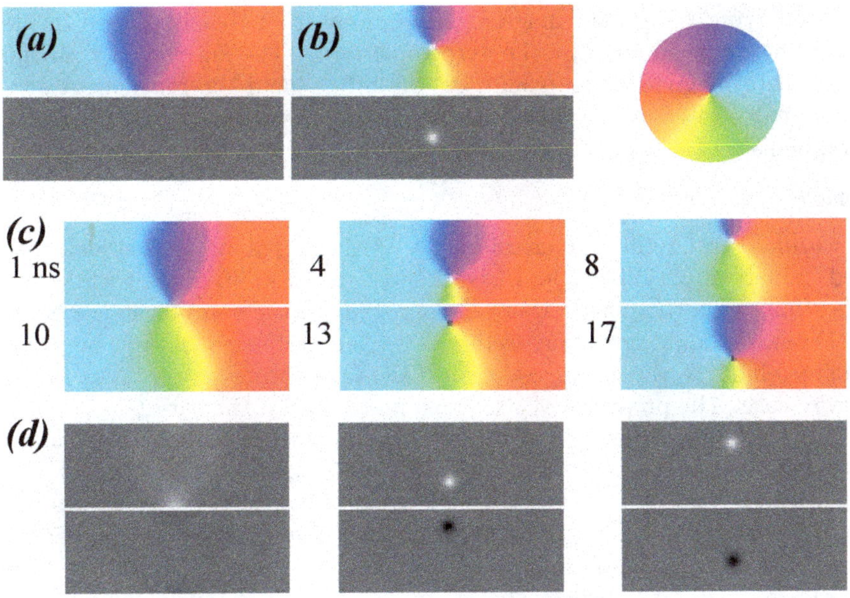

Fig. 14. Pictures of the transverse-wall dynamics in a nanostrip (width 120 nm, thickness 5 nm, mesh size 5 nm) under a field applied along the strip axis of 20 Oe ($\mu_0 H_a = 2\,\mathrm{mT}$), the damping constant being $\alpha = 0.01$. The color wheel on the right indicates the color coding of the inplane magnetization components, whereas the z-component is coded black (-1) to white ($+1$). Structures at rest are **(a)** the transverse wall and **(b)** the unstable antivortex. The dynamics is shown by a series of snapshots at the indicated times, for the inplane **(c)** and out-of-plane **(d)** magnetization components. An antivortex appears, crosses the strip width thus reversing the magnetization of the transverse wall, followed by an antivortex with opposite core appearing and crossing the strip width in the reverse direction, etc. The dynamic images are 320 nm × 120 nm in size

and along the DW, of an antivortex with $+z$ core magnetization. The antivortex (Fig. 14b) is indeed the stable state that is reached when starting from a DW with $+z$ transverse magnetization. Similarly, the conversion from $-y$ to $+y$ orientations occurs by the displacement of an antivortex with $-z$ core magnetization. The strip edge at which the antivortex appears is, at least for not too high fields, that edge where the DW is pinched. Due to the rapid magnetization rotation around this point, an effective field contribution due to the exchange term is present. The direction of the antivortex core magnetization is that favored by precession, around the applied field, of the DW magnetization.

The same argument explains the direction of the motion of the antivortex across the strip width (this was called by *Thiele* the gyrotropic force [53], see Sect. 7.1). Therefore, an alternative view of the antivortex injection and motion is that, at rest, a virtual antivortex exists with its core outside the strip and having the proper magnetization orientation that is pushed inside and then across then strip by the gyrotropic force. The antivortex with the "wrong" core magnetization would, on the other hand, be repelled. A systematic analysis of DW structures in nanostrips, based on topological defect theory, helps to understand this point of view [54].

Comparison to the 1D Model

Can we understand the numerical values of the apparent Walker field obtained numerically? Table 1 and Fig. 7 show that the fitted wall width depends strongly on the magnetization component considered. Thus, the effective 1D model should not apply well, and its parameters should not be properly determined. Nevertheless, we applied the same expressions (42) as in the nanowire case. The predicted Walker field and velocity, that are $H_W = 34\,\mathrm{Oe}$ and $v_W = 2300\,\mathrm{m/s}$ for $\alpha = 0.01$, are much above the numerical values (15 Oe and 600 m/s, respectively), and the effective exchange constant is very large (with the consideration of the effective κ parameter, however, one obtains $H_{\max} = 28\,\mathrm{Oe}$ and $v_{\max} = 1000\,\mathrm{m/s}$ that are closer to the numerical results). This phenomenon is typical of those too complex structures that are no longer well described by the 1D model. Here, the DW magnetization precession appears to be an oversimplified view of the continuous antivortex injection and motion. Note, however, that, despite this obvious shortcoming, the 1D model is off by a factor of no more than 3.

Effect of Roughness

It should be further noted that, as the structure transformation starts at the strip edges, the nature of these edges can affect the DW dynamics. As an example, Fig. 15 shows the effect of rough edges, for the same strip as considered in Fig. 13. The roughness was created here by randomly generating crystallites within the wire, of a given mean diameter D (average grain surface $\pi D^2/4$) and keeping only those entirely located inside a rectangle [29].

It appears clear that the breakdown process is now suppressed, and that the velocity saturates to a value close to that obtained in the perfect sample. However, DW continuous motion is lost at low fields: the roughness also induces a DW propagation field. Moreover, as no thermal fluctuations were introduced in the calculations, the DW motion at low velocity cannot be described [55]. The corresponding snapshots are shown in Fig. 16. They reveal a perturbation of the DW structure due to the roughness, but also the absence of the nucleation of the antivortex (note the locally higher z magnetization component at the tip of the triangle associated with the TW structure).

This introduction of the sample's imperfection has been necessary in order to confront calculated results to experiments [56]. A satisfactory agreement was found [29], using a damping parameter $\alpha = 0.02$ and a roughness parameter $D = 20\,\mathrm{nm}$ for a $120\,\mathrm{nm} \times 5\,\mathrm{nm}$ Permalloy wire. The comparison was based on average velocity data (no real-time measurements nor imaging having been performed).

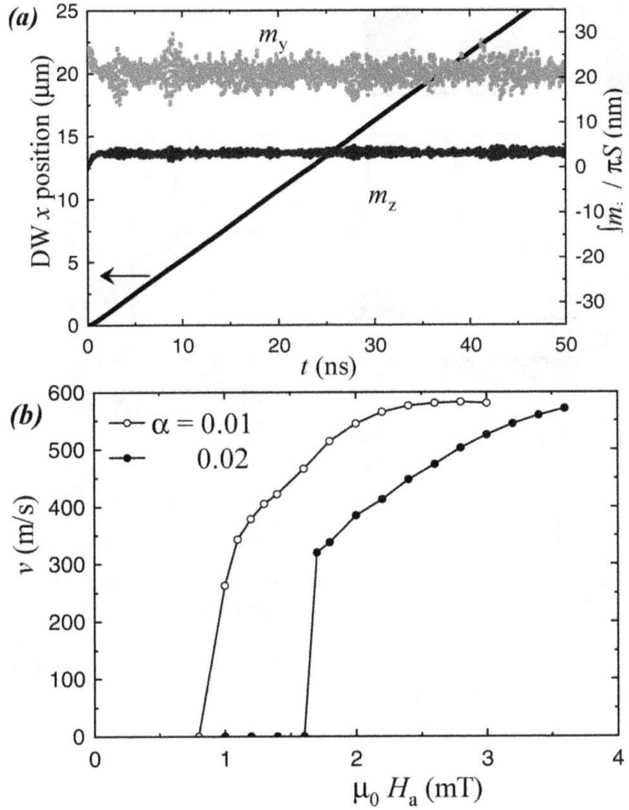

Fig. 15. Transverse-wall dynamics in a rough nanostrip (width 120 nm, thickness 5 nm, mesh size 4 nm × 4 nm × 5 nm, roughness parameter $D = 10$ nm). **(a)** The field H_a applied along the wire axis is 20 Oe ($\mu_0 H_a = 2$ mT), the damping constant is $\alpha = 0.01$. The DW position is plotted as well as the integrated transverse magnetization components. Contrary to Fig. 13, the DW structure does not change much, the DW moves at high velocity and some noise is present. **(b)** DW velocity versus applied field H_a, with the damping coefficient α as a parameter, highlighting the appearance of a propagation field and the suppression of the DW structural transformation as a result of strip edge roughness

5 Dynamics of a Bloch-Point Wall in a Nanowire

We saw in Sect. 2 that for edge sizes above a few exchange lengths (7Λ for a square nanowire without anisotropy), the Bloch-point wall is the stable structure.

In a first step, neglecting the Bloch point itself, we see from Fig. 2 that the precession of the DW moments about an applied axial field will occur differently according to the sign of that field. For a wall bearing a negative charge (tail-to-tail wall) precession around the $a + x$ field will induce a posi-

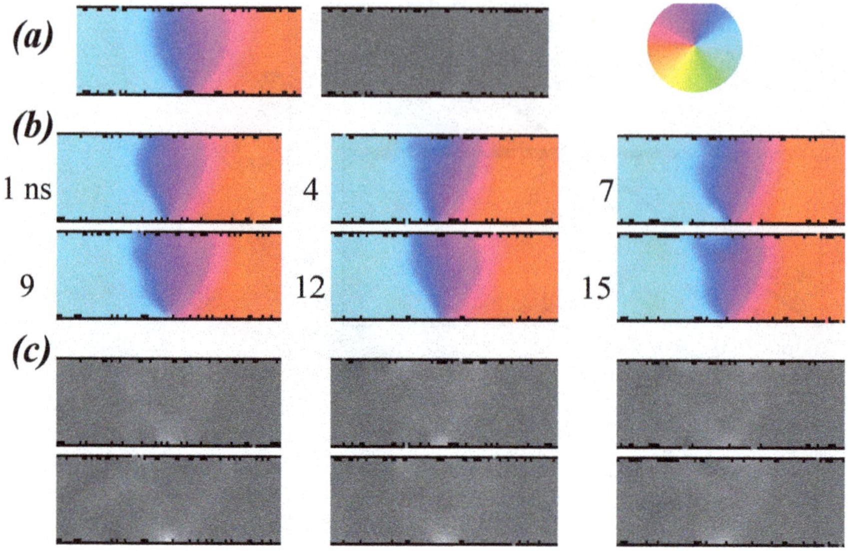

Fig. 16. Transverse-wall dynamics in a rough nanostrip (width 120 nm, thickness 5 nm, mesh size 5 nm, $D = 10$ nm) under a field applied along the strip axis of 20 Oe ($\mu_0 H_\mathrm{a} = 2$ mT), the damping constant being $\alpha = 0.01$. **(a)** Structure at rest. The dynamics is shown by a series of snapshots at the indicated times, for the inplane **(b)** and out-of-plane **(c)** magnetization components. The images are 320 nm wide

tive charge at the center of the wire (Fig. 17a) whereas for the opposite field the negative charge at the BP will increase (Fig. 17b). The former precession should be easier than the latter, considering the magnetostatic energy stored (Fig. 17c). In other words, there should exist two anisotropy fields, thus two Walker fields and Walker velocities, depending on the field sign. As, for a given applied field, (34) shows that velocity increases with transverse anisotropy, we also expect different velocities (in magnitude) for equal and opposite fields.

This description has a very important practical implication: the BPW structure has an intrinsic anisotropy. Taking the circular wire as an extreme case, a huge mobility difference appears between a TW and a BPW, because the TW has no anisotropy hindering the precession of the integrated DW magnetization. Apart from the DW width difference, the mobilities differ by a factor α^2, as seen earlier when comparing the 5 nm × 5 nm and 10 nm × 5 nm wires. This fact has been documented by several authors [31, 32, 35, 36, 44].

The expectations described above are supported by the results of numerical simulations (Fig. 18). At a low field, the steady-state precession angles differ owing to the field sign (Fig. 18a). The positive field, which induces an easier precession (competing with a lower anisotropy), gives rise to a smaller velocity than the negative field. At a higher field (Fig. 18b) a jump of wall-

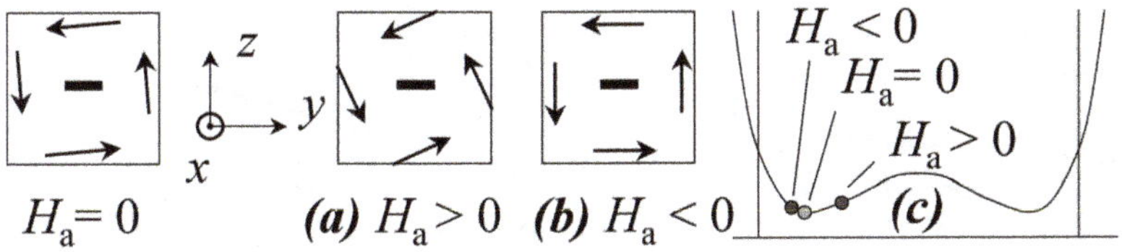

Fig. 17. Schematic of the effect of the magnetization precession under an axial applied field H_a for a tail-to-tail Bloch-point wall of given magnetization circulation sense. **(a)** a positive field reinforces the existing spiral, decreases the negative magnetostatic charge at the wire center and distributes it onto the wire edges. **(b)** a negative field on the contrary concentrates the negative charge at the wire center. **(c)** schematic of energy as a function of the spiraling angle of the DW magnetization (*vertical lines* corresponding to no spiraling, where the angles are 0 or π), with an indication of the effect of the applied field

magnetization angle (by a value a little less than π) occurs for the positive field. The velocity then increases markedly, and becomes the same as that reached under the negative field, as expected because after the jump the wall-magnetization precession is now taking place in the steep side of the potential well.

In order to understand this behavior in greater detail, the profiles of the DW energy and width versus DW "spiraling" angle were computed by the path method [52]. For a BPW, the DW spiraling angle can be defined using the radial and orthoradial average components of its magnetization (these components were plotted in Fig. 5b). The same analysis as in Table 1 was performed in order to extract the effective parameters for the 1D model. The evaluated anisotropy functions $KG(\phi)$ and $K\,\mathrm{d}G/\,\mathrm{d}\phi$ are plotted in Fig. 18c. They are indeed highly nonsymmetrical. From these data and using (44) on the maximum values, one predicts two Walker fields $H_{\mathrm{W}}^{+} = 19\,\mathrm{Oe}$, $H_{\mathrm{W}}^{-} = 75\,\mathrm{Oe}$, and two Walker velocities $v_{\mathrm{W}}^{+} = 635\,\mathrm{m/s}$, $v_{\mathrm{W}}^{-} = 2500\,\mathrm{m/s}$. Comparing with the velocity-field graph (Fig. 18d) shows that this model is only a qualitative description of the low-field phenomena, as it estimates fields and velocities up to a factor of about 2. This was to be expected as the effective exchange constant deduced from the estimation of the effective parameters is $A = 5.5 \times 10^{-11}\,\mathrm{J/m}$. Moreover, the higher Walker field is not seen in the numerical results. The reason for this is already apparent in Fig. 18a, where for the negative field a change starts around $t = 5\,\mathrm{ns}$. Direct examination of the moving DW pictures shows that the wall surface is becoming tilted and curved, leading to a reduction of velocity. The larger the field, the lower the velocity after this transition. It is difficult to ascertain the origin of this transition: it may be an instability of purely numerical origin (remember that the BPW contains a micromagnetic singularity), or real. More work, including a systematic variation of the numerical mesh [34], is necessary in order to clear up this point.

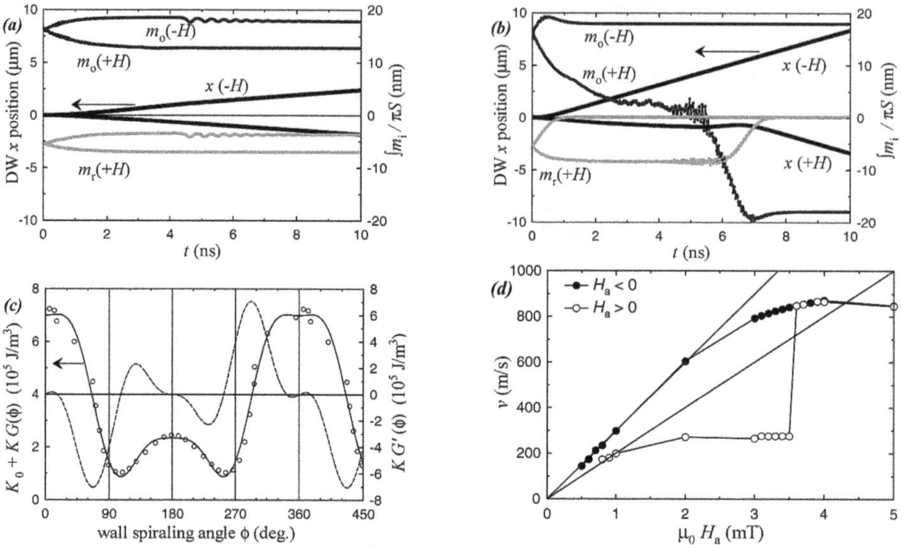

Fig. 18. Dynamics of a tail-to-tail Bloch-point wall in a 40 nm square Permalloy nanowire, for a damping parameter $\alpha = 0.01$. **(a)** For a low field ($\mu_0 H_a = 1\,\text{mT}$), steady-state conditions reached are not symmetrical upon field reversal: the wall moves faster under a negative field. **(b)** At a higher field ($\mu_0 H_a = 4\,\text{mT}$), the positive field causes a transition to a BPW with opposite circulation, that has the same velocity as the structure under the negative field. **(c)** Plot of the estimated effective anisotropy function $KG(\phi)$ deduced from the profiles of the total energy and the wall width associated to the axial component, versus spiraling angle. The full line is a fit of the data by a fourth-degree polynomial in $\cos(\phi)$. The anisotropy derivative was computed from this fit (dashed line, right scale), in order to estimate the Walker fields. **(d)** Relation of velocity to field (absolute values) for both field polarities

An instability at very high fields (200 mT) has also been seen in the calculation by finite elements of a circular nanowire [36]. In that case, magnetic "drops" containing two Bloch points were regularly detached from the DW, as a result of the low DW mobility close to the Bloch point.

At present, no experimental confirmation of this behavior exists. Nanowires (with circular cross section) of this range of diameters do exist [40] but DW dynamics measurements have not been performed yet. The influence of the disorder caused e.g., by a transverse anisotropy with an axis rotating from grain to grain along the wire has also not yet been investigated, so that it is not clear how far the nice calculated behavior will survive in reality.

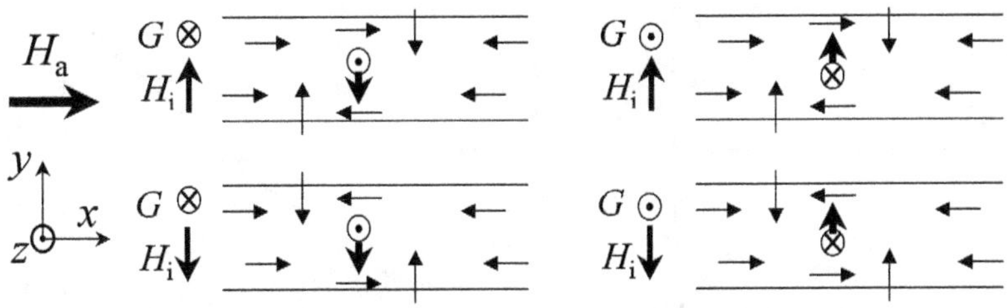

Fig. 19. Displacement of the vortex core inside vortex walls that are all of the head-to-head type, under the same field applied along the nanostrip axis. In addition to the schematic structures, the internal field $\boldsymbol{H}_i$ at the vortex and the gyrovector $\boldsymbol{G}$ (defined in Sect. 7.1) are shown. Both allow us to understand the lateral (along y) motion of the vortex

6 Dynamics of a Vortex Wall in a Nanostrip

We consider finally a 240 nm × 10 nm permalloy nanostrip in which the minimum energy DW structure is the vortex wall (VW, see Fig. 8).

Like the BPW, this structure does not possess a nonzero transverse component (except for that of the vortex core that is oriented perpendicular to the strip plane, but it is small). Therefore, precession inside the DW will be nonuniform, resulting in an evolution of the DW structure. The precession of the vortex core magnetization is equivalent to a motion of the core. As shown in Fig. 19, we expect four different core displacements, under the same field, for the four posible VW structures shown in Fig. 8b. Note that one cannot consider for this purpose the precession around the applied field $\boldsymbol{H}_a$ as, due to the complexity of the VW structure, such a field would not give rise to a motion of the vortex, taken as an isolated entity, along the field direction. That motion corresponds to a field $\boldsymbol{H}_i$, an internal field that will be created by the deformation of the whole DW structure under the applied field.

A y motion means that the field drives the vortex out of the strip. This is again a manifestation of the gyrotropic force (see Sect. 7.1). As the core prefers to sit in the center of the strip as soon as the VW is metastable, there is a restoring force that opposes this motion. Thus we expect that at low fields the VW will move in steady state, with a laterally offset core. Above a certain field, however, the vortex will be expelled and the wall will transform to a TW. The transition field obviously depends on the proximity to the phase boundary displayed in the parameter space of Fig. 9.

These phenomena show up beautifully in the numerical results presented in Fig. 20. The transition from VW to TW occurs with a change of velocity, as shown in Fig. 20d. Snapshots of the DW structure during the high-field motion are presented in Fig. 21. They show the periodic vortex core downwards ($-y$) motion and injection of a new vortex at the top strip edge.

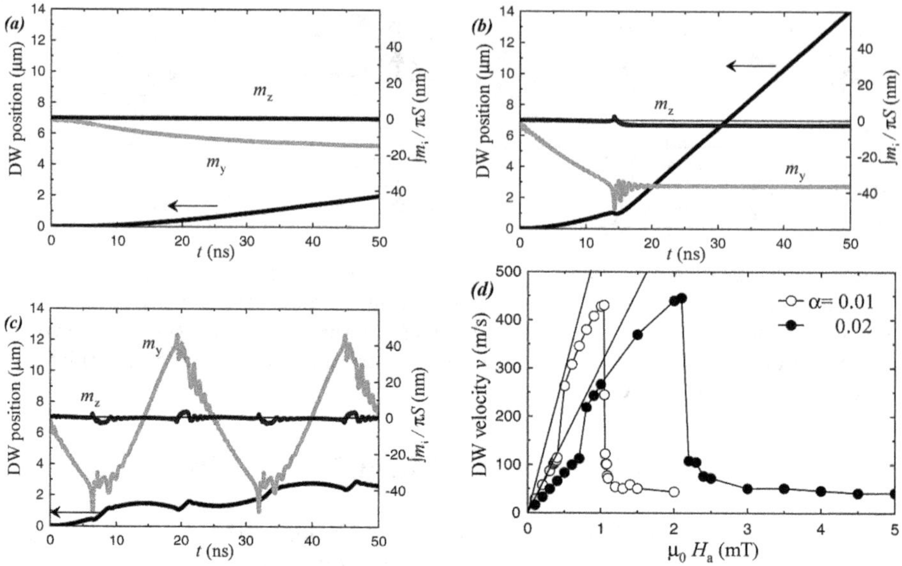

Fig. 20. Vortex-wall dynamics in a nanostrip with a damping parameter $\alpha = 0.01$ (width 240 nm, thickness 10 nm, mesh size 5 nm × 5 nm × 10 nm). **(a)** For a very low field ($\mu_0 H_a = 0.2\,\text{mT}$), the vortex-wall structure is retained, with an offset of the core position shown by the m_y component. **(b)** At a higher field ($\mu_0 H_a = 0.8\,\text{mT}$), the vortex is expelled and the wall transforms into a transverse wall. **(c)** For high fields (here $\mu_0 H_a = 1.5\,\text{mT}$), the Walker instability of the transverse wall (as in Fig. 13) appears and the velocity drops. Note, by comparison to Fig. 13b, the very large anisotropy of the precessing integrated DW magnetization. **(d)** Relation of velocity to field for two values of the damping constant. The three regimes of motion appear clearly, with a jump of mobility when switching to the transverse-wall structure (the lines display the two mobilities for $\alpha = 0.01$)

Comparison to the 1D Model

The estimated parameters of the effective 1D model are given in Table 1. The energy maximum state was obtained by the path technique, it corresponds to a vortex core displaced towards the strip edge and merging with the half-antivortex located there. The energy of this state is well above that of the TW state, that is here metastable. The Walker field at 7.7 Oe (for $\alpha = 0.01$, it is 7.3 Oe with the κ correction) is rather close to the transition field of Fig. 20d. However, the Walker velocity estimated (2435 m/s, 1680 m/s with the κ correction) is 15 times larger than the result of the numerical calculation. The same discrepancy is observed if the mobilities (that do not depend on an instability of the DW structure) are compared.

The solution to this problem was already presented by *Thiele* [53], in his analysis of the stationary motion of arbitrary domain-wall structures, as explained in Sect. 7. This analysis introduces another wall width, de-

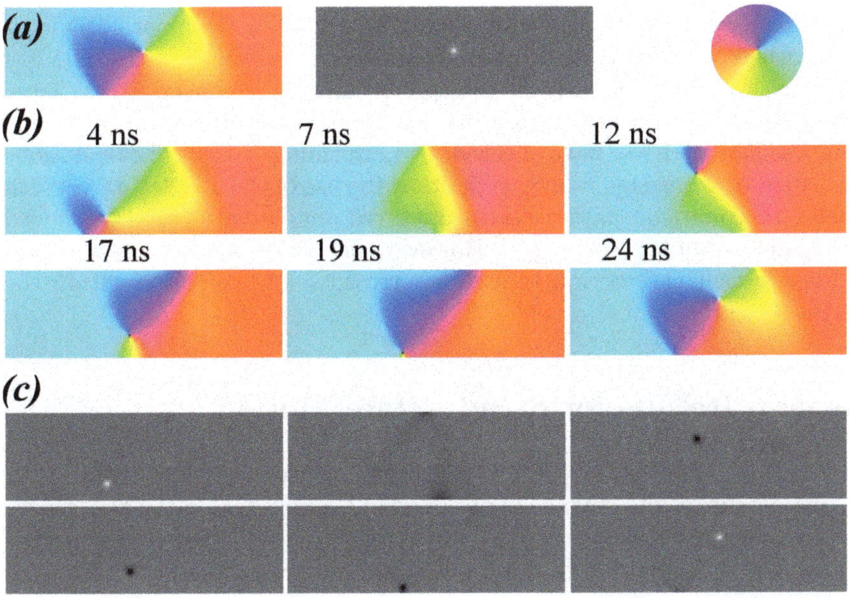

Fig. 21. Pictures of the vortex-wall dynamics in a nanostrip (width 240 nm, thickness 10 nm) under a field applied along the strip axis of 15 Oe ($\mu_0 H_a = 1.5\,\mathrm{mT}$), the damping constant being $\alpha = 0.01$. Color coding of the magnetization components is as in previous figures. **(a)** Structure at rest with color wheel. The dynamics is shown by a series of snapshots at the indicated times, for the inplane **(b)** and out-of-plane **(c)** magnetization components. The images are 750 nm × 240 nm in size

noted Δ_T (Thiele width). As shown in Table 1, it is about 9 times smaller than Δ_x determined by the fit of the Bloch-wall profile. Taking this change of Δ into account for dynamics, we arrive at a Walker velocity of 10[illegible] m/s, reasonably close to the numerical micromagnetics value of 110 m/s. The κ correction would bring that value down to 132 m/s, but one may doubt using it as the numerical data do not show a strong nonlinearity of the velocity-field curve.

Summarizing, we see that a VW can have its own habit of motion only for low fields, and that the DW velocity is well below that predicted by the 1D model. This velocity reduction has been understoodd by the additional magnetization rotation existing around the vortex core, and a new definition of the DW width was introduced to account for this fact. At higher fields the vortex core is expelled and one recovers (dynamically) a TW. This TW undergoes, at still higher fields, a Walker breakdown, as seen previously. The influence of disorder on this process has not yet been completely investigated, and will not be discussed here.

One experimental result exists at present [19], in which the DW motion was measured in real time by the giant magnetoresistance effect, the nanostrip consisting of two magnetic layers separated by a nonmagnetic metallic spacer (strip width $w = 500\,\mathrm{nm}$, thickness of the "free" layer $d = 40\,\mathrm{nm}$). The low DW mobility that was found was interpreted as resulting from a very large damping constant ($\alpha = 0.63$), in a first step where a width parameter $\varDelta = 100\,\mathrm{nm}$ was assumed. We have checked, by performing numerical calculations for the corresponding single-layer strip, that the use of the Thiele's DW width did bring the damping parameter to more reasonable values (0.14 deduced from $\varDelta_{\mathrm{T}}$ instead of 0.29 from $\varDelta_x$). However, this value is very probably still too large, so that the motion that was detected may take place within the overcritical regime.

7 General Results for Steady-State Motion of Domain Walls

Many important results have been obtained in the case where the DW moves without change of its structure, at a constant velocity. Different approaches were used for this purpose, that are described in the next subsections.

7.1 Thiele's Equations

A stationary motion at velocity $\boldsymbol{v}$ implies that

$$\frac{\partial \boldsymbol{m}}{\partial t} = -\left(\boldsymbol{v} \cdot \boldsymbol{\nabla}\right) \boldsymbol{m}\,, \tag{49}$$

so that the LLG equations (9) become

$$\begin{aligned} \sin\vartheta \left[\left(\boldsymbol{v} \cdot \boldsymbol{\nabla}\right) \vartheta + \alpha \sin\vartheta \left(\boldsymbol{v} \cdot \boldsymbol{\nabla}\right) \varphi \right] &= \frac{\gamma_0}{\mu_0 M_{\mathrm{s}}} \frac{\delta E}{\delta \varphi}\,, \\ -\sin\vartheta \left(\boldsymbol{v} \cdot \boldsymbol{\nabla}\right) \varphi + \alpha \left(\boldsymbol{v} \cdot \boldsymbol{\nabla}\right) \vartheta &= \frac{\gamma_0}{\mu_0 M_{\mathrm{s}}} \frac{\delta E}{\delta \vartheta}\,. \end{aligned} \tag{50}$$

The force on the DW is, by definition,

$$\boldsymbol{F} = -\frac{\partial \mathcal{E}}{\partial \boldsymbol{r}_{\mathrm{wall}}}\,, \tag{51}$$

where $\mathcal{E}$ denotes the spatial integral of the energy density E and $\boldsymbol{r}_{\mathrm{wall}}$ the DW position. As the magnetization profile depends on the DW position according to $\boldsymbol{m}\left(\boldsymbol{r} - \boldsymbol{r}_{\mathrm{wall}}\right)$, this transforms into

$$\boldsymbol{F} = +\int \frac{\partial E}{\partial \vartheta} \boldsymbol{\nabla} \vartheta + \frac{\partial E}{\partial \varphi} \boldsymbol{\nabla} \varphi\,. \tag{52}$$

Inserting the values of the energy gradients from (50), one arrives at the Thiele equation [53]

$$\boldsymbol{F} + \boldsymbol{G} \times \boldsymbol{v} + \alpha \overline{\overline{D}} \boldsymbol{v} = \boldsymbol{0}\,, \tag{53}$$

where the gyrovector $\boldsymbol{G}$ is given by

$$\boldsymbol{G} = -\frac{\mu_0 M_\mathrm{s}}{\gamma_0} \int \sin\vartheta \boldsymbol{\nabla}\vartheta \times \boldsymbol{\nabla}\varphi\,, \tag{54}$$

and the dissipation matrix acts on the velocity as

$$\overline{\overline{D}}\boldsymbol{v} = -\frac{\mu_0 M_\mathrm{s}}{\gamma_0} \int \left[(\boldsymbol{v}\cdot\boldsymbol{\nabla})\,\vartheta\right]\boldsymbol{\nabla}\vartheta + \sin^2\vartheta\left[(\boldsymbol{v}\cdot\boldsymbol{\nabla})\,\varphi\right]\boldsymbol{\nabla}\varphi\,. \tag{55}$$

The signs adopted are the same as Thiele's [53]. We have chosen to place the damping coefficient in factor of the dissipation matrix, rather than inside it, for compatibility with extensions of this relation in the presence of a spin-polarized current [57]. The equation means that a stationary motion is a dynamic equilibrium of forces that include, in addition to the intuitive force on the DW, a gyrotropic force and a damping friction.

Thiele's Domain-Wall Width

For a field H_a applied along the x-axis the force is simply

$$\boldsymbol{F} = 2\mu_0 M_\mathrm{s} H_\mathrm{a} S\hat{x}\,, \tag{56}$$

where S is the strip cross-sectional area. In a nanostrip, as the thickness is of the order of the exchange length, the magnetization gradients exist only in the strip plane so that $\boldsymbol{G}$ is perpendicular to that plane. Thus, for a stationary DW displacement along the strip axis Thiele's equation (53) yields

$$v = \frac{\gamma_0}{\alpha}\frac{2S}{\int(\partial\boldsymbol{m}/\partial x)^2}H_\mathrm{a}\,, \tag{57}$$

so that we are led to define a "dynamic" DW width parameter as

$$\Delta_\mathrm{T} = \frac{2S}{\int(\partial\boldsymbol{m}/\partial x)^2}\,. \tag{58}$$

This value is given in the table of parameters (Table 1). For the nanowires considered, it does not differ much from the value estimated in the 1D model, but for nanostrips strong differences can be seen. Using this concept, the mobility that was computed for the VW could be very well understood [42].

Gyrotropic Force and Gyrovector

The Thiele equation explains also, and simply, the vortex sideways motion seen in Figs. 20 and 21: as $\boldsymbol{G}$ is along $-z$ and $\boldsymbol{v}$ along $+x$, the gyrovector crossproduct is along $-y$. As the dissipation matrix has small offdiagonal terms (that are usually neglected), this force has to be balanced by a restoring force along $+y$, corresponding to a core motion along $-y$ (in the direction of $\boldsymbol{G}\times\boldsymbol{v}$).

The gyrovector has a nice geometrical interpretation, as the area covered on the unit sphere by the magnetization distribution [53]. Therefore, one has

$$G_z = \pm 2\pi \frac{\mu_0 M_s}{\gamma_0} d \tag{59}$$

for a vortex wall, and very nearly zero for a TW. Thus, the gyrotropic force exists only for vortices and related structures like antivortices.

7.2 Döring's Kinetic Potential

The Lagrangian of micromagnetics defined in (11) reads, for stationary motions

$$L = E - \frac{\mu_0 M_s}{\gamma_0} \cos\theta \, (\boldsymbol{v} \cdot \boldsymbol{\nabla}) \, \varphi \equiv E - \frac{\mu_0 M_s}{\gamma_0} \varphi \sin\theta \, (\boldsymbol{v} \cdot \boldsymbol{\nabla}) \, \theta \, . \tag{60}$$

The second equality was obtained by subtracting from the original Lagrangian the time derivative of $(\mu_0 M_s / \gamma_0) \, \varphi \cos\theta$, which does not change the variational equations. In terms of the Lagrangian, the LLG system (50) reads

$$\begin{aligned} \frac{\delta L_{\mathrm{int}}}{\delta\theta} &= \frac{\alpha \mu_0 M_s}{\gamma_0} (\boldsymbol{v} \cdot \boldsymbol{\nabla}) \, \theta - \frac{\partial E_{\mathrm{ext}}}{\partial\theta} \, , \\ \frac{\delta L_{\mathrm{int}}}{\delta\varphi} &= \frac{\alpha \mu_0 M_s}{\gamma_0} \sin^2\theta \, (\boldsymbol{v} \cdot \boldsymbol{\nabla}) \, \varphi - \frac{\partial E_{\mathrm{ext}}}{\partial\varphi} \, . \end{aligned} \tag{61}$$

We have here separated the energy density E into internal (E_{int}) and external (E_{ext}) parts, with L_{int} the Lagrangian derived from E_{int}. For the field-driven case, the external energy is simply the Zeeman energy of the applied field, whereas the internal energy gathers the exchange, anisotropy and demagnetizing energies.

The internal Lagrangian is the kinetic potential defined by *Döring* [8]. In his paper, Döring did not consider damping and assumed that the effect of the applied field could be neglected. The system (61), with right-hand sides set equal to zero, means that the kinetic potential is stationary with respect to all variations of the domain-wall structure. This provides a technique by which the structure of the moving DW can be directly obtained, through a calculation similar to that of the DW structure at rest. A very important result of such calculations is that, generally, it is possible to converge towards a DW structure only for velocities smaller than some value. This implies that a maximum velocity exists for stationary motion of any DW structure. As an example, we leave it as an exercise to the reader to solve the 1D model by this technique: the maximum velocity, assuming a constant Δ, is found to be the Walker value and, for a varying Δ, the maximum velocity derived earlier (36).

This formulation also helps understanding intuitively why the maximum velocity obtained in the numerical calculations is, in first approximation, independent of α. Similarly, the dynamic transition from VW to VW in a nanostrip can be understood as a manifestation of the maximum velocity of the VW structure. In fact, that maximum velocity is a measure of the stability of the VW structure.

Justification of the Döring Stationarity Principle

Can we justify rigorously the "principle" of stationary kinetic potential, beyond the 1D model where we have shown that it is correct? We remark first that the spatial integral of the right-hand sides of (61), multiplied respectively by $\dot{\theta}$ and $\dot{\varphi}$ yields

$$-\left[\frac{\mathrm{d}\mathcal{E}_{\mathrm{ext}}}{\mathrm{d}t}+\frac{\alpha\mu_0 M_{\mathrm{s}}}{\gamma_0}\int\left(\frac{\mathrm{d}\boldsymbol{m}}{\mathrm{d}t}\right)^2\right]=0\,. \tag{62}$$

In (62), the last equality means that the energy supplied by the external world is dissipated by damping, so that the internal energy stays constant, as it should for a stationary motion. Thus, the stationarity of the kinetic potential does not result only from the neglect of dissipation and applied field, as we have exhibited one magnetization variation that respects it. However, that both equations in (61) are identically equal to zero seems hard to prove. In fact, calculations performed with very large damping constants ($\alpha \approx 0.1-1$) show that maximum velocities reached differ from the value at low damping, especially for nanostrips with a large cross section. Therefore, it can be concluded that the Döring stationarity "principle" holds only in the limit of small damping, which means, nevertheless, that it applies to the majority of the present experimental situations (samples in fact).

Finally, comparing the two approaches of Thiele and Döring, it can be pictured that the Döring principle allows us to find how the DW structure realizes the balance of forces transverse to the velocity. In other words, (61) separates the terms that do work (on the right) from those that do not (on the left). The energy balance determines the DW velocity, and the balance of internal forces determines how that velocity is sustained by the DW structure.

8 Conclusions and Outlook

In this Chapter, we have described the great variety of the field-driven dynamics of domain walls in nanowires and nanostrips. This diversity has been related to the structure of the domain wall, that in turn depends mainly on the wire/strip transverse dimensions. We have tried to demonstrate that the apparent complexity could be dissipated by the consideration of an effective one-dimensional model. This model has been worked out in detail, both for statics and dynamics, and practical procedures for the estimation of the effective parameters of the model described. Our general philosophy is that this model is very helpful for a qualitative understanding, and that it becomes quantitative at really small sizes (a few exchange lengths Λ). The present state of sample fabrication and measurement is well above this limit, but this might change in the near future.

We have also seen some phenomena that cannot be described by this model, such as the dynamical expulsion of a vortex or the appearance of a very low domain-wall velocity. The question is therefore open about what would be the next-order model. One direction for improvement could include more dynamical parameters in the model. Alternatvely, the connection to Thiele's analysis could be pushed further. We hope that, fueled by new experimental results, research in this field will advance rapidly in the near future.

The spin-transfer-induced displacement of domain walls constitutes another area where we could test our knowledge on domain-wall dynamics. The present situation is that we can test any new term to be included in the magnetization dynamics equation according to its effect on the domain-wall dynamics. Again, the one-dimensional model proves very helpful in this respect, although the structures presently fabricated belong to the class that support complex domain walls with a complex dynamics.

Finally, we have studied briefly the role of imperfections on the dynamical behavior of domain walls. The first results indicate that domain-wall motion may be extremely sensitive to edge roughness. If samples with controlled disorder cannot be produced (but this does not seem to be the case at present), a statistical theory of the motion of domain walls should be developed.

Acknowledgements

The authors would like to thank Jacques Miltat for a critical reading of the manuscript. A. T. thanks Jean-Marie George and Kamel Ounadjela, Albert Fert, and Olivier Fruchart for insightful questions. This work has been partly supported by the Human Potential programme of the European Community, contract number HPRN-CT-2002-00318 ULTRASWITCH.

References

[1] K. J. Sixtus, L. Tonks: Phys. Rev. **37**, 930–958 (1931)
[2] K. J. Sixtus, L. Tonks: Phys. Rev. **42**, 419–435 (1932)
[3] L. R. Walker: Bell Telephone Laboratories Memorandum (1956, unpublished); see, however, the presentation by *Dillon* [4] and the later paper by *Schryer* and *Walker* [6]
[4] J. F. Dillon, Jr.: in G. T. Rado, H. Suhl (Eds.): *Magnetism*, vol. III (Academic, New York 1963) pp. 415–464
[5] F. Bloch: Z. Phys. **74**, 295 (1932)

[6] N. L. Schryer, L. R. Walker: J. Appl. Phys. **45**, 5406–5421 (1974)
[7] L. Landau, E. Lifshitz: Phys. Z. Sowjetunion **8**, 153–169 (1935)
[8] W. Döring: Z. Naturforsch. **3a**, 373–379 (1948)
[9] A. P. Malozemoff, J. C. Slonczewski: *Magnetic Domain Walls in Bubble Materials* (Academic, New York 1979)
[10] A. Hubert: in J. Ehlers, K. Hepp, H. A. Weidenmüller (Eds.): *Theorie der Domänenwände in geordneten Medien*, Lecture Notes in Physics **26** (Springer, Berlin 1974)
[11] A. Hubert, R. Schäfer: *Magnetic Domains* (Springer, Berlin 1998)
[12] U. Ebels, L. D. Buda, K. Ounadjela, P. E. Wigen: in B. Hillebrands, K. Ounadjela (Eds.): *Spin Dynamics in Confined Magnetic Structures I* (Springer, Berlin 2002) pp. 167–216
[13] V. G. Bar'yakhtar, M. V. Chetkin, B. A. Ivanov, S. N. Gadetskii: in G. Höhler (Ed.): *Dynamics of Topological Magnetic Solitons*, Tracts Mod. Phys. **129** (Springer, Berlin 1994)
[14] J. Shen, J. Kirschner: Surf. Sci. **500**, 300–322 (2002)
[15] S. O. Demokritov, B. Hillebrands: in B. Hillebrands, K. Ounadjela (Eds.): *Spin Dynamics in Confined Magnetic Structures I* (Springer, Berlin 2002) pp. 65–92
[16] M. Aprili, T. Kontos, W. Guichard, J. Lesueur, P. Gandit: Physica C **408**, 606 (2004)
[17] W. Wernsdorfer, B. Doudin, D. Mailly, K. Hasselbach, A. Benoit, J. Meier, J.-P. Ansermet, B. Barbara: Phys. Rev. Lett. **77**, 1873–1876 (1996)
[18] R. Ferré, K. Ounadjela, J.-M. George, L. Piraux, S. Dubois: Phys. Rev. B **56**, 14066–14075 (1997)
[19] T. Ono, H. Miyajima, K. Shigeto, K. Mibu, N. Hosoito, T. Shinjo: Science **284**, 468–470 (1999)
[20] Y. Henry, K. Ounadjela, L. Piraux, S. Dubois, J.-M. George, J.-L. Duvail: Eur. Phys. J. B **20**, 35–54 (2001)
[21] Y. Nakatani, N. Hayashi, T. Ono, H. Miyajima: IEEE Trans. Magn. **37**, 2129–2131 (2001)
[22] A. Thiaville, J. M. García, J. Miltat: J. Magn. Magn. Mater. **242–245**, 1061–1063 (2002)
[23] L. Piraux, J. M. George, J. F. Despres, C. Leroy, E. Ferain, R. Legras, K. Ounadjela, A. Fert: Appl. Phys. Lett. **65**, 2484–2486 (1994)
[24] E. Snoeck, R. E. Dunin-Borkowski, F. Dumestre, P. Renaud, C. Amiens, B. Chaudret, P. Zurcher: Appl. Phys. Lett. **82**, 88–90 (2003)
[25] D. A. Allwood, N. Vernier, G. Xiong, M. D. Cooke, D. Atkinson, C. C. Faulkner, R. P. Cowburn: Appl. Phys. Lett. **81**, 4005–4007 (2002)
[26] Y. Nakatani, Y. Uesaka, N. Hayashi: Jpn. J. Appl. Phys. **28**, 2485–2507 (1989)
[27] N. Hayashi, K. Saito, Y. Nakatani: Jpn. J. Appl. Phys. **35**, 6065–6073 (1996)
[28] H. Fukushima, Y. Nakatani, N. Hayashi: IEEE Trans. Magn. **34**, 193–198 (1998)
[29] Y. Nakatani, A. Thiaville, J. Miltat: Nature Mater. **2**, 521–523 (2003)
[30] R. P. Cowburn: J. Phys. D: Appl. Phys. **33**, R1–R16 (2000)
[31] H. Forster, T. Schrefl, W. Scholz, D. Suess, V. Tsiantos, J. Fidler: J. Magn. Magn. Mater. **249**, 181–186 (2002)
[32] R. Hertel: J. Magn. Magn. Mater. **249**, 251–256 (2002)
[33] E. Feldtkeller: Z. angew. Phys. **19**, 530–536 (1965)

[34] A. Thiaville, J. M. García, R. Dittrich, J. Miltat, T. Schrefl: Phys. Rev. B **67**, 094410–12 (2003)
[35] R. Hertel, J. Kirschner: Physica B **343**, 206–210 (2004)
[36] R. Hertel, J. Kirschner: J. Magn. Magn. Mater. **278**, L291–L297 (2004)
[37] A. Aharoni: J. Appl. Phys. **83**, 3432–3434 (1998)
[38] W. F. Brown, Jr.: in E. P. Wohlfarth (Ed.): *Magnetostatic Principles in Ferromagnetism*, Selected Topics in Solid State Physics **I** (North Holland, Amsterdam 1962)
[39] D. G. Porter, M. J. Donahue: J. Appl. Phys. **95**, 6729–6731 (2004)
[40] M. Darques, A. Encinas, L. Vila, L. Piraux: J. Phys. D: Appl. Phys. **37**, 1411–1416 (2004)
[41] R. D. McMichael, M. J. Donahue: IEEE Trans. Magn. **33**, 4167–4169 (1997)
[42] Y. Nakatani, A. Thiaville, J. Miltat: J. Magn. Magn. Mater. **290–291**, 750–753 (2005)
[43] M. Kläui, C. A. Vaz, J. A. Bland, L. J. Heyderman, F. Nolting, A. Pavlovska, E. Bauer, S. Cherifi, S. Heun, A. Locatelli: Appl. Phys. Lett. **85**, 5637–5639 (2004)
[44] R. Wieser, U. Nowak, K. D. Usadel: Phys. Rev. B **69**, 064401 (2004)
[45] A. Kirilyuk, J. Ferré, D. Renard: Europhys. Lett. **24**, 403 (1993)
[46] S. V. Tarasenko, A. Stankiewicz, V. V. Tarasenko, J. Ferré: J. Magn. Magn. Mater. **189**, 19–24 (1998)
[47] T. L. Gilbert: Phys. Rev. **100**, 1243 (1955) (abstract only)
[48] J. Miltat, G. Albuquerque, A. Thiaville: in B. Hillebrands, K. Ounadjela (Eds.): *Spin Dynamics in Confined Magnetic Structures I* (Springer, Berlin, Heidelberg 2002) pp. 1–33
[49] T. L. Gilbert: *Formulation, foundations and applications of the phenomenological theory of ferromagnetism*, Ph.D. thesis, Illinois Institute of Technology, Chicago (1956)
[50] T. L. Gilbert: *Project A059, Supplementary Report*, Technical report, Armour Research Foundation (1956)
[51] W. F. Brown, Jr.: *Micromagnetics* (Wiley Interscience, New York 1963)
[52] R. Dittrich, T. Schrefl, D. Süss, W. Scholz, H. Forster, J. Fidler: J. Magn. Magn. Mater. **250**, 12–19 (2002)
[53] A. A. Thiele: Phys. Rev. Lett. **30**, 230–233 (1973)
[54] O. Tchernyshyov, G. W. Chern: Phys. Rev. Lett. **95**, 1977204 (2005)
[55] J. Ferré: in B. Hillebrands, K. Ounadjela (Eds.): *Spin Dynamics in Confined Magnetic Structures I* (Springer, Berlin, Heidelberg 2002) pp. 127–165
[56] D. Atkinson, D. A. Allwood, G. Xiong, M. D. Cooke, C. C. Faulkner, R. P. Cowburn: Nature Mater. **2**, 85–87 (2003)
[57] A. Thiaville, Y. Nakatani, J. Miltat, Y. Suzuki: Europhys. Lett. **69**, 990–996 (2005)

Index

Domain-Wall Dynamics in Magnetic Logic Devices

Del Atkinson[1], Colm C. Faulkner[1], Dan A. Allwood[2], and Russell P. Cowburn[3]

[1] Department of Physics, University of Durham
South Road, DH1 3LE, Durham, UK
[2] Department of Engineering Materials, University of Sheffield
Portobello Street, Sir Robert Hadfield Building, S1 3JD, Sheffield, UK
[3] Blackett Physics Laboratory, Imperial College London
Prince Consort Road, SW7 2BW, London, UK

Abstract. Domain walls propagate very well through magnetic nanowires and have been used to demonstrate a logic scheme in which the two stable magnetization directions in planar nanowires are used as the two Boolean basis states of a logic system. In this case, magnetic domain walls perform the role of a signal transition edge, and information can be spread throughout an interconnected network of nanowires by the propagation of the magnetic domain walls. The performance of such domain-wall logic is dependent on the dynamics of the domain walls within the magnetic nanowire circuits. Here the issues concerning the dynamic behavior of magnetic domain walls in planar nanowires with inplane magnetization are reviewed. The velocity of domain-wall propagation is a key consideration as it relates ultimately to the operational speed of the logic circuit. Theoretical and experimental work on velocity indicates that domain walls can propagate very rapidly (over 1 km/s) when the wire thickness is small (e.g., 5 nm). The propagation of domain walls through the complex-shaped nanowire components that make up a logic circuit is also discussed. These components include corners, interconnects as well as the logic junctions. Finally, for fully functional magnetic logic the need for overlap in the operational field ranges for domain-wall propagation of the individual circuit components is described.

1 Introduction

Many of the magnetoelectronic devices that have been proposed in recent years for information storage and manipulation are based upon discrete magnetic elements that undergo abrupt magnetic reversal at a critical applied magnetic field. The classic example of such elements is the storage element in a magnetic random access memory (MRAM) cell. Magnetization reversal of such elements either involves the initial nucleation of a reversed domain, or else a quasicoherent rotation of the magnetization of the entire element. Interest continues to grow, however, in alternatively engineered magnetic elements, and, in particular, those that involve the controlled motion of a pre-existing domain wall through the device, perhaps between stable resting

B. Hillebrands, A. Thiaville (Eds.): Spin Dynamics in Confined Magnetic Structures III, Topics Appl. Physics **101**, 207–224 (2006)

positions [1]. One of the advantages of such a scheme is that there is less opportunity for topological complexities in the micromagnetics of the element, such as 360° domain walls and unannihilated vortex cores [2], which can lead to irreproducibility in the magnetic switching or ultimately to failure of the device.

Domain walls often propagate very well through magnetic nanowires and the nanowires can be used to move domain walls from one place to another. In particular, a scheme has recently been proposed [3] in which the two stable magnetization directions in planar nanowires can be used as the two Boolean basis states of a logic system. In this case, magnetic domain walls perform the role of a signal transition edge, and information can be spread throughout an interconnected network of nanowires by the propagation of the magnetic domain walls.

Understanding and controlling the behavior of domain walls as they propagate through complex geometry planar magnetic nanowires is critical for the successful development of magnetic logic circuits. Key factors are the propagation velocity of domain walls and the influence of structural design features, such as junctions and corners, on domain-wall movement. Knowledge of the domain-wall velocity and its dependence on magnetic field when combined with magnetic logic pathlength information allows estimates of device operating times and equivalent clock speeds. Understanding the interaction of domain walls with structural features can lead to improvements in both the dynamic behavior of walls through these structures and the logic functionality (for example, by increasing the operating margin between states).

2 Domain-Wall Propagation in Nanowires – Velocity and Dynamics

The propagation of magnetic domain walls is a major process of the magnetization change for many magnetic materials and as such has been very widely studied. The propagation of domain walls has been shown to depend on a wide variety of factors, including the material composition, structural dimensions and the number and nature of included defects. For the magnetic logic scheme under discussion here the magnetic materials of current interest are alloys in the nickel-iron series with atomic ratios around 80:20 nickel to iron, generically referred to as Permalloy. Direct dynamic visualization of domain walls as they propagate in nanowires is not currently possible due to a combination of their small size and the short time scales of the wall motion. However, insights and understanding of the dynamic behavior of domain walls in magnetic logic nanowires can be obtained from generalized theoretical developments, micromagnetic modeling and experimental investigation of thin films and most importantly nanostructured Permalloy.

2.1 Domain-Wall Dynamics in Continuous and Nanostructured Permalloy Films

The propagation behavior of domain walls has been widely studied both theoretically [4–7] and experimentally for thin-film Permalloy [8–11] but it is only recently that domain-wall dynamics have been studied in magnetic nanostructures [12–14] and particularly the behavior in planar nanowires [15–18]. Models of domain-wall propagation have highlighted the dependence of wall velocity on the domain-wall width and damping of the gyromagnetic precession and support the idea of the wall as a clearly defined entity, which changes relatively little in width as the wall propagates with field. Such modeling, which excludes defects and temperature effects, typically shows an initially linear (or nearly so) magnetic-field dependence of the velocity from zero field with small deviations due to field-dependent changes of the domain-wall width, followed by larger deviations or oscillations due to instability of the wall structure. An important parameter for describing the field dependence of velocity is the domain-wall mobility that is defined as the slope of the velocity against magnetic field curve. For a simple plane wall the field dependence of the velocity is linear and limited by gyromagnetic behavior of the spin reorientation process. In this case, the domain-wall mobility has been expressed by the relation

$$\mu = \gamma \Delta / \alpha, \tag{1}$$

where μ is the domain-wall mobility, γ is the gyromagnetic ratio and α is the phenomenological precessional damping parameter [4]. Recent micromagnetic modeling of the domain-wall mobility in thin films and nanostructured Permalloy showed an initially linear or near-linear field dependence of the velocity, obtaining wall mobilities of 30 m/(s Oe) for a 10 nm thick film [5] and 31 m/(s Oe) in a 5 nm thick, 200 nm wide planar nanowire [19].

2.2 Domain-Wall Velocity in Permalloy Thin Films

Experimental studies of domain-wall dynamics in nickel-iron and cobalt thin films report field-dependent velocity behavior consistent with thermally activated wall motion [8, 9, 20] while other domain observations and magnetization measurements also show more complex behavior in which reversal proceeds via nucleation and growth of reversed magnetization in and around a diffuse domain boundary region sweeping through the film [8, 20].

Domain-wall velocities in Permalloy thin films have been determined from observations of the domain-wall displacement produced by the application of pulses of magnetic field of known duration [9, 21]. Such experimental work has shown that the magnetic-field dependence of the domain-wall velocity may be divided into two regimes. In the high-field regime the domain-wall

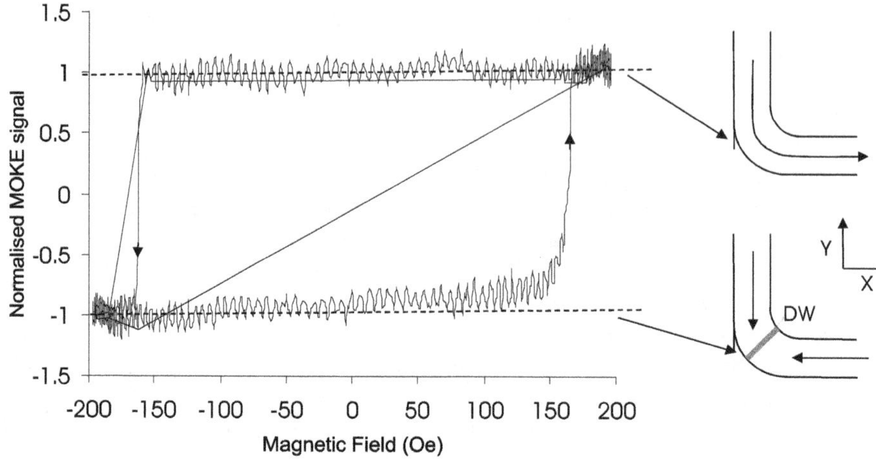

Fig. 1. Magnetization behavior in a 200 nm wide 5 nm thick planar Permalloy nanowire subjected to a horizontal magnetic field. The domain structures at positive and negative remanence are shown schematically. Reversal from negative to positive remanence occurs via propagation of an existing wall and as the field increases the wall is eventually swept out of the wire. Switching the magnetization back to negative requires the nucleation of a reverse domain and domain wall at higher fields

behavior is governed by intrinsic gyromagnetic damping and at lower fields the behavior is affected by thermally activated pinning and depinning of the wall during propagation. In the high-field regime the velocity varies linearly with magnetic field giving a single-valued mobility broadly in agreement with the models. For example, a mobility value of 38 m/(s Oe) was obtained at high field in the thinnest film (31 nm) reported by *Konishi* et al. [9].

2.3 Domain-Wall Behavior in Permalloy Planar Nanowires

Submicrometer patterning of magnetic thin films adds a further degree of structural confinement that can simplify the domain-structure configuration. For planar Permalloy nanowires the magnetization is constrained to lie along the long axis by shape anisotropy and axial domain walls do not occur as the width of the wall is comparable with the scale of the nanostructure width. In these wire structures magnetization reversal takes place by the propagation of a 180° head-to-head domain wall along the axis of the wire.

The basic behavior of domain-wall propagation in a single Permalloy nanowire is illustrated by the magnetization loop in Fig. 1. The magnetization behavior was measured along the horizontal arm of the structure as an average of many hysteresis loops using a continuous-wave laser magneto-optic Kerr effect (MOKE) magnetometer [22]. The structure shown in Fig. 1 allows a domain wall to be positioned at the corner of the structure. The ap-

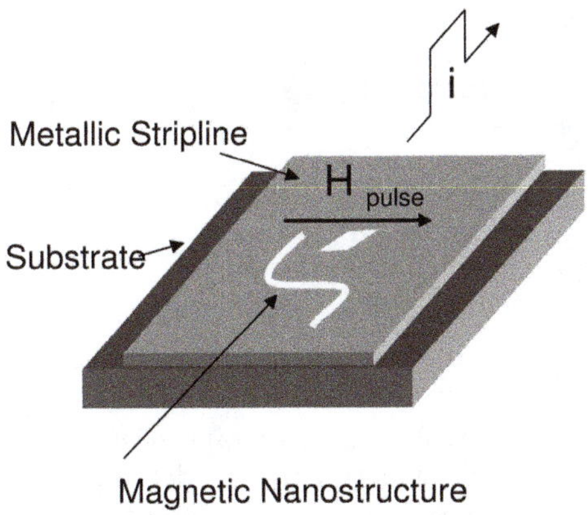

Fig. 2. A schematic of the sample used for investigating domain-wall dynamics in a 200 nm wide Permalloy structure

plication of a positive field along the x-axis sweeps the wall through and out of the horizontal section of the structure. In this case, the horizontal measured magnetization reversal process is distributed over many oersteds, changing gradually at first then more rapidly at higher fields. In contrast, when the magnetization is switched back to the original orientation the switching field is higher and the switching event is sharper, occurring over a very small field range. Similar switching behavior is observed in measurements of single magnetic-field-cycle hysteresis, although the data have higher noise. The first observation is consistent with domain-wall motion in which a pre-existing wall moves through a series of pinning sites during reversal, while a wall nucleated at higher fields propagates essentially unimpeded by the pinning sites.

Studies of domain-wall dynamics in submicrometer-scale structures are limited. *Koch* et al. [15] studied the magnetization reversal of a 5 nm thick $Ni_{60}Fe_{40}$ 800 nm wide and 1.6 μm long patterned element using tunneling magnetoresistance. In this case, the switching was described in terms of viscous (thermally assisted) domain-wall motion and an effective domain-wall mobility of 16 m/(s Oe) was obtained. A study of the magnetization switching in a 500 nm wide, 40 nm thick $Ni_{80}Fe_{20}$ layer that formed part of a trilayer giant magnetoresistive structure measured domain-wall velocity at temperatures between 100 and 160 K. The velocities obtained range between one hundred and three hundred m/s and the mobility obtained was 2.6 m/(s Oe) and constant over the temperature investigated. This mobility is much lower than that obtained from micromagnetic modeling and thin-film measurements described earlier, and has been attributed to a significantly higher gyromagnetic damping due to edge and surface defects [16].

More recently, domain-wall dynamics have been studied in a single-layer Permalloy nanowire with a geometry directly related to the structures used for the current magnetic logic circuits [17, 19]. A planar nanowire structure similar to the one shown in Fig. 1 was used to study the dynamics and specifically the velocity of domain-wall propagation at room temperature.

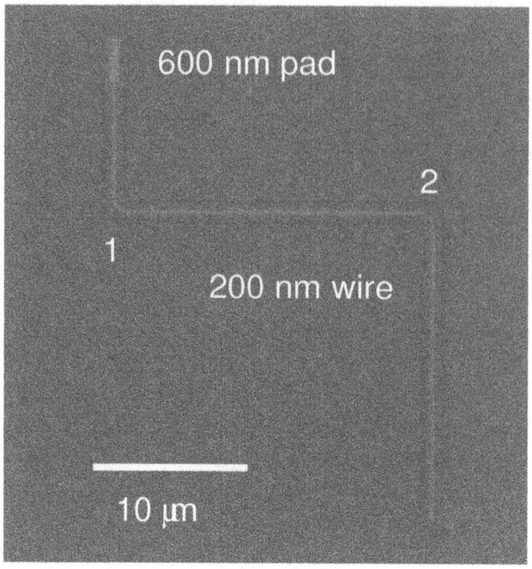

Fig. 3. A micrograph of a 200 nm wide structure used to determine the dynamic behavior of domain walls. The structure was fabricated by thermal evaporation and focused ion beam milling with 30 keV Ga^+ ions

This time, the nanostructure was fabricated directly on top of a thin film of aluminum that formed the conductor in a stripline transmission line. In this experiment, longitudinal MOKE was again used to detect reversals of magnetization in the nanostructure. Domain walls were propagated by applying pulsed magnetic fields to the nanostructure. The temporal resolution in the experiment comes from the controlled duration of the magnetic-field pulses. Figure 2 illustrates schematically the experimental arrangement and Fig. 3 shows an image of the nanowire structure fabricated by focused ion beam milling. This structure also allowed the study of domain-wall propagation into a corner in the nanowire. This aspect will be discussed later.

Domain-wall velocities were obtained from the transit time taken for a wall to travel along a known length of the wire. The transit time was taken from the duration of the pulsed field. A domain wall was first nucleated in the "pad" and moved to the first corner by a field in the y-direction from an electromagnet. This field was removed and a rapid flat-topped magnetic field pulse was applied in the x-direction by a current pulse in the stripline. During this pulse the domain wall propagated along the middle section. A second field from the electromagnet was then applied in the y-direction. The magnitude of this field was such that if a domain wall had reached the second corner (moved during the field pulse) it would be propagated down into the final part of the structure and detected as a switching event. The magnetization state was reset after each sequence. Care was taken to ensure that directly nucleated reversal did not occur in the middle section of the structure. This was achieved by determining the field amplitude needed for reversal in the middle section without a wall being positioned previously at the first corner. Propagation experiments were then carried out with pulsed fields below this amplitude. For each pulse length investigated, measurements were made as a function of pulsed-field amplitude to determine the switching

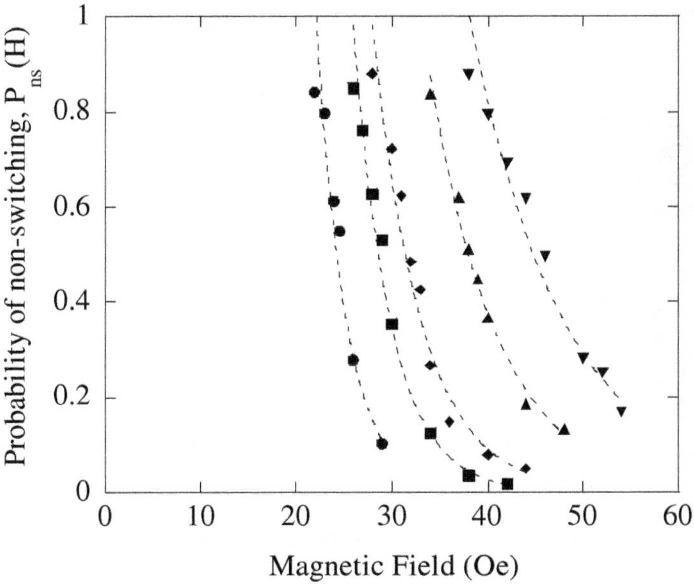

Fig. 4. Nonswitching probabilities as a function of field amplitude for fixed pulse durations (•508 ns, ■150ns, ♦71 ns, ▲45 ns, ▼24 ns). The exponential fits are guides to the eye

behavior. At each combination of pulselength and field amplitude the propagation experiment was repeated many hundreds of times from which the average magnetization reversal signal obtained represented the proportion of times the domain wall propagated along the wire to the detection position. This proportion of switching events observed is described more commonly in terms of the nonswitching probability. Figure 4 shows the effect of increasing the magnetic-field amplitude on the nonswitching probability. Figure 4 reveals that the domain-wall motion is stochastic as the walls do not always travel the same distance under each field pulse. The probability of switching occurring, i.e., the wall traveling through the detection point, increases with both applied field and pulse length. This is qualitatively in agreement with thermally activated wall motion along the nanowire and consistent with other work on domain-wall behavior.

As mentioned earlier the velocity is obtained from the known length of the magnetic wire that the wall has to travel and the duration of the magnetic field pulse that is also known. Figure 5 shows the magnetic-field dependence of the domain-wall velocity as measured from reversals detected in the position shown on the inset image of the structure. In this figure the magnetic-field value for each velocity point was obtained from data shown in Fig. 4 for the probability of nonswitching as a function of field at the level of 50 %. Velocity data was not obtained for magnetic fields higher than 50 Oe because the

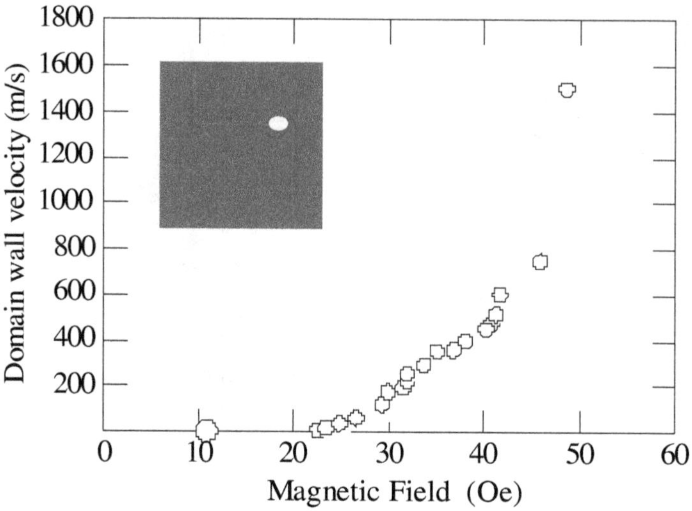

Fig. 5. Domain-wall velocity as a function of magnetic field for a 200 nm wide 5 nm thick planar Permalloy nanowire

switching process occurred without the need for a domain wall initially present, i.e., switching occurred by directly nucleation reversal in the horizontal section of the structure.

Domain-wall motion was observed at fields down to 12 Oe, where the velocity was very low and up to nearly 50 Oe where the velocity was over 1 km/s. The form of the curve is consistent with thermally activated wall motion [17, 23] but has also been shown by micromagnetic modeling to be largely consistent with domain-wall motion that is significantly affected by the roughness of the edges of the structure [18].

The nature of the wall during the reversal process could not be elucidated from these measurements, but across the field range presented switching was only observed when a domain wall was initially present at the first corner. At the highest velocities the structure of the wall may be significantly different as observed for thin films by *Kryder* and *Humphrey* [8].

3 Domain-Wall Dynamics in Complex Structures for Magnetic Logic

3.1 Magnetic Logic Circuits

Magnetic logic circuits are complex, incorporating, in addition to linear nanowire interconnects, corners, constrictions and junction structures that the domain walls must negotiate in order to transmit and transform logic-state information. A scheme has been developed comprising four basic junction structures. These four structures perform, respectively, the Boolean

NOT-function, the Boolean AND-function, the fanning out of one nanowire into two ("Fan-out") and the crossing over of two nanowires in the same plane ("Cross-over"). Figure 6a shows, for example, a magnetic logic circuit that contains at least one example of each of these basic junctions [24]. In addition, one sees in the magnetic circuit linear nanowire interconnects and rounded corners. Since information is represented by the ferromagnetic magnetization direction, processing of that information is achieved by manipulating the topology of the nanowire network.

The magnetic circuit of Fig. 6a works in the following way. The NOT gate is formed within a closed loop, such that its output is connected back around into its input. This closed loop forms a synchronous ring oscillator, in which domain walls circulate under the action of the externally applied rotating magnetic field. Each time the domain wall passes through the NOT gate, it is converted from a head-to-head wall into a tail-to-tail wall and vice versa. Simple analysis of such a system will show that it should oscillate with a frequency of one third of the external field frequency [3]. A fan-out junction is incorporated into the closed loop in order to extract a copy of the circulating information and to feed it into a second fan-out junction. One copy from this second fan-out is routed directly up the right-hand side of the magnetic circuit, while the other undergoes a 360° loop rotation. Although this extra loop does not add any extra logic functionality, it delays the signal by one clock cycle, and, more importantly, it allows the insertion of a properly operated crossover junction. The two signals are then recombined in an AND gate.

Figure 6b shows low-frequency (27 Hz) magneto-optical measurements of the operation of this magnetic circuit. The top two traces show the x- and y-components of the rotating magnetic field that is applied to the entire magnetic circuit. Trace I shows a magneto-optical measurement of the magnetization made locally on one of the nanowire interconnects marked "I" in panel (a). As expected, one sees a trace with a frequency of one third of that of the oscillating field. Traces II and III show schematically the same data, but phase shifted to take account of the synchronous delays associated with the extra 360° loop and the 90° corners and correspond to the expected signals at the two inputs to the AND gate. Trace IV shows a magneto-optical measurement of the magnetization made locally at the output of the AND gate, marked "IV" in panel (a). Here, the short pulses in trace IV correspond to the overlap points of the two input traces, as would be expected for the AND function.

Some of the details of the interconnect and the logic junctions used in this nanocircuit will now be described in greater detail, with particular reference to their dynamical properties.

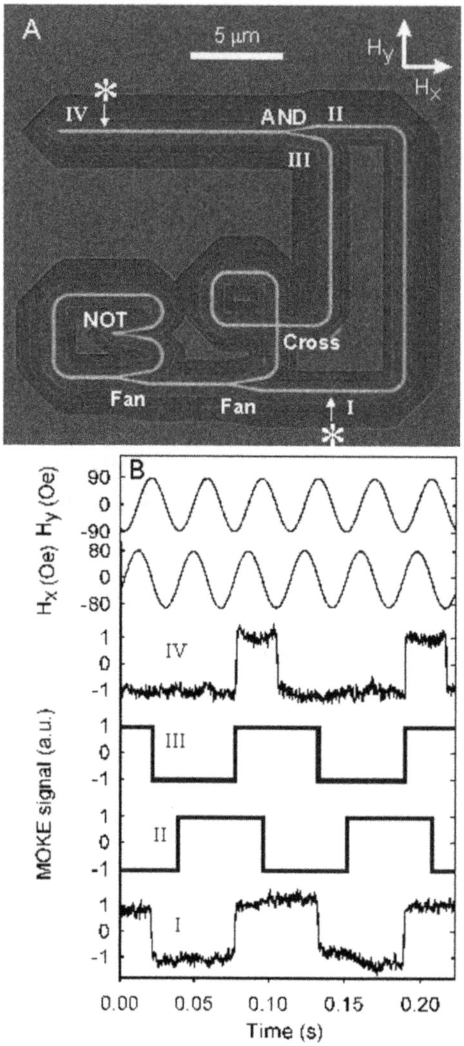

Fig. 6. (**a**) A magnetic circuit containing NOT, AND, crossover and fan-out functions fabricated from 5 nm thick Permalloy nanowires. (**b**) Magneto-optical measurements of the operation of the magnetic logic circuit. The entire circuit is exposed to a rotating magnetic field, described by the top two traces

3.2 Domain-Wall Propagation through Corner Structures

Corners allow the connection of nanowire components with different orientations allowing the movement of magnetic domain walls and hence the flow of logic data within the plane of the logic network. The propagation of domain walls around corners can be achieved with combinations of orthogonal fields (in the plane of the nanowire network) in an appropriate sequence for example as shown by magnetic force microscope images of a NOT gate in a loop in Fig. 7 (where the bright contrast region in the rectangular wire structure is attributed to stray field from the domain wall). The geometry of the corner may affect the dynamics of the wall propagation through the

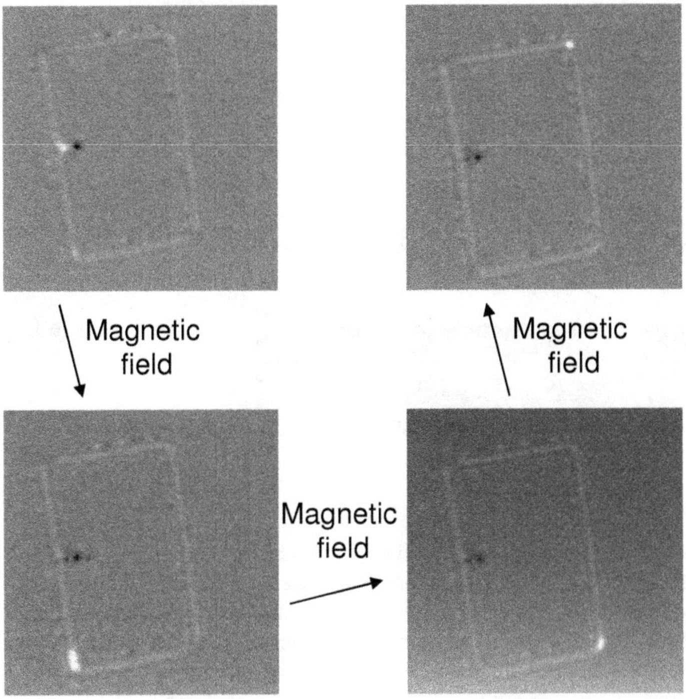

Fig. 7. Magnetic force microscope images showing the location of a domain wall propagated around a planar nanowire loop containing a single NOT gate structure (MFM contrast comes from the stray field of the domain wall and is shown as the white regions denoted with DW). The images were obtained in the remanent state. The arrows indicate the directions of the magnetic field applied to move the wall. The nanowire structure is 200 nm wide and is fabricated by focused ion beam milling from a 5 nm thick thermally evaporated film of Permalloy. The long axis of the rectangular nanowire loop is approximately 20 μm long. Images courtesy of X. Zhu and P. Grütter

corner, for example, sharp corners might well be different from rounded ones since they may form a potential trap for the wall. For practical magnetic logic interconnects the free propagation of domain walls through corners has been directly demonstrated using the structure shown in Fig. 8. In this case magnetization measurements indicate that domain-wall-mediated magnetization reversal is observed to occur at the same field levels before and after the corners in the structure showing no pinning due to the corners. In this case the measurement was undertaken with an alternating magnetic field at 27 Hz. As discussed earlier the velocity of wall propagation increases with magnetic field and it is relevant to enquire as to how the walls behave at corners as their speed increases. As part of the experimental work on wall velocity [17, 19] the domain walls were propagated into the second corner of structure shown in Fig. 3. If a domain wall entered and remained in the

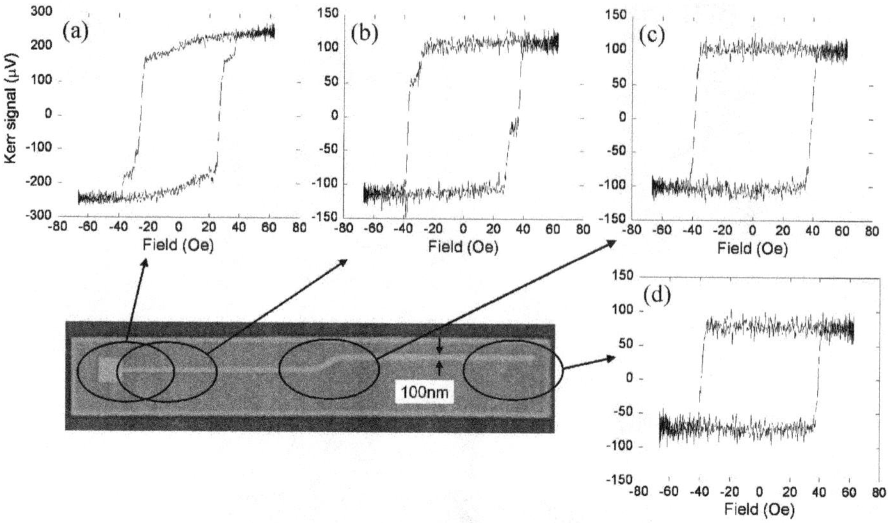

Fig. 8. Comparison of the domain-wall-mediated switching behavior of a 200 nm planar nanowire before and after the section containing two corners. The switching field is the same before and after the wall showing that the corners do not add additional components. The measurement was undertaken with an alternating magnetic field at 27 Hz. The elliptical marks indicate the magnetization measurement position

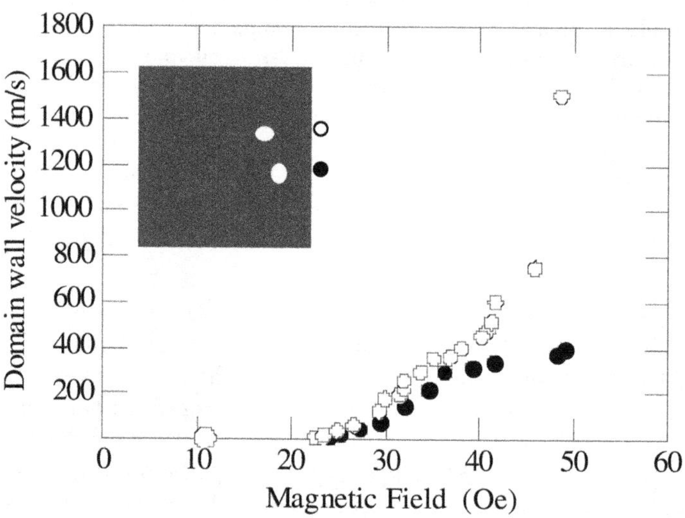

Fig. 9. The influence of the corner on the domain-wall velocity as a function of magnetic field for a 200 nm wide 5 nm thick planar Permalloy nanowire. The inset image shows the relative positions of the magnetization reversal measurement

corner after the pulsed field was applied then a subsequent magnetic field applied in the y-axis would propagate the wall down to be detected by the MOKE magnetometer. Using this method an effective domain-wall velocity was obtained. Figure 9 compares the magnetic-field dependence of the wall velocities for the direct measurements and the measurements into the corner. The two velocity datasets are in reasonable agreement up to 35 Oe and then diverge at higher fields with the effective velocity obtained from propagation into the corner lower than that obtained in the direct measurement. This difference may be attributed to the effect of the change of spin orientation that occurs at the corner that may effectively "reflect" the wall back from the corner, requiring a further increase in field in order to push the wall into the corner and allow it to be detected. This interpretation suggests that fast-traveling walls may not propagate through corners as easily as slower walls although in a logic circuit a combination of appropriately phased orthogonal fields would be used. Alternatively, it has been suggested that the observed divergence in the field dependence of the velocity represents an experimental uncertainty [18], although it should be noted that repeated measurements at field values around the point of divergence between the two velocity datasets gave consistent results.

3.3 Domain-Wall Propagation through Logic Junction Structures

As discussed earlier, logic functionality is obtained by fabricating relatively complex junction structures. In these structures domain walls are required to propagate through in a controlled manner as a function of magnetic field. The logical NOT function has been realised with a cusp or smooth Y-shaped gate structure [3, 26]. The operation of this structure as a NOT gate is realised through a combination of alternating orthogonal magnetic fields phased to produce a rotating magnetic field. The rotating field drives the domain wall into the gate and then a domain wall travels out from the gate at the appropriate phase of the magnetic field cycle. The magnetization dynamics of the gate are critical to the successful operation of the structure as a logic circuit. A wall must propagate into the junction and a wall must emerge from the junction at the correct points in the field cycle. The magnetization process has been studied using micromagnetic simulations. Figure 10 shows the micromagnetic simulations at various points in a magnetic-field cycle. The most striking feature is the change in the domain structure as the wall enters the apex of the cusp. The magnetic structure becomes complex and a simple head-to-head wall can no longer be identified. With further rotation of the applied field a domain wall emerges into the output wire of the gate. For practical applications the logic structure must operate at relatively high frequencies. For experimental reasons during the development of the NOT gate the frequency of the alternating magnetic field was only 27 Hz. Single-axis pulsed-field measurements are not suitable for investigating the fast dynamics of this structure. However, successful operation of the NOT

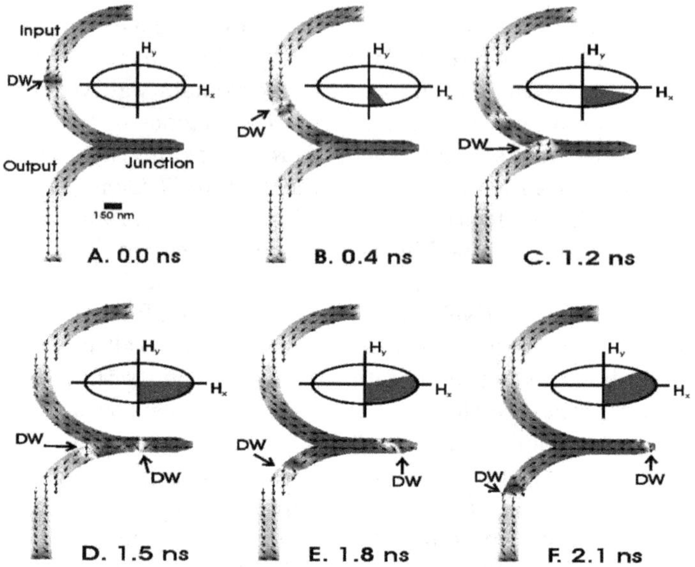

Fig. 10. Micromagnetic simulations of a domain wall propagating in a NOT-gate structure at various points in a rotating magnetic field cycle. Modeled using OOMMF [25]. The width of the structure was 200 nm, the thickness was 5 nm and the mesh size was 5 nm. The color scheme denotes the direction of magnetization with red indicating moments oriented left to right, blue indicating moments pointing right to left and paler colors through to white indicating an increasing vertical component to moment orientation. DW indicates the location of the domain wall within the structure at a given point in the field cycle. The inset ellipses show the magnitude and orientation of the magnetic field applied through one field cycle (the horizontal component of the applied field is larger than the vertical hence the field profile is elliptical). The infilled parts of the ellipses indicate the portion of the magnetic-field cycle elapsed for each snapshot of the simulated magnetization behavior

gate has been obtained with a rotating magnetic field at a frequency of 2 kHz using a ferrite-cored electromagnet.

Three-terminal structures with potential for AND or OR functionality have also been investigated. These structures typically have two input wires that merge at a junction with a single output wire [27]. The operation of these structures requires clear separation in the amplitude of the switching field in the output wire that is dependent on the number of domain walls presented from the input wires. For the AND gate this is translated into domain-wall behavior that leads to a significantly higher pinning energy at the junction when one wall is input compared to when two walls are present. This requirement has been successfully demonstrated in measurements at 27 Hz, as shown in Fig. 11 where the number of domain walls input to the

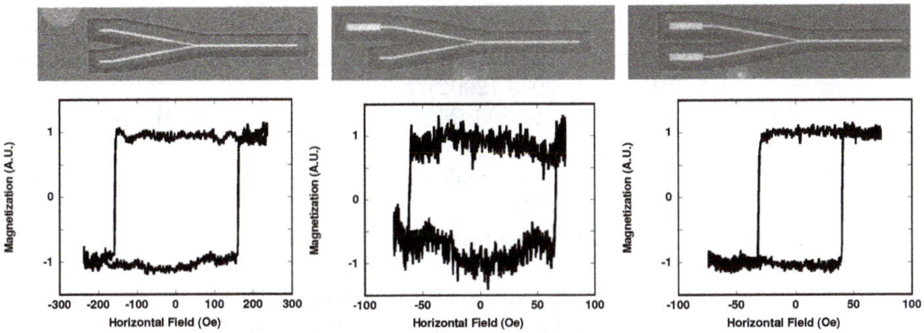

Fig. 11. Micrographs of three-terminal structures with zero, one and two domain-wall injection pads and below them the associated magnetic switching behavior measured at the right-hand side of each structure. These structures illustrate the effect that the number of domain walls incident upon the nanowire junction has on the switching field

junction was controlled by the presence or absence of a "nucleation pad" structure on the the ends of the input arms of the nanowire structure.

The successful implementation of a fully functional logic scheme based on domain-wall propagation in magnetic nanowires requires the integration of several different structural elements including both the logic gates and the circuit-connectivity elements such as corners and crossovers. For integrated operation the individual structural elements must operate reliably within the same range of applied magnetic fields without significant risk of pinning of existing domain walls or nucleation of new walls. To improve integration, significant improvements in the operating field range have been achieved by systematically modifying the gate geometry [26], which may be attributed to subtle changes of the micromagnetic structure of the logic gate.

So far several combinations of logic elements and connective structures have been successfully demonstrated at 27 Hz and the technical prospects for a fully functional magnetic domain-wall logic scheme are encouraging. The operational performance of domain-wall logic will ultimately depend on the size of the logic circuits and the dynamic behavior of the domain walls through these magnetic circuits and this too is encouraging.

References

[1] M. Klaui, C. A. F. Vaz, J. A. C. Bland, et al.: Controlled magnetic switching in single narrow rings probed by magnetoresistance measurements, Appl. Phys. Lett. **81(1)**, 108–110 (2002)

[2] J. G. Zhu, Y. Zheng: The micromagnetics of magnetoresistive random access memory, in B. Hillebrands, K. Ounadjela (Eds.): *Spin Dynamics in Confined Magnetic Structures*, vol. 1 (Springer, Berlin, Heidelberg 2001) pp. 289–324

[3] D. A. Allwood, G. Xiong, M. D. Cooke, C. C. Faulkner, D. Atkinson, N. Vernier, R. P. Cowburn: Submicrometer ferromagnetic NOT gate and shift register, Science **296**, 2003–2006 (2002)
[4] A. P. Malozemoff, J. C. Slonczewski (Eds.): *Magnetic Domain Walls in Bubble Materials* (Academic Press, New York 1979)
[5] C. E. Patton: Nonlinear domain wall motion in magnetic thin films, IEEE Trans. Magn. **9**, 559–562 (1973)
[6] N. L. Schryer, L. R. Walker: The motion of 180° domain walls in uniform dc magnetic fields, J. Appl. Phys. **45**, 5406–5421 (1974)
[7] M. Redjdal, J. Giusti, M. F. Ruane, F. B. Humphrey: Thickness dependent wall mobility in thin permalloy films, J. Appl. Phys. **91**, 7547–7549 (2002)
[8] M. H. Kryder, F. B. Humphrey: Micromagnetic characteristics of transverse diffuse boundaries in permalloy thin films, IEEE Trans. Magn. **6**, 725–728 (1971)
[9] S. Konishi, S. Yamada, T. Kusuda: Domain-wall velocity, mobility and mean-free-path in Permalloy films, IEEE Trans. Magn. **6**, 722–724 (1971)
[10] M. Lambeck, A. Scheibe: Investigation of the thermally activated wall movement in a wide range of temperature, IEEE Trans. Magn. **5**, 569–573 (1969)
[11] M. A. Akhter, D. J. Mapps, Y. Q. Ma, A. K. Petford-Long, R. Doole: Domain walls and magnetic properties of very thin permalloy films for magnetoresistive sensors, IEEE Trans. Magn. **34**, 1147–1149 (1998)
[12] N. Giordano, J. D. Monnier: Magnetization reversal and domain wall motion in thin Ni wires, Physica B **194–169**, 1009–1010 (1994)
[13] W. Wernsdorfer, K. Hasselbach, A. Benoit, B. Barbara, B. Doudin, J. Meier, J. P. Ansermet: Measurements of switching in individual nickel nanowires, Phys. Rev. B **55**, 11552–11559 (1997)
[14] A. Thiaville, J. M. Garcia, J. Miltat: Domain wall dynamics in nanowires, J. Magn. Magn. Mater. **242–245**, 1061–1063 (2002)
[15] R. H. Koch, J. G. Deak, D. W. Abraham, P. L. Trouilloud, R. A. Altman, Y. Lu, W. J. Gallagher, R. E. Scheuerlein, K. P. Roche, S. S. P. Parkin: Magnetization reversal in micron-sized magnetic thin films, Phys. Rev. Lett. **81**, 4512–4515 (1998)
[16] T. Ono, H. Miyajima, K. Shigeto, K. Mibu, N. Hosoito, T. Shinjo: Propagation of a magnetic domain wall in a submicrometer magnetic wire, Science **284**, 468–469 (1999)
[17] D. Atkinson, D. A. Allwood, G. Xiong, M. D. Cooke, C. C. Faulkner, R. P. Cowburn: Magnetic domain-wall dynamics in a submicrometre ferromagnetic structure, Nature Mater. **2**, 85–87 (2003)
[18] Y. Nakatani, A. Thiaville, J. Miltat: Faster magnetic walls in rough wires, Nature Mater. **2**, 521–523 (2003)
[19] D. Atkinson, D. A. Allwood, C. C. Faulkner, G. Xiong, M. D. Cooke, R. P. Cowburn: Magnetic domain wall dynamics in a Permalloy nanowire, IEEE Trans. Magn. **39**, 2663–2665 (2003)
[20] D. Atkinson, D. A. Allwood, M. D. Cooke, R. P. Cowburn: Nanosecond pulsed field magnetization behaviour of NiFe thin films, J. Phys. D: Appl. Phys. **34**, 3019–3023 (2001)
[21] R. V. Telesnin, E. N. Ilyicheva, N. G. Kanavina, N. B. Stepanova, A. G. Shishkov: Domain-wall motion in thin Permalloy films in pulsed magnetic field, IEEE Trans. Magn. **5**, 232–235 (1969)

[22] D. A. Allwood, G. Xiong, M. D. Cooke, R. P. Cowburn: Magneto-optical Kerr effect analysis of magnetic nanostrucures, J. Phys. D: Appl. Phys. **36**, 2175–2182 (2003)

[23] A. Kirilyuk, J. Ferre, D. Renard: Domain walls in ultrathin ferromagnetic films: velocity and fractal dimension, Europhys. Lett. **24**, 403–408 (1993)

[24] D. A. Allwood, G. Xiong, C. C. Faulkner, D. Atkinson, D. Petit, R. P. Cowburn: Magnetic domain wall logic, Science **309**, 1688–1692 (2005)

[25] The object oriented micromagnetic framework (OOMMF) project at ITL/NIST
URL `http://math.nist.gov/oommf/`

[26] D. A. Allwood, G. Xiong, M. D. Cooke, C. C. Faulkner, R. P. Cowburn: Characterization of submicrometer ferromagnetic NOT gates, J. Appl. Phys. **95**, 8264–8270 (2004)

[27] C. C. Faulkner, D. A. Allwood, M. D. Cooke, G. Xiong, D. Atkinson, R. P. Cowburn: Controlled switching of ferromagnetic wire junctions by domain wall injection, IEEE Trans. Magn. **39**, 2860–2862 (2003)

Index

Spin-Transfer Torque and Dynamics

Mark D. Stiles[1] and Jacques Miltat[2]

[1] National Institute of Standards and Technology, Electron Physics Group, Gaithersburg, MD, 20899-8412, USA
mark.stiles@nist.gov

[2] Université Paris-Sud & CNRS, Laboratoire de Physique des Solides, 91405 Orsay, France
miltat@lps.u-psud.fr

Abstract. The currents in magnetic multilayers are spin polarized and can carry enough angular momentum that they can cause magnetic reversal and induce stable precession of the magnetization in thin magnetic layers. The flow of spins is determined by the spin-dependent transport properties, like conductivity, interface resistance, and spin-flip scattering in the magnetic multilayer. When an electron spin carried by the current interacts with a magnetic layer, the exchange interaction leads to torques between the spin and the magnetization. The torque that results from this interaction excites the magnetization when the current is large enough. The qualitative features of the dynamics that result from current-induced torques are captured by a simple model in which the magnetization of the layer is assumed to be uniform. Even greater agreement results when finite temperature effects are included and the magnetization is allowed to vary throughout the film.

1 Introduction

Ferromagnets are stable in multiple states, a property that can be used to store information, as in magnetic tapes, magnetic disk drives, and more recently magnetic random access memory (MRAM). In these magnetic devices, information is stored by switching the magnetic state to the desired configuration. The switching is accomplished by a magnetic field that overcomes the coercive force of the material. In many cases, in MRAM for example, the magnetic fields are generated by electrical currents, in which case they are referred to as Oersted fields. In 1996, *Slonczewski* [1, 2] and *Berger* [3, 4] predicted that the current flowing through magnetic multilayers could have a more direct effect on the magnetic state. When a current flows through a ferromagnet, it becomes spin polarized and hence carries angular momentum. The current remains polarized in neighboring nonmagnetic layers so that the angular momentum carried by the current can interact with the magnetization in subsequent magnetic layers. The spin current exerts a *spin-transfer torque* on the magnetizations in the device. For large enough currents this torque leads to precession and reversal.

In addition to the intrinsic scientific interest in this phenomena, it has several possible applications [5]. As mentioned above, in existing implementations of MRAM, the bits are switched by Oersted fields. The long range

B. Hillebrands, A. Thiaville (Eds.): Spin Dynamics in Confined Magnetic Structures III, Topics Appl. Physics **101**, 225–308 (2006)

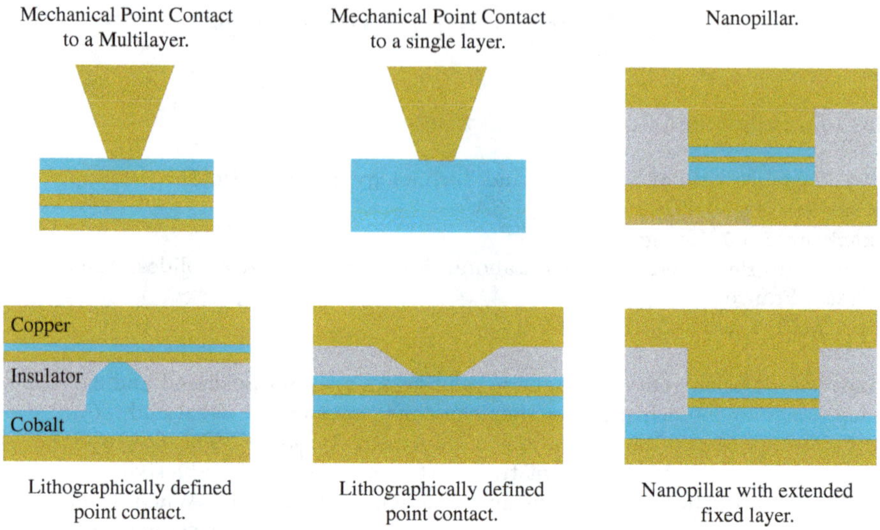

Fig. 1. Schematic experimental geometries. In each geometry, copper layers are tan, cobalt layers are blue, and insulating layers are light gray, as labeled in the *lower left* panel. Other ferromagnetic materials may be substituted for Co and nonmagnetic materials for Cu

of these fields means that reduced fields are applied to neighboring bits. The requirement that only the correct bits switch, and those bits switch reliably, places severe manufacturing constraints on the devices. The possibility that this spin-transfer torque could replace the Oersted fields for reversing otherwise stable magnetic configurations in MRAM has inspired a significant amount of the research on this effect. In MRAM, the information is stored in lithographically defined devices. A related technique could be used to store information in media like a magnetic hard disk. In this case, a probe would write magnetic bits with current pulses while being scanned over the media. Another possible application of spin-transfer torques comes from the precessional behavior observed in other regimes of operation. Here, the precessional motion converts a DC input current into an AC output voltage. Such behavior might be useful for making current-controlled oscillators.

Since the prediction of the effect in 1996, spin-transfer-induced magnetization dynamics have been observed in a number of laboratories [6–28]. These observations have been made in a number of device geometries, some of which are illustrated in Fig. 1, including mechanical point contacts [6, 21], lithographically defined point contacts [7, 17, 18], electrochemically grown nanowires [8–10], manganite junctions [11], lithographically defined nanopillars [12–16, 19, 20, 22–25], tunnel junctions [27, 29–31], and semiconductor structures [28]. These devices all share two characteristics, magnetoresistive readout of the magnetic state and small cross-sectional area.

In all of these devices, the resistance of the device depends on its magnetic state, typically through the giant magnetoresistance effect, so that measuring the resistance of the structure allows the magnetic state to be inferred. Frequently, the layers are fabricated so that one layer, referred to as the free layer, responds to the current-induced torques and another, referred to as the fixed layer, does not. This layer stays fixed either because it is exchange biased by being coupled to an antiferromagnet or simply because it is thicker than the free layer and less susceptible to spin-transfer torques. Typical measurements report the resistance V/I or differential resistance $\mathrm{d}V/\mathrm{d}I$ as a function of the current. The signature of a spin-transfer effect is a change in the resistance of the device that is asymmetric in the current. The change in resistance indicates that the magnetic configuration has changed. The asymmetry in current indicates that the cause is the spin-transfer effect rather the the Oersted field. Any changes in the magnetic configuration due to the Oersted field alone are expected to be symmetric in the current, whereas the spin-transfer torque is explicitly asymmetric.

All of the devices have small cross-sectional areas for two reasons. First, it turns out that fairly high current densities are required to transfer enough angular momentum to affect the magnetization. The heat generated by the current would destroy the device if it were not concentrated into a small area that is in good thermal contact to a large mass. Second, the relative effect of the Oersted field as compared to the direct spin transfer decreases as the cross-sectional area decreases. For a uniform current density and uniform magnetization, the spin-transfer torque is uniform. The spin-transfer torque then couples strongly to the uniform precession mode. The torque per area is independent of the area for constant current density. The Oersted fields on the other hand are nonuniform and do not couple to the uniform precession of a uniform magnetization. As the cross-sectional area decreases, the magnetization tends to become more uniform and it becomes harder to excite nonuniform modes. In addition, the maximum Oersted field decreases as the area decreases for a constant current density. Inside a wire with a circular cross section, for instance, the Oersted field increases linearly from the origin. Thus, the smaller the diameter, the smaller the maximum Oersted field.

The simplest way to get a small cross-sectional area is through a mechanical point contact. In these devices, a sharp tip is lightly pressed into a sample. By monitoring the resistance of the junction while the tip is pressed into the sample, it is possible to stop when the contact area is small. The first observation of a spin-transfer effect by *Tsoi* et al. [6] used such a mechanical point contact to a magnetic multilayer. They observed peaks in the differential resistance for only one polarity of the current, see Fig. 2, and interpreted them as evidence of a spin-transfer-torque-induced transition into a precessing state. The current at which this peak occurred increased linearly with increasing magnetic field. Subsequent measurements on this system [32], showed that as the current was increased to values greater than that of the

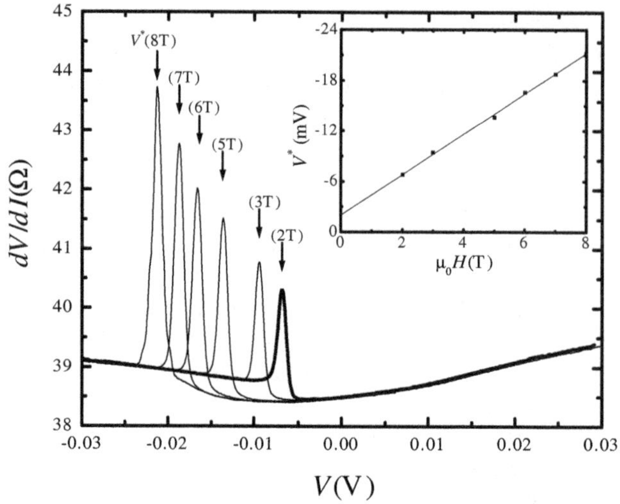

Fig. 2. Differential resistance of a mechanical point contact as a function of current for various applied magnetic fields. The peak in differential resistance that occurs for only one current direction indicates a spin-transfer effect. The *inset* shows that the current at the peak in the differential resistance increases linearly with the applied field. Used by permission from *Tsoi* et al. [6]

peak, the system made a transition into a state that coupled strongly to externally applied microwaves.

More controlled devices can be made through nanolithography. *Myers* et al. [7] fabricated lithographically defined point contacts and *Katine* et al. [12] fabricated nanopillars. In high magnetic fields, both of these systems exhibited a similar peak in the differential resistance as seen in the mechanical point contact experiment. However, in lower fields, both showed hysteretic switching between two stable states, see Fig. 3. The resistivity of these states was equal to the resistivity of the parallel and antiparallel magnetic alignments measured in more standard magnetoresistance measurements. These experiments demonstrated that spin-transfer torques could be used to switch the configuration of bistable magnetic devices. Subsequent experiments have shown that the hysteresis loops can be quite square, showing clean switching between fully parallel and fully antiparallel alignments. Applied fields greater than the coercive field inhibit a stable antiparallel state and the system switches from a parallel configuration into more complicated states including precession.

Establishing the rest of the phenomenology for the behavior of these devices requires measurement of the time or frequency response of the resistance. Two processes dominate these measurements, two-level switching and precession. Hysteresis, as seen in Fig. 3, is always a time-dependent phenomenon. Bistability is observed for a range of currents because the different

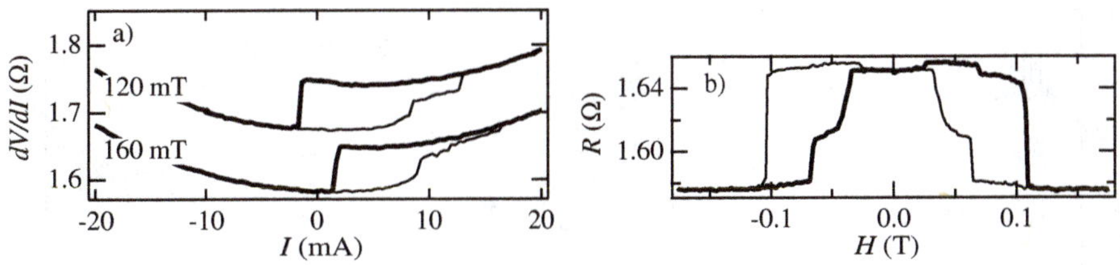

Fig. 3. Differential resistance. (**a**) $\mathrm{d}V/\mathrm{d}I$ of a pillar device exhibits hysteretic jumps as the current is swept. The current sweeps begin at zero; *light* and *dark lines* indicate increasing and decreasing current, respectively. The traces lie on top of one another at high bias, so the 120.0 mT trace has been offset vertically. (**b**) Zero-bias magnetoresistive hysteresis loop for the same sample. Used by permission from *Katine* et al. [12]

states of the system are metastable. A system at finite temperature will find its ground state over a long enough measurement time. On even longer time scales, the system would exhibit two-level switching between the two states with the relative time spent in each state dependent on the energy difference between them compared to the temperature. The rates for two-level switching depend strongly on external parameters, and the switching can be quite prominent for the right choices of parameters [6, 19, 33–35].

Two-level switching is one possible source for the low-frequency noise frequently observed in these systems [36]. The frequency distribution of the noise is typically described by the power spectrum density. Sometimes the noise has a Lorentzian power spectrum density characteristic of true two-level switching, but these systems also exhibit a variety of more general power spectral densities.

In the magnetic nanopillars, at low applied fields, the switching time is much longer than the measurement time for a range of currents and the system displays hysteresis. The width of the hysteretic region depends on the measurement time and temperature [33]. As the applied field is increased, the lifetime of the two states decreases and eventually crosses over with the measurement time, see Fig. 4. At this point, the hysteretic steps in the resistance combine and become a broadened, reversible transition between the two states. *Urazhdin* et al. [19] showed that the derivative of this transition gives the peak seen in the differential resistance.

A topic of significant controversy in the interpretation of the two-level switching is the degree to which there is heating of the magnetic degrees of freedom above the temperatures of the electronic and lattice degrees of freedom. Much of the experimental results can be understood using models based on magnetic heating [19, 34]. However, careful investigations of the temperature dependence at low temperatures by *Krivorotov* et al. [37] rule this out, at least for the system studied in that experiment.

The precessing state was first established through peaks in the power-spectrum density in frequency-dependent measurements by *Kiselev* et al. [36]

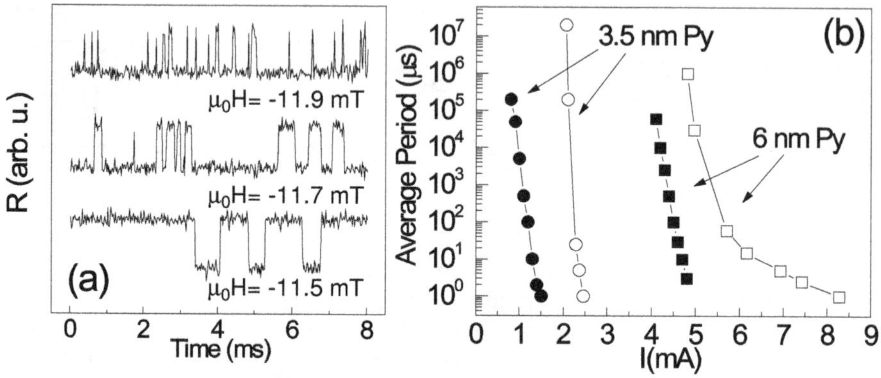

Fig. 4. Two-level switching. (**a**) Time traces of the static resistance $R = V/I$ at $T = 295\,\mathrm{K}$ at different applied fields H. *Top two curves* are offset. (**b**) Current dependence of the average telegraph noise period. The applied field along the in-plane easy-axis H was adjusted approximately linearly with current I, so that the average dwell times in antiparallel and parallel states were equal. *Solid circles*: $T = 295\,\mathrm{K}$, $\mu_0 H = 9.3\,\mathrm{mT}$ to $12.1\,\mathrm{mT}$; *open circles*: same sample at $4.2\,\mathrm{K}$, $\mu_0 H = 30.0\,\mathrm{mT}$ to $31.5\,\mathrm{mT}$; *solid squares*: different sample at $295\,\mathrm{K}$, $\mu_0 H = 11.3\,\mathrm{mT}$ to $12.3\,\mathrm{mT}$; *open squares*: same sample at $4.2\,\mathrm{K}$, $\mu_0 H = 30.0\,\mathrm{mT}$ to $45.0\,\mathrm{mT}$. Used by permission from *Urazhdin* et al. [19]

and *Rippard* et al. [38]. Precession was observed for currents near the peak in the differential resistance previously attributed to precession, but also for a range of currents that extended below the peak. In these measurements, a DC input current generates a high-frequency precessing state. Since the magnetization varies rapidly, the resistance does as well. While the precession is fast on human time scales, on the order of GHz, it is slow on electronic time scales, so the time-varying resistance leads to a time-varying voltage, which is what is measured, see Fig. 5. In some geometries, the resonance can be extremely narrow as a function of frequency, Q values as large as 18 000 have been observed [39]. These narrow resonances raise the possibility of using spin-transfer effects to make current-controlled oscillators for use in high-speed electronic applications.

A number of experiments have tested the predictions of Slonczewski's model for spin-transfer torques. Several authors [12, 40, 41] have compared predictions of macrospin simulations, see Sect. 4, with measured results, for example. The computed phase diagrams, the behavior as a function of applied magnetic field and current, qualitatively agree with those measured experimentally, see Fig. 6 for a measured phase diagram. Fully quantitative agreement is still difficult because many of the necessary parameters are not independently measurable.

Measurements of the thickness dependence [42] show that the critical currents increase as the thickness of the free layer increases. This behavior is consistent with the torque being an interface effect as opposed to an effect

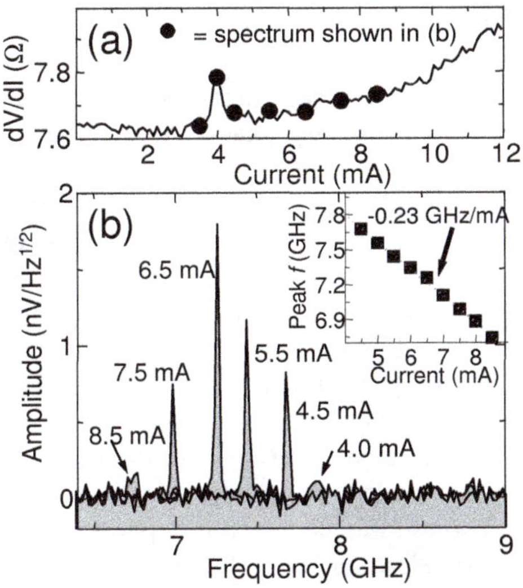

Fig. 5. Precession. (**a**) Differential resistance $\mathrm{d}V/\mathrm{d}I$ vs. current I for $\mu_0 H = 0.1\,\mathrm{T}$ along the in-plane easy axis. (**b**) High-frequency spectra taken at several different values of current through the device, corresponding to the symbols in (a). Variation of frequency f with current (*inset*). Used by permission from *Rippard* et al. [38]

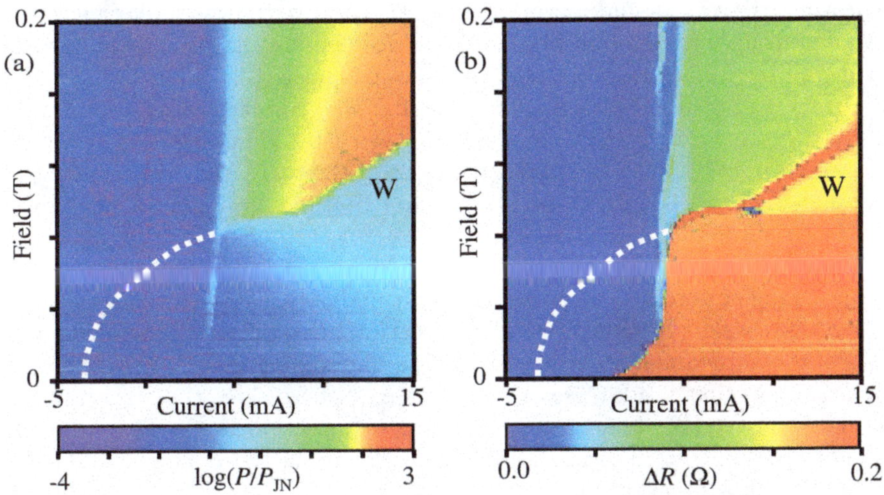

Fig. 6. Phase diagram. (**a**) Microwave power measured in a nanopillar plotted in color scale versus current I and applied field H along the in-plane easy axis. Current is swept from negative to positive values. The *dotted white line* shows the position of the antiparallel to parallel transition when current is swept in the opposite direction. (**b**) Differential resistance plotted in *color scale*. A smooth current-dependent, H-independent background has been subtracted. Resistance changes are measured relative to the parallel state. The "W" region is discussed in the text. Used by permission from *Kiselev* et al. [36]

that is active throughout the bulk of the free layer. Measurements of the dependence of the critical current on the initial angle between the free and fixed layers [43] show the expected increase in critical current as the fixed-layer magnetization becomes perpendicular to the easy axis of the free layer. While asymmetry in the current is taken as an indication of spin-transfer effects, such asymmetry only holds if the fixed layer and free layer are inequivalent. In one experiment [44], a symmetric structure showed symmetric behavior in current.

Several experiments have studied the changes in behavior due to changes in material properties. It is established that it is possible for ferromagnetic layers to polarize the currents with different signs, that is either majority or minority carriers can dominate the current. Measurements [45, 46] show that the sign of the current that leads to an instability in the free layer depends on the spin-dependent conductivity of the fixed layer, but not the free layer. The model for spin transfer described in Sect. 3 shows that such a dependence is expected for typical geometries. Other experiments [17, 18] showed that the dependence of the critical current on the saturation magnetization and the polarization of the current are consistent with this model for spin-transfer torques.

A set of related experiments grew out of the idea of repeating the original mechanical point contact experiment [6] but with a single film of Co rather than a multilayer. *Ji* and coworkers [21, 47] observed very similar peaks in the differential resistance as were observed for the multilayer. Nanopillar experiments by *Özyilmaz* et al. [48] also showed peaks that were asymmetric in current and evolved with magnetic field. These results are somewhat surprising because the models based on the work by Slonczewski depend on non-collinearity between the magnetizations of the two layers to generate the dynamics. Section 3.7 discusses how these experiments might be explained by closely related models.

The rest of this Chapter is organized to give a pedagogical understanding of spin-transfer torques, in particular, the understanding that has evolved from Slonczewski's original work. Each section gives a result that forms the basis of the subsequent section, providing a hierarchical structure for the explanation. Section 2 addresses the question, "where does the spin-transfer torque come from?" The answer is found in quantum-mechanical calculations of the behavior of electrons scattering from magnetic interfaces. Given the torque as a function of the spin current raises the question, addressed in Sect. 3, "how big is the spin current?" Determining the spin current for different magnetic configurations is the domain of semiclassical calculations of the transport. The results of Sect. 2 enter these calculations as boundary conditions. Having found the spin-transfer torque as a function of the magnetic configuration and current, leads to the next question, "how does the system respond?" Micromagnetic calculations of various types address this question, ranging from stability analysis of simple models to full micromagnetic simulations. These are discussed in Sect. 4.

2 Quantum Scattering

In this section, we describe how a spin current incident on an interface with a ferromagnet exerts a torque on the magnetization. We start by considering the equations of motion for the spin density in the regions near an interface between a nonmagnetic material and a ferromagnet. A variety of effects lead to a transfer of angular momentum from the flowing electrons to the magnetization. For many systems, there is essentially complete transfer of the transverse incident spin current. The transferred angular momentum can be interpreted as a torque.

The ultimate driving force for magnetic effects is the exchange interaction. In atomic physics, the exchange interaction causes most atoms to have magnetic moments. Hund's first rule describes the energetic considerations of arranging the electrons in a partially filled atomic level. It states that electrons should have as many spins as possible parallel. To understand why this is so, recall that the electronic wavefunction of an atom must be antisymmetric. A consequence of this antisymmetry is that the wavefunction must go to zero whenever two parallel spins are close to each other. This Pauli repulsion means that the Coulomb energy for parallel spins is lower than that for antiparallel spins. Within a partially filled level, where all states have the same kinetic energy in the absence of interactions, any cost in kinetic energy is smaller than the gain in Coulomb energy. In solids, where the electron states form bands, the situation is more complicated because it always costs kinetic energy when a minority electron is promoted into a majority electron. In most solids, the cost is too great for a polarization to develop. Nonetheless, in the transition-metal ferromagnets, Fe, Co, and Ni, this intra-atomic exchange is large enough to cause a ferromagnetic ground state. The resulting net polarization of the electronic spins means that majority (spins parallel to the majority of spins) and minority (spins antiparallel to the majority) electrons have different properties. These differences drive the effects described in the rest of this Chapter.

In solid-state physics, there is an additional consequence of the exchange interaction that is commonly discussed, interatomic (as opposed to intra-atomic) exchange. In transition-metal ferromagnets, this is the interaction that drives the spins on neighboring atoms to be parallel to each other. A common representation of this interaction is the Heisenberg form, $-J\boldsymbol{S}_i \cdot \boldsymbol{S}_j$, where $\boldsymbol{S}_i$ and $\boldsymbol{S}_j$ are the net spins on neighboring sites and J is the strength of the interaction. In micromagnetics, a mesoscale, continuum model for magnetism, the interatomic exchange interaction has the form

$$\frac{A_{\mathrm{ex}}}{M_{\mathrm{S}}^2} \int d^3r \Big\{ \left[\boldsymbol{\nabla} M_x(\boldsymbol{r})\right]^2 + \left[\boldsymbol{\nabla} M_y(\boldsymbol{r})\right]^2 + \left[\boldsymbol{\nabla} M_z(\boldsymbol{r})\right]^2 \Big\}, \tag{1}$$

where A_{ex} is the exchange constant, M_{S} the saturation magnetization, and M_x is the x-component of the vector magnetization $\boldsymbol{M}$. This energy has its minimum when all gradients are zero and the magnetization is uniform. This form of the exchange interaction will be important in Sect. 4, which discusses the time evolution of the magnetization.

A spin current consists of spins moving. Since both spin and velocity are vectors, the spin current is a tensor quantity. Classically, it is the outer product of the electron's spin and its velocity, $\boldsymbol{Q} = \boldsymbol{S} \otimes \boldsymbol{v}$. In a quantum-mechanical description of an ensemble of spins, the spin density and the spin current are

$$\begin{aligned} \boldsymbol{s}(\boldsymbol{r}) &= \sum_{i\sigma\sigma'} \psi^*_{i\sigma}(\boldsymbol{r})\, \boldsymbol{S}_{\sigma,\sigma'}\, \psi_{i\sigma'}(\boldsymbol{r}), \\ \boldsymbol{Q}(\boldsymbol{r}) &= \sum_{i\sigma\sigma'} \mathrm{Re}\left[\psi^*_{i\sigma}(\boldsymbol{r})\, \boldsymbol{S}_{\sigma,\sigma'} \otimes \hat{\boldsymbol{v}}\, \psi_{i\sigma'}(\boldsymbol{r})\right], \end{aligned} \tag{2}$$

where $\hat{\boldsymbol{v}}$ is the velocity operator, $\boldsymbol{S}_{\sigma,\sigma'}$ are the spin operators ($\hbar/2$ times Pauli spin matrices), and the sum over i is over all occupied states. While the spin current for a collection of electrons cannot generally be factored into the outer product of a net spin direction and a net velocity vector, for one-dimensional problems it usually can. In this case it is useful to factor the spin current into the outer product of a dimensionless spin direction $\hat{\boldsymbol{s}}$ and the number current $\boldsymbol{j}$ times a dimensionless, scalar polarization P

$$\boldsymbol{Q} = (\hbar/2)\, P\, \hat{\boldsymbol{s}} \otimes \boldsymbol{j}. \tag{3}$$

At various points in this Chapter we will discuss the direction associated with a spin current even though the spin current is a tensor. The direction we mean is the direction of the spin taken from this factorization. In addition, transverse (or perpendicular) spin currents $\boldsymbol{Q}_\perp$ are the components of $\boldsymbol{Q}$ such that $\hat{\boldsymbol{M}} \cdot \boldsymbol{Q}_\perp = 0$, or

$$\boldsymbol{Q}_\perp = (\hbar/2)\, P\, [\hat{\boldsymbol{s}} - \hat{\boldsymbol{M}}(\hat{\boldsymbol{s}} \cdot \hat{\boldsymbol{M}})] \otimes \boldsymbol{j}, \tag{4}$$

in terms of the factorization in (3).

To develop the equation of motion for the spin density, it is useful to first consider the related expression for the electron number density n. The equation of motion for the number density is found by taking its commutator with the Hamiltonian. The number density commutes with each term in the Hamiltonian except the kinetic energy. The result is the familiar continuity equation

$$\frac{\partial n}{\partial t} = -\boldsymbol{\nabla} \cdot \boldsymbol{j}. \tag{5}$$

This equation simply states that the time rate of change in the density is given by the net flow of electrons into a region.

When this procedure is repeated for the spin density, the kinetic energy produces a similar term, but there are other terms in the Hamiltonian that do not commute with the spin density

$$\frac{\partial \boldsymbol{s}}{\partial t} = -\boldsymbol{\nabla} \cdot \boldsymbol{Q} + \boldsymbol{n}_{\mathrm{ext}}\,, \tag{6}$$

where $\boldsymbol{\nabla} \cdot \boldsymbol{Q} = \partial_k Q_{ik}$, summing over the repeated index k. The other terms in the Hamiltonian, which do not commute with the spin density, like the Zeeman interaction and the magnetocrystalline anisotropy, give rise to $\boldsymbol{n}_{\text{ext}}$, the external torque density. The interpretation of (6) is similar to the continuity equation. The time rate of change in the spin density is given by the net flow of spins into a region plus the torques that tend to rotate the spin density.

There are two contributions to the spin current. One contribution $\boldsymbol{Q}_{\text{ex}}$ is the mediator of the exchange interaction and is carried by all of the electrons contributing to the magnetization. In general, each electronic eigenstate in a solid carries a spin current. When a ferromagnet has a spatially uniform magnetization and there is no applied current, the spin current carried by the right-moving states exactly cancels the spin current carried by the left-moving states. However, when the magnetization direction in a ferromagnet is nonuniform, the spin currents carried by the eigenstates do not cancel, leaving a net spin current even in the absence of an applied current. The gradient of this spin current gives a torque that tends to rotate the inhomogeneous magnetization $\boldsymbol{n}_{\text{ex}} = -\boldsymbol{\nabla} \cdot \boldsymbol{Q}_{\text{ex}}$. This torque is the microscopic torque corresponding to the micromagnetic exchange interaction described in (1).

The contribution to the spin current of interest to this Chapter, $\boldsymbol{Q}_{\text{trans}}$, is the transport spin current, which is carried by an imbalance in the populations of the states near the Fermi energy. This strictly nonequilibrium spin current comes from an imbalance in the populations of the right-moving and left-moving states rather than a net spin current carried by equal populations of both. For the rest of this Chapter we treat $\boldsymbol{n}_{\text{ex}}$ as an explicit torque, grouping it into $\boldsymbol{n}_{\text{ext}}$ and we refer to the transport spin current as $\boldsymbol{Q}$ dropping the "trans" label. This is the normal description for micromagnetics.

In transition metals, in which the orbital moment only makes a small contribution to the magnetization, multiplying the spin density by $g_{\text{e}}\mu_{\text{B}}/\hbar$ gives the magnetization density. It is unfortunate that since $g_{\text{e}} \approx -2.002319$, the electron's spin and moment are opposite to each other. This leads to confusion when discussing spins in ferromagnets [49]. To minimize confusion in this discussion of transport, we will typically assume that the magnetization $\boldsymbol{M}$ is in the $-\hat{\boldsymbol{x}}$-direction so that the *ferromagnetic spin density* $\boldsymbol{s} = -\hbar \boldsymbol{M}/(|g_{\text{e}}|\mu_{\text{B}})$ is in the $\hat{\boldsymbol{x}}$-direction. Thus, majority electrons have their spins along $\hat{\boldsymbol{x}}$ and can be referred to as "spin up". Their moments are along $-\hat{\boldsymbol{x}}$, parallel to the magnetization. The subscript $\uparrow$ refers to majority electrons, which have moments parallel and spins antiparallel to the magnetization. The subscript $\downarrow$ refers to minority electrons.

If the spin density is converted to magnetization, (6) becomes the Landau–Lifshitz–Gilbert equation without damping [50]. Damping is missing because (6) has been written for a single-particle Hamiltonian. A more complete Hamiltonian that included the many-body interactions with other parts of the system would lead to additional terms. Some of these terms can be combined to give phenomenological contributions to the right-hand side

$$\frac{\partial \boldsymbol{s}}{\partial t} = -\boldsymbol{\nabla} \cdot \boldsymbol{Q} + \boldsymbol{n}_{\text{ext}} - \alpha \hat{\boldsymbol{s}} \times \frac{\partial \boldsymbol{s}}{\partial t} - \hat{\boldsymbol{s}} \frac{\delta s}{\tau_{\text{sf}}}, \tag{7}$$

where $\hat{\boldsymbol{s}}$ is a unit vector in the direction of the spin density and δs is the spin density in excess of the equilibrium value (spin accumulation). The third term on the right-hand side is the phenomenological damping with damping parameter α. The last term leads to relaxation of any deviation of the ferromagnetic spin density from its equilibrium value. It is characterized by a spin-flip scattering time, τ_{sf} and is discussed at greater length in Sect. 3.

It turns out that the spin-transfer torque of interest in this article is an interface effect [1, 2, 51–53], so we need to consider the equation of motion for the spin density near the interface. We consider a pillbox of area A that extends from just inside the nonmagnet to just inside the ferromagnet, see Fig. 7. If we integrate the equation of motion for the spin density, (6) over the volume of the pillbox, the divergence of the spin current becomes a surface integral of the flux. Note that the spin flux is a vector because it is given by a tensor spin current dotted into the interface normal. For a uniform spin current incident from the nonmagnet, the spin flux through the interface in the nonmagnet has a contribution from the incident and the reflected spin currents $(\boldsymbol{Q}_{\text{in}} + \boldsymbol{Q}_{\text{ref}}) \cdot (-\hat{\boldsymbol{z}})A$. For this interface, the interface normal is $-\hat{\boldsymbol{z}}$. For the interface in the ferromagnet, the transmitted spin current in $\boldsymbol{Q}_{\text{tr}} \cdot \hat{\boldsymbol{z}}A$. The total change in spin flux acts like a torque on the magnetization in the pillbox

$$\boldsymbol{N}_{\text{c}} = (\boldsymbol{Q}^{\text{in}} - \boldsymbol{Q}^{\text{tr}} + \boldsymbol{Q}^{\text{ref}}) \cdot A\hat{\boldsymbol{z}} \approx \boldsymbol{Q}_{\perp}^{\text{in}} \cdot A\hat{\boldsymbol{z}} \,. \tag{8}$$

The equality in this expression holds for all systems. In the rest of this section we show that the approximate form, where $\boldsymbol{Q}_{\perp}^{\text{in}}$ is the part of $\boldsymbol{Q}^{\text{in}}$ with its spin direction perpendicular to $\hat{\mathbf{s}}$, holds for transition-metal interfaces.

To proceed, it is instructive to consider the electronic structure of typical materials used to study the spin-transfer effect. The electronic structure of transition-metal ferromagnets is complicated. Ferromagnetism in transition metals is driven by atomic-like intra-atomic exchange and correlation effects in the partially filled d-electron shells. The atomic-like effects suggest a localized description of this part of the electronic structure. However, the d orbitals are strongly hybridized with both d and s-p orbitals on neighboring atoms. The strong hybridization suggests an itinerant description of the electronic structure. Reconciling these aspects of the physics is an ongoing area of research, and the resulting models are not simple [54–56]. Here, we adopt the approach used in most calculations of transition-metal electronic structure, the local spin density approximation (LSDA) [57–60]. This approximation accurately describes the itinerant aspects of the electronic structure, while treating the atomic-like exchange and correlation effects in mean field theory. This approximation was derived for computing the ground-state properties

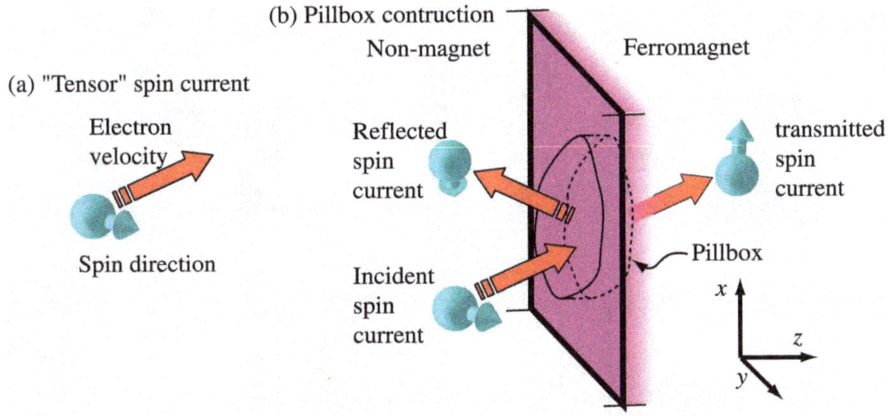

Fig. 7. (**a**) An electron moving in one direction with its spin in another illustrating a tensor spin current. (**b**) A pillbox around an interface for computing the interfacial torque

of materials. For transition-metal ferromagnets, it works quite accurately for properties like the cohesive energy, equilibrium lattice constant, and the magnetic moment [61]. Formally, it is not justified for computing the electronic structure, but it is a good combination of simplicity and accuracy even for this case.

The Fermi surfaces of Cu and face-centered cubic (fcc) Co calculated using the LSDA are shown in Fig. 8. The Fermi surfaces for the Cu and Co majority electrons are close to free-electron like, but that of the Co minority electrons is not. The complicated Fermi surface for the minority electrons is a consequence of the strong hybridization between the d-levels and also between the d-levels and the s-p levels. The differences in the Fermi surfaces lead to different properties for majority and minority electrons. Two important spin-dependent properties are the conductivity and the interface reflection. The former will be discussed in the next section and the latter will be of importance below.

Other models that are used to describe magnetic multilayers are the s-d model and the related local moment model. In these models, the d electrons are responsible for the magnetism and their itinerancy is ignored. The s electrons are responsible for transport, and they are only weakly spin dependent. In these models, the Fermi surfaces are largely spin independent. However, these models include spin-dependent scattering rates so they are able to capture most of the details of transport in these systems. However, when the details of the states at the Fermi surface are important, caution is required.

The behavior of spin currents at interfaces is dominated by the spin-dependent reflection of electrons. The exchange interaction in the ferromagnet gives the electrons at the Fermi energy different properties depending on whether their moments are parallel or antiparallel to the magnetization. As a

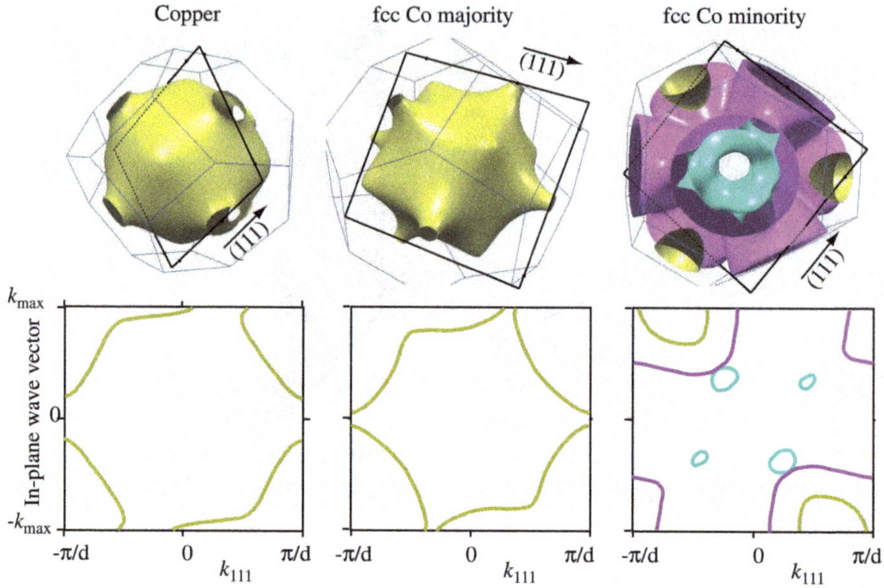

Fig. 8. The Fermi surfaces for Cu and fcc Co. The *top three panels* show 3-dimensional representations of the three Fermi surfaces. For the Co minority states, three bands intersect the Fermi energy, each is indicated in a separate color. Slices through the Fermi surfaces are indicated for each. The *bottom three panels* show these slices and the intersection of the slices with the Fermi surfaces. At (111) interfaces, electron states with the same in-plane wavevector, that is, along a horizontal line in the lower panels, couple across an interface. In Co, the Fermi energy is above the majority d-bands, but in the middle of the minority d-bands, so the majority Fermi surface is very similar to that for Cu, but the minority surface is very different

consequence, the reflection amplitudes for electrons coming from a nonmagnet and scattering from the interface with a ferromagnet depend on whether the electron's moment is parallel, $R_\uparrow$, or antiparallel, $R_\downarrow$, to the magnetization. The transmission amplitudes, $T_\uparrow$ and $T_\downarrow$, are similarly spin dependent. The spin-dependent reflection gives rise to a spin-dependent interface resistance even for ideal interfaces [62–65]. These resistances will be discussed further in the next section, which deals with transport.

The reflection and transmission of electrons with moments at arbitrary angles to the magnetization (along $-\hat{\boldsymbol{x}}$) determines the torque. The behavior of these electrons is found with little additional work. Since electrons are spin-1/2 particles, the state of an electron with a spin at a polar angle θ and azimuthal angle ϕ is a coherent superposition of the spin-up state and the spin-down state

$$|\theta,\phi\rangle = \cos(\theta/2)\,\mathrm{e}^{-\mathrm{i}\phi/2}\,|\uparrow\rangle + \sin(\theta/2)\,\mathrm{e}^{\mathrm{i}\phi/2}\,|\downarrow\rangle\,. \tag{9}$$

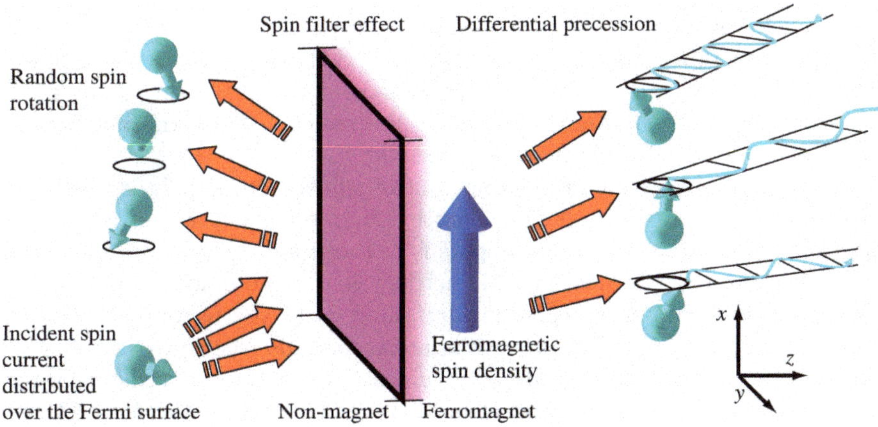

Fig. 9. Mechanisms contributing to absorption of incident transverse spin current. Electrons incident from the nonmagnet (*lower left*) are distributed over a distribution of states represented by three different incident directions. All of these electrons are in the same spin state, which is transverse to the ferromagnetic spin density (*blue arrow*). The reflected electron spins have predominantly minority character and their transverse components are distributed over many directions (random spin rotation) because of the variation over the Fermi surface of the phases of the reflection amplitudes. The transmitted electron spins precess as they go into the ferromagnet because the wavevectors for the majority and minority components are different. Electrons with different initial conditions precess at different rates, leading to classical dephasing (differential precession)

Note that although polar coordinates are usually defined with respect to the z-axis, here we define them with respect to the x-axis for consistency with the coordinate choice made in the rest of this Chapter. Thus, the reflection amplitudes for majority and minority electrons determine the reflection amplitudes for spins at arbitrary angles. When an electron starting in the nonmagnet in the state $\mathrm{e}^{\mathrm{i}kz}\,|\theta,\phi\rangle$ reflects, it comes back in a different state

$$\mathrm{e}^{-\mathrm{i}kz}\left[R_\uparrow \cos(\theta/2)\,\mathrm{e}^{-\mathrm{i}\phi/2}\,|\uparrow\rangle + R_\downarrow \sin(\theta/2)\,\mathrm{e}^{\mathrm{i}\phi/2}\,|\downarrow\rangle\right] . \tag{10}$$

The reflected electron is rotated with respect to the incident electron. The new polar angle θ' is determined by the relative magnitudes of the reflection amplitudes, $\tan(\theta'/2) = |R_\downarrow/R_\uparrow|\tan(\theta/2)$. The new azimuthal angle is determined by the relative phases of the reflection amplitudes $\phi' = \phi + \mathrm{Im}[\ln(R_\uparrow^* R_\downarrow)]$. For cartoons of these rotations, see Fig. 9.

Similar considerations apply to the transmitted electrons but with an additional complication. When the electrons are in the ferromagnet, the different spin components have different wavevectors, $k_\uparrow$ and $k_\downarrow$. The differences between the majority and minority Fermi surfaces seen in Fig. 8 illustrate the general difference between the wavevectors for the two different states. The consequence of this difference is that the phase between the up-spin and down-spin amplitudes, $e^{i(k_\uparrow - k_\downarrow)z}$ changes as the electron penetrates further into the ferromagnet. As can be seen from (9), a change in the relative phase corresponds to a change in the azimuthal angle at which the spin is directed. This changing azimuthal angle is simply precession around the magnetization[1]. For cartoons of this precession, see Fig. 9.

To see the consequences of noncollinear spin scattering, it is useful to consider a simple model first described by *Waintal* et al. [51] in which the reflection probability for minority spins is one and for majority spins is zero. While this situation is not typical, neither is it uncommon [66]. Imagine an electron with its spin perpendicular (along $\hat{\boldsymbol{z}}$) to the ferromagnetic spin density (along $\hat{\boldsymbol{x}}$) scattering from this interface. This electron is a coherent superposition of majority and minority amplitudes. After scattering from the interface, all of the majority amplitude is transmitted into the ferromagnet and all of the minority amplitude is reflected back into the nonmagnet. The incident, reflected and transmitted spin currents are

$$\begin{aligned}\boldsymbol{Q}_{\text{in}} &= \frac{\hbar}{2}\hat{\boldsymbol{z}} \otimes v\hat{\boldsymbol{z}},\\ \boldsymbol{Q}_{\text{refl}} &= \frac{\hbar}{4}(-\hat{\boldsymbol{x}}) \otimes (-v\hat{\boldsymbol{z}}),\\ \boldsymbol{Q}_{\text{trans}} &= \frac{\hbar}{4}\hat{\boldsymbol{x}} \otimes v\hat{\boldsymbol{z}}.\end{aligned} \tag{11}$$

Computing the torque on the ferromagnetic spin density inside the pillbox of Fig. 7 as in (8) gives $\boldsymbol{N}_{\text{c}} = \hat{\boldsymbol{z}}Av\hbar/2$. This simple model illustrates two key aspects of the scattering process. The first, which holds in all cases, is that the spin current along the magnetization is conserved. In this example, there is no spin current along the magnetization in the incoming state, and no net spin current along the magnetization in the outgoing states. The torque is always perpendicular to the magnetization. The second aspect is that the reflected and transmitted spin currents have no transverse components. This feature is only approximately true in general, as is discussed below. In combination, these two aspects give the simple result that the spin-transfer torque is approximately given by the absorption of the transverse part of the incident spin current.

This model simply illustrates one of the ideas behind *Slonczewski's* original model [1, 2] for this effect. Here, a spin-filter effect spatially separates the majority and minority spin currents from each other. Since the two components do not overlap, they no longer interfere with each other so that there is no transverse spin current in the outgoing states. The transverse part of the incident spin current has been absorbed by the magnetization. The mecha-

[1] Spatial precession occurs when a coherent superposition of up and down spin states have the same energy (E_{F} in this case) but different wave vectors. The more familiar case of temporal precession occurs when a coherent superposition of up and down spin states have the same wave vector but different energies. In both cases, an accumulation of phase changes the orientation of the spin vector.

nism for this transfer of angular momentum is the exchange interaction felt by electrons in the ferromagnet. The exchange interaction exerts a torque on the electron spins. In turn, there is a reaction torque on the magnetization.

Even though the assumptions of this simple model do not hold for realistic cases, the conclusion holds approximately for transition-metal interfaces like Co/Cu [52, 53]. Two other effects contribute to its approximate validity; both are illustrated schematically in Fig. 9. In general, the reflected spins have a transverse component. However, as shown in (10), the transverse component will be rotated with respect to the incident state. It turns out that the relative phase of the reflection amplitudes varies widely over the Fermi surface, so the azimuthal rotation does as well. In fact, when the distribution of electrons carrying the current is summed over the Fermi surface, the reflected transverse spin largely cancels out and ends up close to zero. This random spin rotation is an example of classical dephasing.

The same dephasing occurs for transmitted electrons, but it is not as effective at eliminating the transverse component. However, as discussed above, the electron spins precess as they penetrate the ferromagnet [1, 2, 67]. Electrons on different parts of the Fermi surface precess at different rates. So, while the dephasing was not complete just after transmission into the ferromagnet, the differential precession increases the cancellation the greater the distance into the ferromagnet. For free-electron models with different spherical Fermi surfaces for majority and minority electrons, the magnitude of the transverse spin current decays as one over this distance into the ferromagnet. For transition-metal interfaces the transverse spin current is reduced to 10% of its incident value after a distance of roughly 1 nm [53].

In calculations of spin transfer based on the local-moment model, such as those by *Heide* et al. [68–70], the similarity of all the Fermi surfaces greatly reduces or even eliminates the mechanisms for spin transfer described above. In this model, the transverse spin current is not transferred to the magnetization at the interface, so that a transverse spin accumulation develops in the ferromagnet. This spin accumulation precesses relatively slowly around the magnetization giving a reaction torque on the magnetization. However, when scattering is included [71–73] the spin accumulation decays exponentially from the interface on a short length scale so that the transverse spin current is effectively absorbed at the interface.

Finally, it is important to note that the descriptions of spin transfer given above treat the magnetization of the ferromagnet as a classical quantity. It can be excited in arbitrarily small increments of energy and angular momentum. In fact, the excitations of the magnetization, spin waves, have a finite minimum energy and a discrete angular momentum. *Urazhdin* [74] shows that a quantum-mechanical description of the magnetization changes the transfer dynamics. However, this model ignores the quantum-mechanical aspects of the electron motion so it is difficult to determine how the changes will survive in a calculation that includes all of the quantum-mechanical details.

Several groups [3, 4, 68–70, 75–79] have developed a model originally proposed by Berger in which the current-induced torque is due to spin-flip scattering creating a coherent excitation of the uniform spin wave. This model requires that the critical current for excitation is set by the spin accumulation (see the next section) being large enough that a spin can flip from minority to majority and excite a spin wave while conserving energy and angular momentum. Support for this model comes from the observation that experiments frequently show a critical voltage. However, a critical voltage is equivalent to a critical current density. Further, *Tserkovnyak* et al. [80] showed that a model without direct magnon excitation can give the same critical current density.

3 Semiclassical Transport

The previous section describes how a spin current gives rise to a torque. Here we address the issue of how the spin current arises in the first place. This involves solving for the charge and spin transport throughout the whole device. The results of the previous section enter this section as boundary conditions.

Transport in magnetic multilayers occurs in three different regimes. In the most perfect samples and at low temperatures, electron transport can be quantum-mechanically coherent. That is, when an electron reflects from several interfaces, the propagating wave interferes with the incident wave. This interference can then modify the transport through the structure. This type of interference is the origin of the oscillatory interlayer exchange coupling that has been extensively studied in magnetic multilayers [81–84]. However, even in multilayers showing interlayer exchange coupling, such effects have not been identified in their transport properties. To a large extent, the interference features of transport get averaged out because electrons from most of the Fermi surface contribute to the transport. On the other hand, tunnel barriers can restrict the states that contribute to the transport to a small fraction of the Fermi surface. The restriction allows for the observation of interference effects on transport properties in multilayers with tunnel barriers [85, 86].

Even if the transport is not coherent, it can still be ballistic. In ballistic transport, the electrons do not scatter when they propagate from one interface to another. This limit is frequently achieved for thin layers. Finally, electrons that scatter multiple times when propagating from interface to interface are in the diffusive limit.

Most magnetic multilayers do not fall cleanly into one limit or the other. Even in a single structure, some layers may be thin enough to be in the ballistic limit, while others are thicker or more disordered and in the diffusive limit. Frequently, some layers are of a thickness where neither limit applies. This range of behaviors has led to a number of different approaches to compute the transport properties; see *Bauer* et al. [87] for a discussion. *Edwards* et al. have treated the coherent limit [88] using the Keldysh formalism. There have been several calculations using the Boltzmann equation [72, 89] that can treat both the ballistic limit and the diffusive limit. The Boltzmann equation describes the evolution of the distribution function for states near the Fermi energy. It is based on the assumption that the distribution function can be defined locally. This assumption is consistent with ignoring quantum-mechanical coherence between different parts of the device. If the layers are thin, the distribution function does not change much from one interface to the other, describing ballistic transport. On the other hand, if the layers are thick, there is a lot of scattering and the distribution function evolves diffusively. In this case many of the details of the scattering can be averaged over.

There are several approaches that can be used to derive equations of motion in which the details of the distribution are averaged over. One such approach to describe the transport through disordered bulk regions that does not depend on assuming a well-defined distribution function in the disordered region is the random-matrix theory as used by *Waintal* et al. [51]. Another approach is the drift-diffusion equation, which is an approximation to the Boltzmann equation that is valid in the diffusive limit [13–15, 90–92]. This approach has been used extensively in the context of giant magnetoresistance, where it is known as the Valet–Fert theory [93]. Here the behavior of individual electron states is summed over and the theory is described in terms of currents and densities. A slightly different approach, which also sums over the behavior of individual electrons states, is the circuit theory of *Brataas* et al. [94, 95]. *Slonczewski* has used a model that grafted circuit theory to an approximate Boltzmann equation to generate a simple analytic expression for the torque as a function of the relative angles of the magnetization [96, 97].

In each of these approaches, the behavior of the current-carrying electrons as they cross interfaces is described by their transmission and reflection probabilities. In the Boltzmann equation, these probabilities enter directly into boundary conditions for the distribution functions. For the drift-diffusion equation, particular combinations of these transmission and reflection probabilities give rise to interface resistances that relate the currents across the interface to the electrochemical potential differences across them. Circuit theory is formulated in terms of interface conductances, which are slightly different combinations of the same transmission and reflection probabilities.

3.1 Boltzmann Equation and Drift-Diffusion Approximation

In this section, we describe the Boltzmann equation and how it can be manipulated to give a drift-diffusion description of the transport. In Sects. 3.4 and 3.5, we describe the transport through magnetic multilayers in terms of the drift-diffusion approach. The variables in the drift-diffusion approach can be interpreted as sums over the distribution function used in the Boltzmann equation, so this description applies to most aspects of the latter approach.

The connection to circuit theory is then made in Sect. 3.3. For a description of the same physics from the point of view of circuit theory, see [98].

The Boltzmann equation is based on the approximation that the electronic system can be described by the local occupation probability, $f_s(\boldsymbol{r},\boldsymbol{k},t)$, of well-defined bulk states of spin $s=\uparrow,\downarrow$. See [93] for the form of the Boltzmann equation and a formal reduction of it to the drift-diffusion equation. Strictly speaking, this form is only valid when the structure is slowly varying, so the bulk states can be defined. This approximation also ignores any coherence between states. In metallic systems, only the occupancies of the electrons close to the Fermi energy are changed due to transport. So it is common to linearize the distribution function

$$f_s(\boldsymbol{r},\boldsymbol{k},t) \approx f_0\left[\epsilon_s(\boldsymbol{k})\right] + g_s(\boldsymbol{r},\boldsymbol{k}_s^{\mathrm{F}},t) f_0'\left[\epsilon_s(\boldsymbol{k})\right], \tag{12}$$

where f_0 is the Fermi function, f_0' its energy derivative, and $\boldsymbol{k}_s^{\mathrm{F}}$ is a wave-vector restricted to the Fermi surface for spin s. In the following discussion, we use the full form of the distribution function, but the idea that only the occupancies of the states near the Fermi surface vary plays an important role in describing the transport.

In a system with noncollinear magnetizations, the distribution function needs to be modified. In the nonmagnet, the electron spin can point in any direction, so it is necessary to account for the coherence between the up- and down-spin components at each point on the Fermi surface. Thus the real distribution functions are generalized to the 2×2 Hermitian distribution matrices $f_{s,s'}(\boldsymbol{r},\boldsymbol{k},t)$. While this is a generalization of the distribution function used in the Boltzmann equation, it is a reduction of the full density matrix that includes the coherence between different parts of the Fermi surface. In the ferromagnet, this construction of a distribution matrix could not work because majority and minority spins have different Fermi surfaces. These surfaces only intersect along lines. Physically, this picture is consistent with the point made in the previous section – any transverse magnetization will rapidly dephase to zero. For a different point of view see [71, 72, 99–102].

To generate the drift-diffusion equation in the nonmagnet, it is useful to use an equivalent formulation for the distribution matrix, which can be generated from the Pauli spin matrices

$$f_{s,s'}(\boldsymbol{r},\boldsymbol{k},t) = f(\boldsymbol{r},\boldsymbol{k},t)\delta_{s,s'} + \sum_{\alpha=x,y,z} f_\alpha(\boldsymbol{r},\boldsymbol{k},t)[\sigma_\alpha]_{s,s'}\,, \tag{13}$$

where σ_α is a Pauli spin matrix. The Boltzmann equation describes the evolution of f and can be written in terms of either form. The drift-diffusion equation is generated by taking moments of the Boltzmann equation and ignoring some of the details near interfaces. Its basic variables are moments of the distribution function. In the normal metal, these are the change in density δn, the number current $\boldsymbol{j}$, the spin accumulation $\boldsymbol{s}$, and the spin current $\boldsymbol{Q}$

$$\begin{aligned} n_0 + \delta n &= \sum_{\boldsymbol{k}} f(\boldsymbol{k}), & j_\beta &= \sum_{\boldsymbol{k}} f(\boldsymbol{k}) v_\beta(\boldsymbol{k}), \\ s_\alpha &= \frac{\hbar}{2}\sum_{\boldsymbol{k}} f_\alpha(\boldsymbol{k}), & Q_{\alpha\beta} &= \frac{\hbar}{2}\sum_{\boldsymbol{k}} f_\alpha(\boldsymbol{k}) v_\beta(\boldsymbol{k}), \end{aligned} \tag{14}$$

where v_β is a component of the velocity of the state at $\boldsymbol{k}$, and n_0 is the equilibrium density. The spin-independent moment of the Boltzmann equation gives the drift-diffusion equation in the nonmagnet

$$j_\beta = (\sigma/e)E_\beta - D\nabla_\beta \delta n. \tag{15}$$

The first term is the drift term, with conductivity σ, and the second is the diffusion term, with diffusion constant D. An equivalent formulation is to use the density of states $(\partial n/\partial\mu)$ to write the number accumulation as a chemical potential $\mu = \delta n/(\partial n/\partial\mu)$. Then, using the Einstein relation, $e^2\sigma = D\partial n/\partial\mu$, the effect of both the local electric field and the diffusion of the charge accumulation is captured by the local electrochemical potential, $j_\beta = -(\sigma/e)\nabla_\beta(\mu + V)$. In fact, a local excess charge $-e\delta n$ creates a local electric field so that E and δn are related through Poisson's equation. However, only the electrochemical potential $\mu + V$ couples to the transport so it is not necessary to actually solve Poisson's equation and determine the charge accumulation and the electric field separately.

The spin moments of the Boltzmann equation give

$$Q_{\alpha\beta} = -D\nabla_\beta s_\alpha. \tag{16}$$

This expression simply states that for nonmagnets in the diffusive limit, any spin current is due to the diffusion of a spatially inhomogeneous spin accumulation. In this equation, the spin current is given by the gradient of the spin accumulation. In the spin-continuity equation, (7), for time-independent magnetization, the gradient of the spin current is proportional to the spin accumulation. Combining these two equations gives a diffusion equation that has exponential solutions characterized by the spin diffusion length $l_{\mathrm{sf}} = \sqrt{D\tau_{\mathrm{sf}}}$. This is the length scale over which a steady-state spin accumulation decays to zero, the mean distance a spin diffuses before it flips. The spin accumulation is frequently written as a spin chemical potential, $(\hbar/2)\Delta\mu_\alpha = s_\alpha/(\partial n/\partial\mu)$. In the bulk, the two are equivalent. At the interface between two materials with different chemical potentials, either formulation can be used, but the differences in the chemical potentials across the interface are directly related to the current and the spin currents.

In ferromagnetic materials, it is most natural to adopt the two-current formulation of the Boltzmann equation as done by *Valet* and *Fert* [93] provided all of the ferromagnet magnetizations are collinear. The majority and minority electrons are treated as two different species so that the variables in the drift-diffusion equation are

$$n_{s0} + \delta n_s = \sum_{\boldsymbol{k}} f_s(\boldsymbol{k}), \quad j_{s\beta} = \sum_{\boldsymbol{k}} f_s(\boldsymbol{k}_s) v_\beta(\boldsymbol{k}), \tag{17}$$

where n_{s0} is the equilibrium spin density for each spin direction. The drift-diffusion equations are then

$$\boldsymbol{j}_s = (\sigma_s/e)\boldsymbol{E} - D_s \boldsymbol{\nabla} \delta n_s . \tag{18}$$

The sum and the difference of the equations for the two spins give the equations for the total current and the spin current, respectively. The relative size of the conductivities for each spin, $\sigma_\uparrow$ and $\sigma_\downarrow$ is complicated. The ratio depends on the relative density of states, the effective velocities of the states and the scattering rates. In Co, the material principally discussed in this Chapter, the differences in the scattering rates dominate, giving a higher conductivity for majority electrons.

If the magnetization is uniform within each layer, the two-current model can be used with appropriate boundary conditions. If the magnetization within a layer is not uniform, the model needs to be generalized to treat a spatially varying magnetization direction. The simplest approach, which we will adopt, is to assume that the direction of the ferromagnetic spin density $\hat{\boldsymbol{u}}$ (opposite the magnetization) is sufficiently slowly varying that the spin direction of the spin current and the accumulated spin adiabatically follow the magnetization direction [103, 104]. In this case, the spin current is simply $(\hbar/2)\hat{\boldsymbol{u}} \otimes (\boldsymbol{j}_\uparrow - \boldsymbol{j}_\downarrow)$. When this form is inserted into the continuity equation (7) it is clear that a polarized current flowing through a spatially varying magnetization direction gives rise to a torque. There is presently a large body of research on this topic [105–111], building on the original work by *Berger* [112, 113]

As the interfaces play important roles in typical spin-transfer structures, so do the boundary conditions. The longitudinal boundary conditions are most simply written in terms of spin-dependent electrochemical potentials, rather than number and spin accumulations. Working in the frame of reference of the magnetization, the longitudinal current for each spin is conserved across the interface (ignoring spin-flip scattering localized to the interface) and is proportional to the difference in the electrochemical potentials across the interface

$$R_s \boldsymbol{j}_s \cdot \hat{\boldsymbol{n}} = \mu_s^{\mathrm{NM}} - \mu_s^{\mathrm{FM}}, \tag{19}$$

where $\hat{\boldsymbol{n}}$ is the interface normal pointing from the nonmagnet into the ferromagnet. The spin-dependent interface resistance R_s is determined by an appropriate sum over the transmission and reflection probabilities for individual states.

In the previous section, we argued that there is no transverse spin accumulation or spin current in the ferromagnet. In this case, the transverse spin current and spin accumulation in the nonmagnet must be proportional to each other

$$\boldsymbol{Q}_{\perp} \cdot \hat{\boldsymbol{n}} = B \boldsymbol{s}_{\perp}. \tag{20}$$

B is the constant of proportionality and is defined in (22). The $\perp$ subscript indicates the transverse component. This boundary condition is actually simple to understand, given that transverse spin currents get absorbed at the interface. If there is an accumulation of transverse spins, they diffuse in all directions. However, if they are absorbed at the interface there is no backflow of transverse spins giving a net transverse spin current toward the interface. This form can be derived by detailed consideration of the boundary conditions in the Boltzmann equation. The boundary conditions (19) and (20) are used in calculations of the spin transport through an interface in Appendix A. If the magnetization is time dependent, there is another term in this boundary condition due to spin pumping.

3.2 Spin Pumping

Spin pumping, an effect originally proposed by *Berger* [3,4] and developed by *Tserkovnyak* and coworkers [114–116], comes from the spin dependence of the reflection amplitudes. The presence of the ferromagnet induces in the nonmagnetic layer a small, decaying, oscillatory magnetization, which is closely related to the Friedel oscillations in the charge density near a surface. This magnetization arises from the partial reflection of electrons as they scatter from the interface. The incoming and outgoing components of the electron states interfere with each other. Each such state gives rise to an oscillatory density in the nonmagnet. This density oscillation is also a spin-density oscillation because the reflection amplitudes are spin dependent. Summing over all electron states gives rise to a net induced spin density that decays into the nonmagnet.

When the magnetization in the ferromagnet rotates, the induced magnetization rotates as well. Since the change in the induced magnetization is driven by the rotating magnetization in the ferromagnet, it must arise from spin currents "pumped" through the interface between the two materials. Since the spin density that rotates is a superposition of contributions from electron states at all energies below the Fermi energy, the spin current is carried by the electrons, not just those at the Fermi level. The perturbation to the electronic system, the rotating magnetization, is time dependent, so it allows for inelastic scattering. However, for the states well below the Fermi energy, all of the states are occupied and there is no possibility for energy transfer. Thus, the spin current carried by the electrons with energies well below the Fermi level is *reversible*. That is, if the magnetization returns to its initial configuration, no energy (or net spin current) has been transferred to the electronic states in the nonmagnet. However, near the Fermi energy, electrons that interact with the rotating magnetization can be scattered from just below the Fermi energy to just above it. This inelastic scattering means that the component of the spin current that is carried by the electrons with energies near the Fermi energy is *irreversible*, that is, when the magnetization returns to its original configuration, there has been a net flow of energy (and spin current) into the nonmagnet. These two components of the spin current, reversible and irreversible, are conceptually related to the real and imaginary parts of frequency-dependent susceptibilities.

While the idea that energy is transferred from the magnetic system into the electronic system of the nonmagnet is relatively straightforward, it is not so obvious that angular momentum is also transferred. However, lowering the energy of the magnetic system requires increasing the alignment of the magnetization with the instantaneous internal field direction. Since the absolute magnitude of the magnetization is largely fixed, increasing the alignment of the magnetization with its internal field is equivalent to rotating the magnetization in a direction that is perpendicular both to the magnetization direction $\hat{\boldsymbol{u}}$ and to the direction in which the magnetization is precessing $\dot{\hat{\boldsymbol{u}}}$. The change in magnetization in the ferromagnet is pumped into the nonmagnet through a spin current. Thus the boundary condition becomes

$$\boldsymbol{Q}_{\perp} \cdot \hat{\boldsymbol{n}} = B\boldsymbol{s}_{\perp} + B'\hat{\boldsymbol{u}} \times \dot{\hat{\boldsymbol{u}}}. \tag{21}$$

Since the spin-pumping term has exactly the same form as Gilbert damping, it acts as an additional source of damping that is effective at the interfaces. This hand-waving argument for (21) is confirmed by calculations using several different approaches for different models of the electronic structure [114, 115, 117–119].

These same calculations further show that in the approximation that the transverse spin current is absorbed at the interfaces, the coefficients in (21) are simply related to the projected area of the Fermi surface in the nonmetal A_{FS}

$$B' = \frac{A_{\mathrm{FS}}}{(2\pi)^3}\frac{\hbar}{2} = \frac{\hbar^2}{2}\frac{\partial n}{\partial \mu}B. \tag{22}$$

This contribution to the damping has been studied extensively in ferromagnetic resonance [120–125]. The measured behavior of the resonance linewidth as a function of layer thicknesses and frequency quantitatively confirms that spin pumping acts as an additional source of damping. Note that the pumped spin current only functions as an additional source of damping if it does not return to the precessing ferromagnet. Thus the angular momentum carried by the spin current needs to be absorbed by spin-flip scattering in the bulk, or by scattering from another ferromagnetic layer with a noncollinear magnetization.

3.3 Circuit Theory

The circuit theory of *Brataas* et al. [94, 95, 98] is based on the approximation that devices can be divided into nodes and regions that connect them.

In the nodes, the voltages (chemical potentials), and spin accumulation are assumed to be spatially uniform. The nodes are connected by conducting channels, across which the chemical potential differences drive charge and spin currents. The conductances of the channels are determined by the same transmission and reflection probabilities that enter as boundary conditions in the Boltzmann equation or determine the interface resistances in the drift-diffusion approximation. Since the spin accumulation in a nonmagnetic node need not be collinear with the magnetization direction in a ferromagnetic node, describing the transport requires both the conductances for majority $g_\uparrow$ and minority $g_\downarrow$ electrons

$$g_\sigma = \frac{e^2}{h} \sum_{n \in \mathrm{NM}} \sum_{m \in \mathrm{FM}} |t_\sigma^{nm}|^2 , \tag{23}$$

as well as the "mixing conductance"

$$g_{\uparrow\downarrow} = \frac{e^2}{h} \sum_{n \in \mathrm{NM}} \left[1 - \sum_{m \in \mathrm{NM}} r_\uparrow^{nm} \left(r_\downarrow^{nm} \right)^* \right] . \tag{24}$$

The majority and minority conductances describe electrons going from one material into another and hence involve sums over channels in both materials. The mixing conductance on the other hand, describes a spin current that is absorbed by the ferromagnet. Thus, it only depends on a sum over channels in the nonmagnet. The mixing conductance describes the behavior of spins in the nonmagnet that are perpendicular to the magnetization in the ferromagnet. The real part gives the spin current that is aligned with the perpendicular part of the chemical potential. The imaginary part gives the spin current that is perpendicular both to the magnetization and to the chemical potential in the nonmagnet. Both of these components are absorbed at the interface giving rise to a torque on the magnetization. As described in the previous section, the sum over the reflection amplitudes goes to zero because of spin filtering and dephasing. Thus, in transition-metal systems the real part of the mixing conductance is roughly proportional to the number of conducting channels and the imaginary part is close to zero.

Circuit theory differs from the drift-diffusion approach in two ways. First, it is formulated in terms of conductances rather than resistances. This difference is minor, because the drift-diffusion approach could be as well. The larger difference is the neglect of the spatial dependence of the chemical potential within the layers (nodes in the language of circuit theory). This approximation works well in devices in which the total resistance is dominated by the interface resistances. However, in devices that are beyond this limit, it is possible to augment circuit theory with a drift-diffusion calculation within a layer. Thus, the differences between the two approaches are largely conceptual.

Circuit theory does not naturally apply to structures where the chemical potentials vary in space, although it can be generalized to treat such systems. The real strength of circuit theory is that it is quite straightforward to treat complicated structures in which more than two ferromagnetic regions are attached to a nonmagnetic node. The proposed spin-torque transistor of *Brataas* et al. is an example [94, 95]. The Boltzmann equation and drift-diffusion approaches can also treat such cases, but solving the equations for geometries that are not essentially one-dimensional is much more difficult, see Sect. 3.6.

3.4 Collinear Transport

The transport equations and boundary conditions of the previous section allow us to determine the current flow through magnetic multilayers. It is useful to consider the transport through magnetic multilayers in terms of simpler components of the structure, starting with a single interface. Early theoretical studies of the behavior of spin currents and accumulation at interfaces have been made by *Aronov* [126], *Johnson* and *Silsbee* [127], *van Son* et al. [128], and *Valet* and *Fert* [93]. Figure 10 illustrates the spin current and the spin accumulation for electrons flowing into a ferromagnet from a nonmagnet and the reverse. Much of the history of this topic is reviewed in [129]. Far away from the interface in either material, the spin current just depends on the spin dependence of the conductivity. In the nonmagnet, the conductivity is unpolarized, so there are an equal number of majority and minority spins flowing. In the ferromagnet on the other hand, the conductivity is spin dependent, and there is a finite spin current. For the case of electrons flowing from the nonmagnet on the left, there are equal numbers of majority and minority electrons flowing toward the interface, but more majority electrons flowing away from it. This imbalance leads to an accumulation of minority spins near the interface. These minority spins diffuse away from the interface into both materials. A diffusion of minority spins away from the interface in the $-\hat{z}$-direction in the nonmagnet gives rise to a positive spin current. On the other hand, the diffusion of minority spins in the $\hat{z}$-direction into the ferromagnet gives a negative correction to the spin current flowing into the bulk. Interestingly, the diffusion of minority spins away from the interface means that the current in the nonmagnet is spin polarized before it enters the ferromagnet.

The system reaches a steady state because of spin-flip scattering. As more nonequilibrium spins accumulate, they relax faster toward equilibrium. Eventually, the change in the magnetization due to spin-flip scattering exactly cancels the net arrival of excess spins. The length scale over which the spin current recovers its bulk values is the spin diffusion length. Here the spin diffusion length is shorter in the ferromagnet than in the nonmagnet. Spin-flip scattering is a result of the coupling of the electrons to the lattice and so serves as a source of angular momentum in the electron subsystem. The electrons in the nonmagnet carry no angular momentum into the interface region,

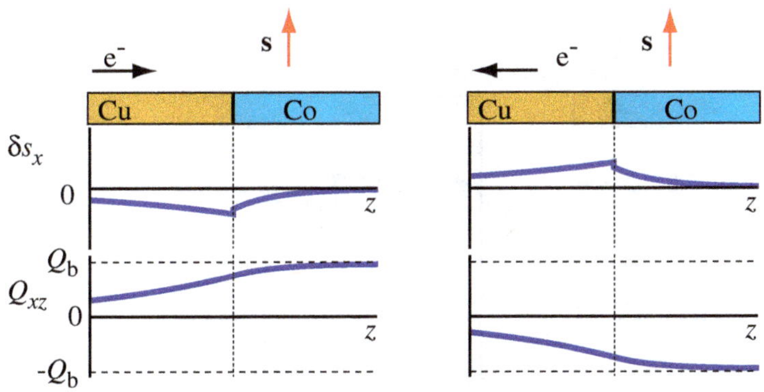

Fig. 10. Spin accumulation (δs_x) and spin current (Q_{xz}) for electron flow from a nonmagnet into a ferromagnet (*left side*) and vice versa (*right side*). $Q_{\rm b}$ is the magnitude of the spin current in the bulk ferromagnet far from any interfaces due to the spin-dependent conductivity. The magnetization in the ferromagnet, $\boldsymbol{M}$ is along the $-\hat{\boldsymbol{x}}$-direction so that the ferromagnetic spin density $\boldsymbol{s}$ is along the $\hat{\boldsymbol{x}}$-direction

but the electrons in the ferromagnet carry some away. Spin flip scattering allows for this apparent nonconservation. The change in the spin-dependent conductivities in the structure plus the spin-flip scattering leads to a transfer of angular momentum from the lattice into the flowing electrons.

There is a discontinuity in the spin chemical potential (spin accumulation) across the interface because the interface resistance is spin-dependent. To see the origin of the discontinuity, note that whenever the polarization of the current P (see (3)), is different from the polarization of the *bulk* conductivity $P_\sigma = (\sigma_\uparrow - \sigma_\downarrow)/(\sigma_\uparrow + \sigma_\downarrow)$, there must be a gradient in the spin accumulation to provide a diffusive component. Similarly, whenever the polarization of the current is different from the polarization of the *interface* conductivity $P_\Sigma = (\Sigma_\uparrow - \Sigma_\downarrow)/(\Sigma_\uparrow + \Sigma_\downarrow)$, there must be a discontinuity in the spin chemical potential. On the other hand, the spin current is constant across the interface because there is no spin-flip scattering localized to the interface (in this model).

The detailed forms of the currents and densities in Fig. 10 are found from the bulk solutions in the leads and the matching conditions at the interface. The details are given in Appendix A.

In the right-hand set of panels in Fig. 10, we see that for current flowing in the opposite direction, more majority electrons flow into the interface region than flow out, so there is an accumulation of majority spins near the interface, changing the sign of the spin accumulation. These majority electrons diffuse away from the interface changing the sign of the spin current.

The behavior of a current flowing through a finite thickness ferromagnetic layer embedded in a nonmagnet can roughly be constructed by superposing the accumulation and spin current for single interfaces. The result is shown in Fig. 11. An equal number of majority and minority electrons flow into and

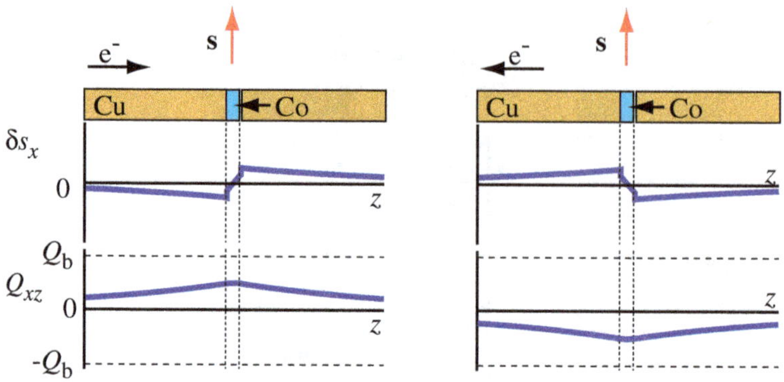

Fig. 11. Spin accumulation (δs_x) and spin current (Q_{xz}) for electron flow through a ferromagnetic layer embedded in a nonmagnetic host. The current flow is in the positive $\hat{z}$-direction on the *left-hand* side and in the negative direction on the *right*. Q_b is the magnitude of the spin current in the bulk ferromagnet far from any interfaces due to the spin-dependent conductivity. The ferromagnetic spin density in the ferromagnet, $\boldsymbol{M}$ is along the $-\hat{x}$-direction so that the ferromagnetic spin density $\boldsymbol{s}$ is along the $\hat{x}$-direction

out of the region near the interface, but more majority electrons flow through the ferromagnet. This leads to an accumulation of minority electrons before the ferromagnetic layer and an accumulation of majority electrons after the layer. These diffuse away from the interface giving rise to a current that is equally spin polarized before and after the layer. For this system, angular momentum is coupled into the electron system from the lattice in the left lead, is carried through the ferromagnetic layer and deposited in the lattice in the right lead.

The spin current and spin accumulation for two ferromagnetic layers with a thin spacer layer can be seen as a superposition of the values for two single layers provided the two magnetizations are collinear with each other. Figure 12 illustrates this superposition for the cases of parallel and antiparallel alignment of the magnetizations. The case of parallel alignment looks very much like a single ferromagnetic layer. There is accumulation of minority spins before the layers, majority spins after the layers, and almost none in the spacer layer. A substantial spin current flows through the whole trilayer. The case of antiparallel alignment is just the opposite. There is almost no spin current and almost no spin accumulation in the leads. There is a large majority spin accumulation in the spacer layer. The spacer-layer spin accumulation is a combination of majority spin accumulation after the first layer and minority (with respect to the reversed magnetization) accumulation before the second layer. This spatially varying spin accumulation provides the diffusive contribution necessary to cancel the natural bulk spin current in each ferromagnetic layer, Q_b.

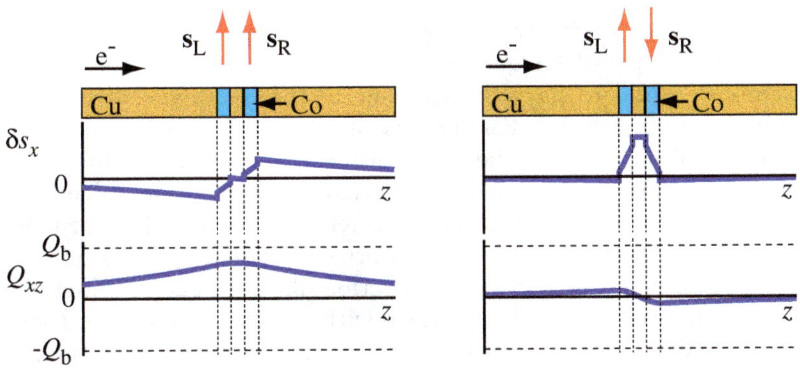

Fig. 12. Spin accumulation (δs_x) and spin current (Q_{xz}) for electron flow through two ferromagnetic layers embedded in a nonmagnetic host and separated by a thin nonmagnetic spacer layer. The magnetizations of the two layers are parallel along the $\hat{x}$ direction on the left side and antiparallel on the right. Q_b is the magnitude of the spin current in the bulk ferromagnet far from any interfaces due to the spin-dependent conductivity

The difference in the spin current between the cases of parallel and antiparallel alignment is the origin of the giant magnetoresistance. In the case of parallel alignment, more of the current flows through the lower-resistance majority channel, lowering the average resistance of the multilayer.

One of the important points of this section is that the polarization of the current and the spin polarization are not local properties. They depend on everything in the device within a few spin diffusion lengths in both directions. In Sect. 2, we showed that the torque depends on the transverse spin current. The arguments in this section show that the spin current depends on the properties of the whole device, not simply the materials in the immediate vicinity of the interface. Finding the torque will also require solving for the transport throughout the whole device.

3.5 Non-Collinear Transport and Torque

When the magnetizations of the two layers are not collinear, the direction of the spin accumulation and spin direction of the spin current (see (3)) vary throughout the structure. Still, the transport can be understood in very much the same terms as the transport for collinear magnetizations with the addition of the absorption of the transverse component of the spin current at the interfaces. For example, consider the spins along $\hat{x}$, collinear with the magnetization of the left layer, see Fig. 13. More majority electrons go through the layer, so there is an accumulation of minority electrons in the left lead. The majority electrons going through the layer go through the thin spacer relatively unchanged. However, when they reach the interface with the right ferromagnet, they are transverse and the transverse spin current goes to zero inside the ferromagnet. This gives a torque on the right ferromagnet that tends to rotate it into parallel alignment with the left ferromagnet. Note that while it is more natural to have the spins lie in the y–z plane, it is more difficult to illustrate in a figure. For this reason, we have chosen to have the magnetizations and spins lie in the xz-plane. Now, consider the spins polarized along the $\hat{\boldsymbol{z}}$-direction, parallel to the right magnetization. When the transverse spin current hits the right ferromagnet and gets absorbed, primarily minority electrons get reflected and majority electrons get transmitted, giving a positive spin current for a spin direction along $\hat{\boldsymbol{z}}$. The reflected minority electrons diffuse back to the interface with the left ferromagnet (note that minority electrons diffusing backwards also give a positive spin current along $\hat{\boldsymbol{z}}$), where they are transverse to the magnetization and get absorbed. Here, the absorption of the transverse spin current gives a torque that tends to rotate the left magnetization to be antiparallel to the right magnetization. The directions of the spin currents can be difficult to visualize from these graphs, so a cartoon of the spin currents is given at the bottom.

The torques described above tend to make the magnetizations "pinwheel" after each other rather than rotate toward or away from each other or even precess around each other. This apparent violation of the conservation of angular momentum derives from the fact that angular momentum is being provided by the coupling to the lattice through spin-flip scattering. The angular momentum coupled into the right magnetization comes from the lattice in the left lead and the magnetization coupled into the left ferromagnet comes from the right lead. As can be seen from the right-hand panels of Fig. 13, the current direction determines the direction of the pinwheeling. In most systems of experimental interest, in which the magnetization of one of the layers is held fixed, electrons flowing from the fixed layer into the other layer tend to bring the second-layer magnetization into alignment with that of the fixed layer and electron flow from the second layer into the fixed layer tends to antialign the magnetization of the second layer with that of the fixed layer.

The calculation illustrated in Fig. 13 can be repeated as a function of the angle between the magnetizations. This calculation gives two results: the torque as a function of angle and the magnetoresistance as a function of angle. Both have been studied by a number of authors [89, 96, 97, 130–134]. The torque per unit area has the form

$$\frac{\boldsymbol{N}_{\mathrm{c}}}{A} = \frac{\hbar}{2}(\boldsymbol{j}\cdot\hat{\boldsymbol{z}})\ \mathcal{P}\ [\hat{\boldsymbol{s}} \times (\hat{\boldsymbol{s}}_0 \times \hat{\boldsymbol{s}})]\,, \tag{25}$$

where $\hat{\boldsymbol{s}}$ and $\hat{\boldsymbol{s}}_0$ are the free- and fixed-layer ferromagnetic spin density directions, respectively. If the spin current reaching the free layer were independent of the relative orientation of these directions, $\mathcal{P}$ would be a constant. The double crossproduct is the part of $\hat{\boldsymbol{s}}_0$ that is transverse to $\hat{\boldsymbol{s}}$, i.e., $\hat{\boldsymbol{s}}_0-(\hat{\boldsymbol{s}}_0\cdot\hat{\boldsymbol{s}})\hat{\boldsymbol{s}}$. This equation is simply the absorption of the transverse spin current, which has been assumed to be fixed. This form has been used in a number of mi-

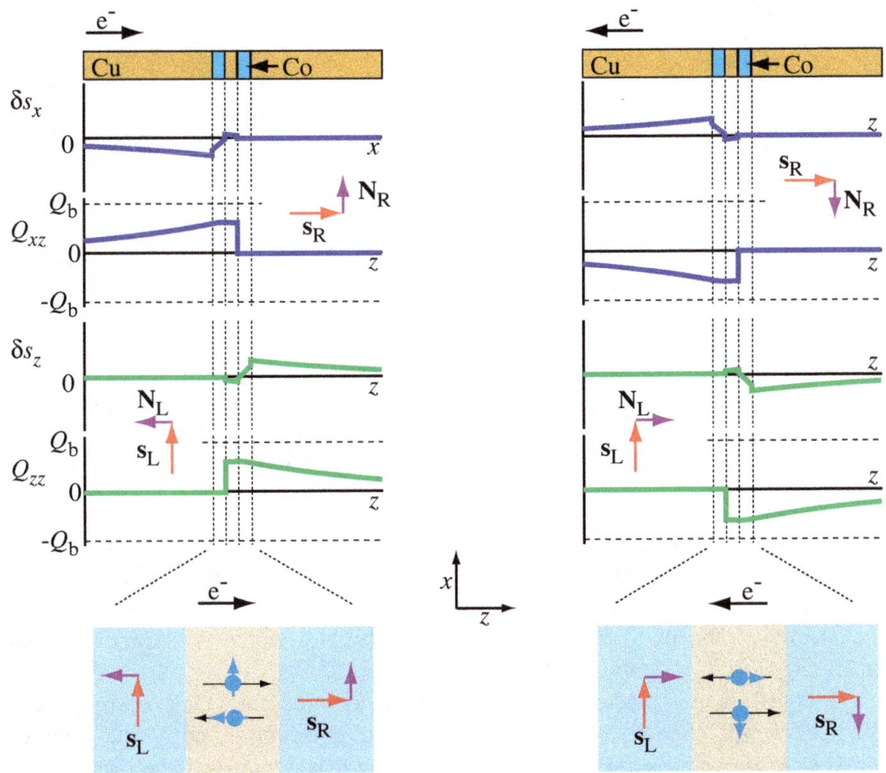

Fig. 13. Spin accumulation (δs_x and δs_z) and spin current (Q_{xz} and Q_{zz}) for electron flow through two ferromagnetic layers embedded in a nonmagnetic host and separated by a thin nonmagnetic spacer layer. The magnetizations of the two layers are perpendicular to each other, for purposes of illustration both are in the plane of the figure. On the *left side* the current flows in the $\hat{z}$-direction and on the *right* in the $-\hat{z}$ direction. Q_{b} is the magnitude of the spin current in the bulk ferromagnet far from any interfaces. The transverse component of the spin current is discontinuous at each interior interface giving rise to torques on the magnetizations of each layer. The directions of the torques are indicated for each of the interfaces. The *bottom panel* gives a cartoon of the spin current in the spacer layer. The *blue arrows* give the direction of the electron spins for electrons moving in the directions given by the *black arrows* (recall that the spin current for an electron with a spin in one direction moving to the left is the same as the spin current from an oppositely directed spin moving to the right). The *purple arrows* repeat the torques from above.

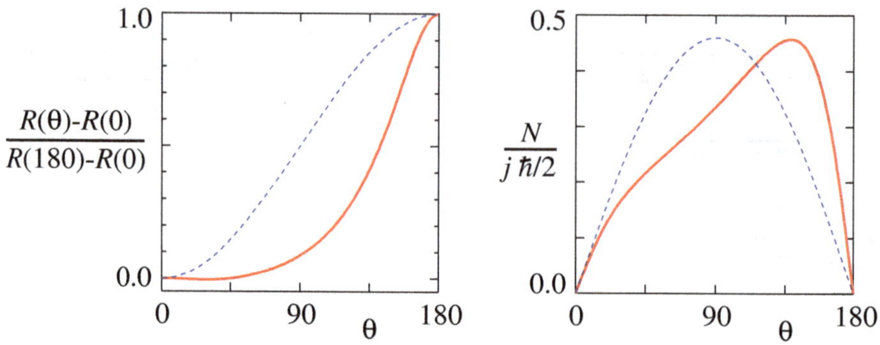

Fig. 14. The angular dependence of the torque and magnetoresistance as a function of the relative angle of the two magnetizations. The *dotted curves* show the simple forms frequently used in micromagnetic simulations, $\sin^2(\theta/2)$ and $\mathcal{P}_0 \sin\theta$ for the relative magnetoresistance and the torque, respectively. Here, $\mathcal{P}_0$ is a constant

cromagnetic simulations for the torque. However, one of the main points of this section is that the spin current is *not* independent of the relative orientations of the two magnetizations. The dependence of the spin current on the orientations of the magnetization has a substantial effect on the torque, as seen in Fig. 14.

In general, it is necessary to solve a transport equation (Boltzmann, drift-diffusion, circuit theory) to compute the angular dependence of the torque. However, *Slonczewski* [96] has derived an analytic formula based on a calculation that combines a density-matrix description of the spacer layer with a circuit theory [94, 95] description of the remainder of the structure. This formula has been extended by several authors [97, 133, 135, 136]. *Xiao* et al. [97] have tested it by comparing it with the results of calculations using the Boltzmann equation. In general, the torque is given by (25) with

$$\mathcal{P} = \left(\frac{q_+}{B_0 + B_1 \cos\theta} + \frac{q_-}{B_0 - B_1 \cos\theta} \right), \tag{26}$$

where $\cos\theta = \hat{\boldsymbol{s}}_0 \cdot \hat{\boldsymbol{s}}$. The parameters B_0, B_1, q_+, and q_- depend on the geometry of the device and the physical properties of the layers and interfaces. In particular, they depend on effective spin-dependent resistances from the spacer layer in each direction over a cumulative spin diffusion length. A similar expression can be written for the angular dependence of the magnetoresistance [133]. Fits to the curves in Fig. 14 with these formulae are indistinguishable from the full calculations.

Calculated values of q_-/q_+ and B_1/B_0 capture some of the important results of the various transport calculations that have been done. The ratio q_-/q_+ depends on the asymmetry of the structure. If the device is symmetric, the ratio goes to zero, giving the limit originally derived by *Slonczewski* [96]. Calculations show that this ratio becomes important when lead lengths become comparable to spin diffusion lengths, a situation not normally found in actual devices. The ratio B_1/B_0 describes the asymmetry of the slope of the torque between angles close to parallel and those close to antiparallel. All transport calculations find a substantial asymmetry. The ratio only becomes small when the spin-current scattering from the free layer is independent of the direction of the free-layer magnetization. This current is only independent of the free magnetization direction when the effective resistances approach the Sharvin resistance. Unfortunately, the torque vanishes in this limit. For some geometries, there are simple analytic formulas for the various parameters [96, 97, 133, 135, 136].

All transport calculations done to date give results consistent with (26). However, there is only limited experimental confirmation of this form. In most cases there is enough uncertainty about important aspects of the samples, as indicated by large sample-to-sample variation, that it is difficult to make a compelling test of the angular dependence of the torque. However, *Smith* et al. [137] have recently shown that critical currents they measured are best described when the angular dependence of the polarization and hence torque is described as in the first term of (26). For samples with short leads, q_- and hence the second term in that equation, is generally negligible. There have been a few tests of the angular dependence of the magnetoresistance [138,139]. These show clear deviation from simple $\sin^2(\theta/2)$ behavior, but less deviation than would be expected from the theories that lead to (26).

The spin-transfer torque is effective near the interface. In most simulations that treat the torque, the practitioners assume that the ferromagnetic layer is thin enough that the magnetization is constant throughout the thickness of the layer and so treat the spin-transfer torque as spread out through the layer. In fact, it is an interfacial torque and needs to be balanced by another interfacial torque. In micromagnetic calculations, interfacial exchange torque is zero when the normal derivative of the magnetization is zero $(\hat{\boldsymbol{n}} \cdot \boldsymbol{\nabla})\boldsymbol{s}(\boldsymbol{r}) = 0$. In the presence of an interfacial spin-transfer torque, this normal derivative becomes non-zero so that the exchange interaction gives rise to an interfacial torque that compensates the spin-transfer torque. The finite derivative at the interface means that the magnetization direction is spatially varying as a function of position in the ferromagnet. The spatially varying magnetization is subject to an exchange torque that effectively spreads out the interfacial spin-transfer torque. In other words, even though the spin-transfer torque is an interfacial torque, the spin-transfer torque plus the exchange interaction together act as though the spin-transfer torque is uniformly spread out over layers, provided those layers are thin enough.

These balancing torques have an interesting physical understanding. Recall from Sect. 2 that the micromagnetic exchange torque can be written as a divergence of a spin current that is carried by all of the electrons. In the non-magnet, the spin current is carried by the nonequilibrium electrons at the Fermi surface. In the ferromagnet, there is no transverse spin current, but there is an exchange torque, which is a spin current carried by all of the electrons. At the interface, these two spin currents are converted into each other.

3.6 Leads/Reservoirs

The entire preceding discussion was based on a one-dimensional approximation for the transport. However, as can be seen in Fig. 1, this is far from the actual situation. Possibly the most important difference is that leads become very wide close to the sample. In fact, this widening is important experimentally to remove the heat generated by the large current densities through the sample.

The simplest treatment of the leads is to consider them as reservoirs. In this context, a reservoir has an infinite density of states, so that the spin chemical potential is zero. This boundary condition is natural in circuit theory and is readily implemented into a drift-diffusion approach. In both of these approximations, the incoming and outgoing electrons are lumped together into a single spin density and a single spin current. In a Boltzmann equation, it is more complicated and the equivalent boundary condition is harder to implement. In this case, the boundary conditions are that the electrons leaving the reservoir must have a bulk-like distribution, but the electrons entering the reservoir can be in whatever state is dictated by the sample. This implementation can give very different results from the drift-diffusion approach.

Even more meaningful results are given by calculations of the transport that treat all three dimensions [140, 141]. These calculations show that accounting for the geometry is important and not as simple as is assumed in treating them as reservoirs. These papers show that the variation of the spin chemical potential in the transition region is quite important. This can be understood from the argument that the effective width of the lead does not go abruptly from small to large, but rather the current spreads out into the leads. These calculations imply that care should be taken in interpreting any one-dimensional calculations for which the results depend sensitively on the length of the leads.

3.7 Lateral Inhomogeneity

If the magnetization in the interface is not uniform, diffusion of spins parallel to the interface will lead to spin-transfer torques [92, 142]. Consider the case of electron flow from the nonmagnet into the ferromagnet. The electrons that are reflected into the nonmagnet and that accumulate there, are *on average* antiparallel to the magnetization direction.

If the magnetization is uniform across the interface, the accumulated spins are aligned. On average, these electrons diffuse away from the interface, but since their motion is diffusive, there is a significant probability that they

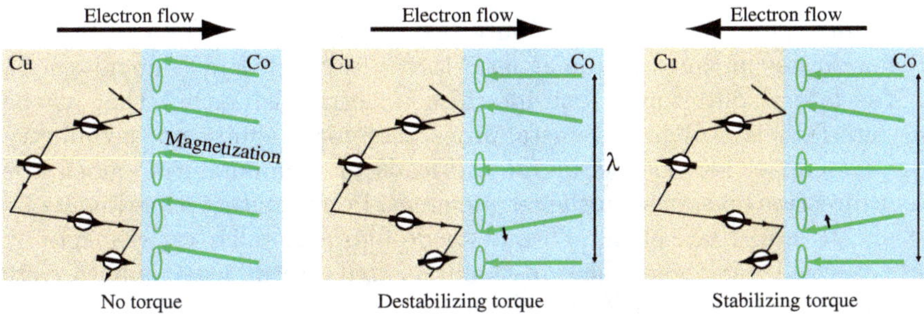

Fig. 15. Spin-transfer torque due to lateral diffusion. Each panel shows an electron diffusing in a nonmagnet and reflecting twice from the interface with a ferromagnet. After it scatters, it is oriented *on average* either parallel or antiparallel to the magnetization depending on the direction of the electron flow, see Fig. 10. In the *first panel*, the ferromagnetic magnetization is uniform laterally so that when the electron scatters the second time, it is aligned with the magnetization and there is no reorientation of either. The electron flow is from the nonmagnet into the ferromagnet, so the accumulated spins are minority spins. In the *second panel*, there is a nonuniform magnetization, and the diffusing minority spin is not aligned with the magnetization the second time it scatters. The magnetization exerts a torque on it and it exerts a torque on the magnetization that tends to amplify the spin wave. In the *third panel*, the electron flow is in the opposite direction so the diffusing spins are majority spins. In this case, the torque in the second scattering event tends to reduce the amplitude of the nonuniform spin wave

return and scatter from the interface in a different location. One such trajectory is illustrated in Fig. 15. The Fermi velocity of the electrons is sufficiently high that the magnetization does not significantly rotate when the electron scatters from the interface. Even if the electron diffuses laterally, it is still antiparallel to the magnetization when it scatters from the interface.

If, on the other hand, the magnetization varies slowly across the interface, the electron spins locally are antiparallel to the local magnetization direction after they reflect. When they diffuse laterally, they are no longer antiparallel to the local magnetization direction when they rescatter from the interface. Since they tend to be antiparallel to the average magnetization after they scatter, they are rotated by this scattering event, exerting a reaction torque on the ferromagnet.

When the net electron flow is from the nonmagnet into the ferromagnet, this lateral diffusion and rescattering tends to amplify deviations away from the average magnetization direction. On the other hand, when the electron flow is from the ferromagnet into the nonmagnet, the diffusing spins are now parallel to the magnetization on average, and the torque due to rescattering tends to suppress fluctuations away from the average direction.

Polianski and *Brouwer* [142] and *Stiles* et al. [92] have shown that these torques can be large enough to lead to instabilities in the magnetization of single thin films or single interfaces when the current is large enough. For a single interface, the instability only occurs for one direction of current flow. For finite-thickness ferromagnetic layers embedded in a nonmagnetic host, the lateral diffusion at one interface is destabilizing, but that at the other interface is stabilizing. Instabilities therefore require an asymmetry in the system, either the details of the leads or a nonuniform (along the current direction) response in the ferromagnet. Point-contact experiments by *Ji* et al. [21] and *Chen* et al. [47] and single-film nanopillar experiments by *Özyilmaz* et al. [48] both show instabilities that are at least qualitatively similar to what is expected from the models.

4 Micromagnetics

Ideally, the models described in the previous sections should be tested directly. Unfortunately the probes available to investigate these systems are rather limited. Generally, only the current and voltage through the structure are measured. This provides a check on the transport calculations, but the test is stringent only when enough of the system parameters are known accurately, which is not generally the case. The torque cannot be directly investigated at all. Only the resulting behavior of the magnetization can be inferred from measurements of the resistance. However, the behavior can be measured as a function of current, applied field, and the geometry of the device. The agreement between calculations and measurements can be quite compelling, albeit not without a few free parameters. In this section we describe the tools used to model the magnetization dynamics and give some results found for these systems.

We start this section with a consideration of the geometry of the devices of interest and their static energetics. Then, we introduce the equation of motion that describes the dynamics of the system – the Landau–Lifshitz–Gilbert equation (see, e.g., [50]). Before considering the full dynamics of the system that results from this equation, we consider the macrospin approximation, a simple limit that illustrates much of the essential physics. In this approximation, the magnetization of the layer undergoing dynamical evolution is considered to remain uniform throughout its motion. This approximation illustrates the way in which these systems exhibit precessional instability for large enough current flows. Further, it allows analytic determination of the regimes of stability of different states. After determination of the zero-temperature "phase diagram" of such systems, we consider the effect of temperature on the dynamics through the introduction of thermally driven fluctuating fields into the equation of motion and finish with some examples of full micromagnetic simulations.

4.1 Geometry and Energetics

A typical GMR pillar geometry for the study of spin-transfer effects is sketched in Fig. 16. It contains a soft thin layer (the top layer in Fig. 16) and a

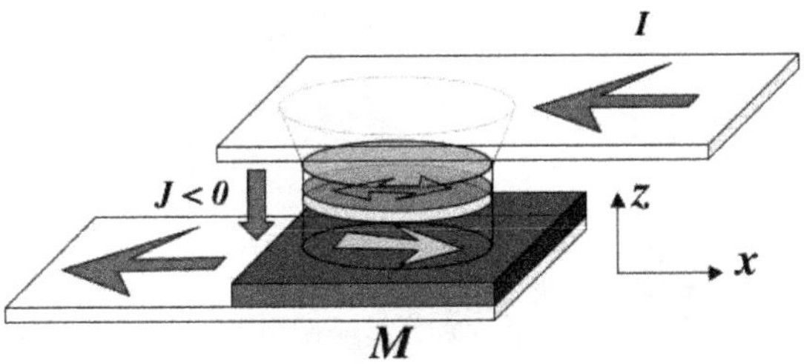

Fig. 16. Axes and sign conventions: note that the axes definition conforms with the convention adopted by most experimentalists

thick, ideally hard, uniformly magnetized layer, often referred to as the fixed or reference layer (bottom layer in Fig. 16). Most authors assume the current density J to be positive for electrons flowing along the $-\boldsymbol{z}$-direction. Under steady current, the soft, thin layer is thus flooded with electrons that become spin polarized through transmission across (reflection from) the thick layer for $J < 0$ ($J > 0$). Without loss of generality, one may safely assume that the magnetization of the fixed layer points along the $+\boldsymbol{x}$ direction. Starting from a parallel (P) magnetization configuration (both magnetization directions close to $+\boldsymbol{x}$), the spin-transfer torque (28) tends to move the magnetization of the soft layer away from its initial direction for $J > 0$. Conversely, starting from an antiparallel (AP) configuration, spin transfer promotes parallel alignment for $J < 0$.

Pillars like those used in the pioneering experiments of *Albert* et al. [143] have typical lateral dimensions in the 100 nm range with an in-plane aspect ratio close to 2 : 1 and soft layers with a thickness of a few nanometers. Because these stacks are nanofabricated from multilayers, we assume that the thickness of those elements remains constant across the element. Such elements are thus best described as flat cylinders with a close to elliptical cross section. As a first approximation, it is convenient to think of the samples as pure ellipsoids because ellipsoids have uniform magnetizations and demagnetizing fields in equilibrium. However, the nanopillars do not behave entirely as pure ellipsoids. Even though the magnetization distribution only weakly departs from a uniform magnetization distribution at rest, the effective field is not uniform, as is seen in Fig. 17.

Figure 17a displays the typical magnetization distribution within such an isolated thin element as a result of the competition between exchange and magnetostatic interactions. Due to demagnetizing effects, the effective easy magnetization axis coincides with the long ellipse principal axis. The color scheme highlights the weak deviations from uniformity close to the apices of this elliptical element. Clearly, these deviations help in reducing boundary magnetic charges at the expense of more distributed volume charges and

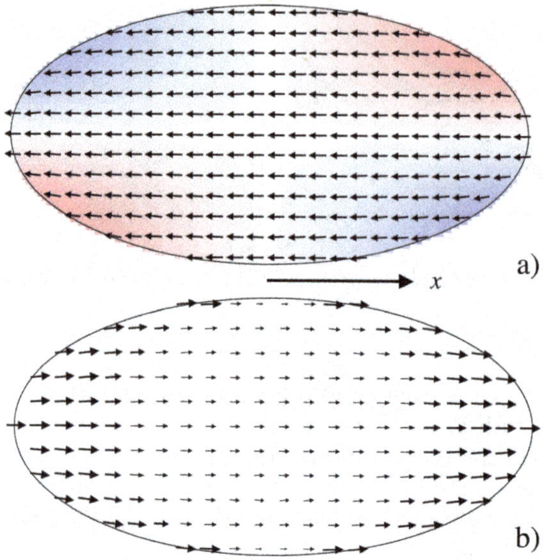

Fig. 17. Magnetization (**a**) and effective field (**b**) distributions typical of flat elliptical elements with long axis in the 130 nm range and 2 : 1 aspect ratio. $Ni_{80}Fe_{20}$, thickness $d = 2.5$ nm.

exchange interactions. At equilibrium, there should be no torque acting on the magnetization [50, 144]:

$$\boldsymbol{M} \times \boldsymbol{H}_{\text{eff}} = \boldsymbol{M} \times \left(-\frac{1}{\mu_0} \frac{\delta \epsilon}{\delta \boldsymbol{M}} \right) \equiv \boldsymbol{0} \,, \tag{27}$$

where $\boldsymbol{H}_{\text{eff}}$ is the so-called effective field and ϵ the energy density, here including just exchange and demagnetizing field contributions. Figure 17b displays a map of the effective field corresponding to the magnetization distribution in Fig. 17a, exhibiting large variations of the modulus of the demagnetizing field with a predictable influence on magnetization dynamics, be it under the action of an applied field or a current-induced spin-transfer torque. These variations lead to nonuniform behavior in the dynamics of such systems, limiting the validity of the macrospin approximation considered in Sect. 4.3.3.

The difference in energy between elliptical cylinders uniformly magnetized either along the long or the short ellipse axis is a good indicator of the thermal stability of such elements. The energy of the magnetization in either direction ($+\boldsymbol{x}$ and $-\boldsymbol{x}$) along the long axis is close to the ground-state energy and that along the short axis is a first approximation to the energy barrier between the two equivalent ground states. An example of the results of such calculations is shown in Fig. 18. Assuming that a $40\, k_{\mathrm{B}}T$ barrier height is sufficient for long-term stability, micromagnetics predicts that elliptical pillars should be stable at room temperature down to $L \approx 100$ nm for the typical thicknesses and magnetic parameters of $Ni_{80}Fe_{20}$. Coming back to Fig. 16, it is clear

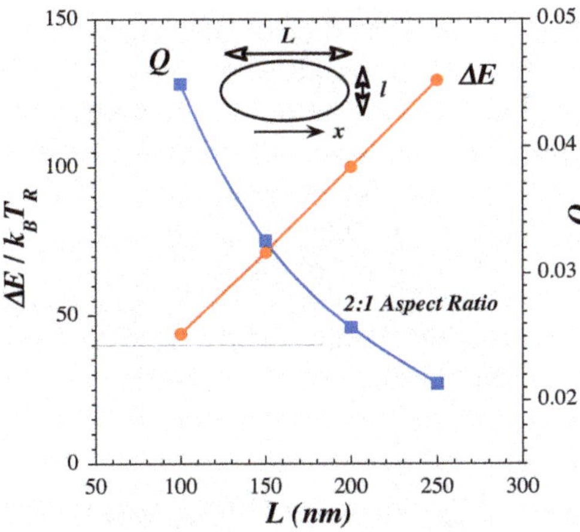

Fig. 18. Energy barrier between long and short axis uniformly magnetized elliptical elements ($Ni_{80}Fe_{20}$, $d = 2.5\,\mathrm{nm}$; room temperature, $T_\mathrm{R} = 300\,\mathrm{K}$). The shape-induced anisotropy $Q = 2K/(\mu_0 M_\mathrm{S}^2) = H_\mathrm{K}/M_\mathrm{S}$ is also plotted

that dipolar coupling between the thick and the thin layers does favor an antiparallel alignment of the magnetization directions (AP state).

4.2 Landau–Lifshitz–Gilbert Equation

We start by converting the spin densities $\mathbf{s}$ and number current densities j describing the torque (25) into magnetizations $\boldsymbol{M}$ and charge current densities J. The magnetization and spin density are related through $\boldsymbol{M} = -(|g|\mu_\mathrm{B}/\hbar)s$, where μ_B is the Bohr magneton ($\mu_\mathrm{B} > 0$). By convention, g_e is negative for free electrons. Since the orbital moment in transition metals is small, the g-factor for transition metals is close to -2. We assume that the current flows normal to the interfaces of any of those devices sketched in Fig. 1, then, $\boldsymbol{j} \cdot \hat{\boldsymbol{z}} = -J/e$, J being the current density and e the charge of the electron ($e < 0$). Lastly, because the free layers of spin-injection devices need to remain ultrathin for an optimized sensitivity to spin-transfer torques (see above), it is assumed that the spin-transfer torque is distributed over the thickness d of the soft element within the stack. The spin-transfer torque is then

$$\left.\frac{\mathrm{d}\boldsymbol{M}}{\mathrm{d}t}\right|_\mathrm{st} = -\frac{|g|}{2}\frac{\mu_\mathrm{B}}{M_\mathrm{S}^2}\frac{1}{d}\frac{J}{e}\mathcal{P}\left[\boldsymbol{M} \times (\boldsymbol{M} \times \boldsymbol{p})\right] , \tag{28}$$

where J is the current per unit area, $\boldsymbol{M}$ the magnetization, $\boldsymbol{p}$ a unit vector in the direction of the fixed-layer magnetization, $\mathcal{P}$ a polarization function that contains all the information stemming from the stack geometry and materials properties within the stack and depends on the relative orientations of the magnetizations. In most samples studied to date, $\mathcal{P}$ is generally positive, but it can be made negative by judicious choice of materials [19].

Starting with a magnetization distribution close to equilibrium under zero current, switching on the current allows the spin-transfer torque to shift the magnetization away from its initial distribution. A new pattern of magnetic charges will appear that generates a new map of the demagnetizing field. Since there is no reason for the latter to be uniform, exchange interactions will also be modified. Summing up, the magnetization will be subject to the micromagnetic effective field, giving rise to a torque proportional to $\boldsymbol{M} \times \boldsymbol{H}_{\text{eff}}$ and the spin-transfer torque. Both are anticipated to be a function of position within the soft element. The total torque reads:

$$\frac{\mathrm{d}\boldsymbol{M}(\boldsymbol{r})}{\mathrm{d}t} = -\gamma_0 \left[\boldsymbol{M}(\boldsymbol{r}) \times \boldsymbol{H}_{\text{eff}}(\boldsymbol{r})\right] - \frac{|g|}{2}\frac{\mu_{\text{B}}}{M_{\text{S}}^2}\frac{1}{d}\frac{J}{e}\mathcal{P}(\boldsymbol{r})\left[\boldsymbol{M}(\boldsymbol{r}) \times (\boldsymbol{M}(\boldsymbol{r}) \times \boldsymbol{p})\right] + \alpha\left[\boldsymbol{M}(\boldsymbol{r}) \times \frac{\mathrm{d}\boldsymbol{M}(\boldsymbol{r})}{\mathrm{d}t}\right] , \quad (29)$$

where γ_0 is the gyromagnetic ratio. The last term is the phenomenological Gilbert damping, which we assume without further proof may still be applied to magnetization dynamics in the presence of a spin-polarized current. The Gilbert damping constant α is typically of the order of 0.01. Equation (29) is the Landau–Lifshitz–Gilbert (LLG) equation of magnetization motion augmented with the spin-transfer torque term (28).

For the rest of the discussion, we ignore the spatial variation of the polarization function $\mathcal{P}$ and that of the magnetization direction of the nominally fixed layer $\boldsymbol{p}$. In reality, both of these vary. In a first approximation, the spatial variation of the polarization function could be treated using a laterally varying one-dimensional transport theory as outlined in Sect. 3. A better approximation would include lateral transport building on the discussion in Sect. 3.7. The spatial variation of the fixed-layer magnetization could be treated explicitly through an equation like (29). However, these effects are complications that should be added to an understanding built on simpler models. For instance, the inclusion of the dependence of the polarization function $\mathcal{P}$ on the magnetization direction, illustrated in Fig. 14, significantly complicates the use of generalized perturbation techniques, see Appendix B. For pedagogical purposes, the ability to use such techniques outweighs the quantitative differences in the results.

Up to this point, all equations in Sect. 4 have been written in the SI system of units. We can derive concise expressions for the LLG equation of magnetization motion with the appropriate variable reductions. Reduced variables include fields, magnetization, energy density and time, i.e., $\boldsymbol{H}$, $\boldsymbol{M}$, ϵ, t. Defining M_{S} as the saturation magnetization, SI and cgs units transform in dimensionless units as

$$\text{SI}: \quad \boldsymbol{m} = \boldsymbol{M}/M_\text{S}, \quad \boldsymbol{h} = \boldsymbol{H}/M_\text{S}, \quad w = \epsilon/(\mu_0 M_\text{S}^2), \quad \tau = \gamma_0 M_\text{S} t; \tag{30}$$

$$\text{cgs}: \quad \boldsymbol{m} = \boldsymbol{M}/M_\text{S}, \quad \boldsymbol{h} = \boldsymbol{H}/(4\pi M_\text{S}), \quad w = \epsilon/(4\pi M_\text{S}^2), \quad \tau = 4\pi\gamma M_\text{S} t. \tag{31}$$

The dimensionless LLG equation then reads:

$$\frac{\mathrm{d}\boldsymbol{m}}{\mathrm{d}\tau} = -\left(\boldsymbol{m} \times \boldsymbol{h}\right) - \chi\left[\boldsymbol{m} \times \left(\boldsymbol{m} \times \boldsymbol{p}\right)\right] + \alpha\left(\boldsymbol{m} \times \frac{\mathrm{d}\boldsymbol{m}}{\mathrm{d}\tau}\right), \tag{32}$$

where we have suppressed the spatial dependence of all the quantities. The prefactor of the spin-transfer term

$$\chi = \frac{\hbar}{2}\frac{1}{\mu_0 M_\text{S}^2}\frac{1}{d}\frac{J}{e}\mathcal{P}: \qquad [\text{Js}]\left[\frac{1}{\text{Jm}^{-3}}\right]\left[\frac{1}{\text{m}}\right]\left[\frac{\text{Cs}^{-1}\text{m}^{-2}}{\text{C}}\right][1] \tag{33}$$

is dimensionless as verified to the right of this equation by the units of each factor. This equation describes the magnetization dynamics in spin-torque systems.

Manipulating (32) further gives an expression that characterizes the energy flow within these system. We take the crossproduct of the left- and right-hand sides of this equation with $\mathrm{d}\boldsymbol{m}/\mathrm{d}\tau$, use the triple crossproduct formula and factor out a common factor of $\boldsymbol{m}$ to give

$$\left(\boldsymbol{h} \cdot \frac{\mathrm{d}\boldsymbol{m}}{\mathrm{d}\tau}\right) + \chi\left[\left(\boldsymbol{m} \times \boldsymbol{p}\right) \cdot \frac{\mathrm{d}\boldsymbol{m}}{\mathrm{d}\tau}\right] = +\alpha\left(\frac{\mathrm{d}\boldsymbol{m}}{\mathrm{d}\tau}\right)^2. \tag{34}$$

Equation (34) underlies the whole of magnetization dynamics, including spin torque, in the classical limit. It expresses the tradeoff between dissipation, the work per unit time (power) of the effective field, and power of the spin torque for which the (dimensionless) equivalent field is $\chi(\boldsymbol{m} \times \boldsymbol{p})$. In the next section, we consider the macrospin approximation in detail. Before we do so, it is worthwhile to examine (34) in this limit. We see that when $\chi = 0$ ($J = 0$), the magnetization may not precess along a closed orbit in the absence of any time-periodic driving field, unless the damping is also zero. On the other hand, when $\chi \neq 0$, closed magnetization orbits are allowed even for nonzero damping. Integrating (34) around a closed path expresses this point mathematically. Micromagnetic fields are conservative so that integrating them along a closed path gives zero, that is $\oint \boldsymbol{h}^{\mu\text{Mag}} \cdot d\boldsymbol{m} \equiv 0$. For a stable trajectory, the dissipated energy must equal the work due to the spin transfer over one period. Mathematically, this amounts to saying that for any closed magnetization trajectory Γ, the following relation should (see (34)) be satisfied:

$$\int_\Gamma \left[\chi\left(\left(\boldsymbol{m} \times \boldsymbol{p}\right) \cdot \frac{\mathrm{d}\boldsymbol{m}}{\mathrm{d}\tau}\right) - \alpha\left(\frac{\mathrm{d}\boldsymbol{m}}{\mathrm{d}\tau}\right)^2\right] \mathrm{d}\tau = 0. \tag{35}$$

Trajectories that satisfy this condition are called precessional states. The existence of such closed magnetization orbits is a key prediction of spin-transfer induced magnetization dynamics.

4.3 The Single or Macrospin Approximation

Despite the words of caution above, the macrospin approximation remains essential in deciphering the main features of magnetization dynamics under the action of the spin-transfer torque. One needs first to define an energy functional for the system under study. In a minimal approach, due to the large lateral dimension/thickness ratio, the demagnetizing field may be approximated by that of a thin elliptical cylinder. Since the thickness is so small compared to the in-plane dimensions, the dominant term is approximately the demagnetizing field of a uniformly magnetized thin film with infinite lateral dimensions, namely $\boldsymbol{H}_{\mathrm{D}} = [0, 0, -M_{\mathrm{S}} m_z]$. The associated energy density amounts to $\epsilon = -(1/2)\,\mu_0 \boldsymbol{M} \cdot \boldsymbol{H}_{\mathbf{D}} = -(1/2)\,\mu_0 M_{\mathrm{S}}^2 m_z^2$. The remaining terms describe demagnetizing effects linked to the shape of the soft element and are equivalent to a shape anisotropy of type $K\left(1 - m_x^2\right)$. Including the Zeeman energy, the overall energy density reads:

$$\epsilon = K\left(1 - m_x^2\right) + \frac{1}{2}\mu_0 M_{\mathrm{S}}^2 m_z^2 - \mu_0 \boldsymbol{M} \cdot \boldsymbol{H}^{\mathrm{a}}\,, \tag{36}$$

where $\boldsymbol{H}^{\mathrm{a}}$ is the applied field. In the following, we assume that the applied field has a nonzero component only along $\boldsymbol{x}$, the in-plane easy axis (see Fig. 16). When pulled out of equilibrium, the magnetization is subject to an effective field

$$\boldsymbol{H}_{\mathrm{eff}} = -\frac{1}{\mu_0 M_{\mathrm{S}}} \frac{\delta\epsilon}{\delta\boldsymbol{m}} = \left[H_x^{\mathrm{a}} + H_{\mathrm{K}} m_x,\ 0,\ -M_{\mathrm{S}} m_z\right]\,, \tag{37}$$

where $H_{\mathrm{K}} = 2\mathrm{K}/(\mu_0 M_{\mathrm{S}})$ is the anisotropy field.

In dimensionless form, (36) and (37) become:

$$\begin{aligned} w &= \frac{1}{2} Q\left(1 - m_x^2\right) + \frac{1}{2} m_z^2 - \boldsymbol{m} \cdot \boldsymbol{h}^{\mathrm{a}}, \\ \boldsymbol{h}_{\mathrm{eff}} &= -\frac{\delta w}{\delta \boldsymbol{m}} = \left[h_x^{\mathrm{a}} + Q m_x, 0, -m_z\right], \end{aligned} \tag{38}$$

with $Q = 2K/(\mu_0 M_{\mathrm{S}}^2) = H_{\mathrm{K}}/M_{\mathrm{S}}$. Q provides a measure of the strength of the shape anisotropy versus stray field energy.

The magnetization vector $\boldsymbol{m}$ has a fixed length, which suggests that we convert the equations of motion into spherical coordinates, namely:

$$\begin{aligned} \frac{\mathrm{d}\theta}{\mathrm{d}\tau} &= h_\phi - \alpha \sin\theta \frac{\mathrm{d}\phi}{\mathrm{d}\tau} \\ \sin\theta \frac{\mathrm{d}\phi}{\mathrm{d}\tau} &= -h_\theta + \alpha \frac{\mathrm{d}\theta}{\mathrm{d}\tau}\,, \end{aligned} \tag{39}$$

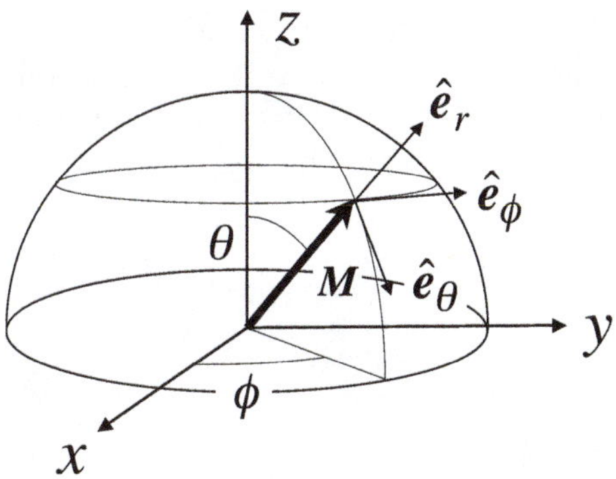

Fig. 19. Spherical coordinates: definition of angles. The usual Cartesian magnetization components read: $m_x = \cos\phi \sin\theta$, $m_y = \sin\phi \sin\theta$, $m_z = \cos\theta$

where θ and ϕ are the polar and azimuthal angles, as defined in Fig. 19, and h_θ, h_ϕ the reduced field components stemming both from the micromagnetic response of the system and spin transfer

$$h_\theta = h_\theta^{\mu\mathrm{Mag}} + h_\theta^{\mathrm{ST}}; \qquad h_\theta^{\mu\mathrm{Mag}} = -\frac{\partial w}{\mathrm{d}\theta}; \qquad h_\theta^{\mathrm{ST}} = -\chi p_\phi \tag{40}$$

$$h_\phi = h_\phi^{\mu\mathrm{Mag}} + h_\phi^{\mathrm{ST}}; \qquad h_\phi^{\mu\mathrm{Mag}} = -\frac{1}{\sin\theta}\frac{\partial w}{\mathrm{d}\phi}; \qquad h_\phi^{\mathrm{ST}} = +\chi p_\theta\,, \tag{41}$$

p_θ and p_ϕ being the components of the fixed-layer magnetization direction in spherical coordinates. It is also convenient to convert the LLG equation from the Gilbert form, in which the damping depends explicitly on $\mathrm{d}\boldsymbol{m}/\mathrm{d}\tau$ to the Landau–Lifshitz form in which it depends on the effective field. The vector LLG equation (32) becomes a set of two first-order differential equations

$$\begin{aligned}(1+\alpha^2)\frac{\mathrm{d}\theta}{\mathrm{d}\tau} &= h_\phi + \alpha h_\theta\,,\\ (1+\alpha^2)\sin\theta\frac{\mathrm{d}\phi}{\mathrm{d}\tau} &= -h_\theta + \alpha h_\phi\,.\end{aligned} \tag{42}$$

4.3.1 Onset of Precessional States

Numerical integration of an equation similar to (29), or, preferably (42) gives stable precessional states, as first demonstrated by *Sun* [145]. Within the macrospin approximation, the current threshold for the establishment of a stable magnetization trajectory may be simply derived from standard perturbation theory. Clearly, in the absence of current and under the action of any applied field $h_x^{\mathrm{a}} > 0$, the stable magnetization direction satisfies $m_x = 1$, or, $\theta = \pi/2, \phi = 0$. The damping constant α is generally small. In the trajectories we are interested in, the spin-transfer torque roughly balances the damping so that χ is of the same order of magnitude as α and both can be treated as small parameters. To investigate the onset of stable precessional states, we focus on trajectories in which the magnetization is close to its equilibrium direction. This suggests the replacement $\theta = \pi/2 + \xi$, so that ξ and ϕ can be treated as small. Taking $1+\alpha^2 \cong 1$ leads to the following linearized equations of magnetization motion

$$\begin{aligned}\frac{\mathrm{d}\xi}{\mathrm{d}\tau} &= h_\phi + \alpha h_\theta,\\ \frac{\mathrm{d}\phi}{\mathrm{d}\tau} &= -h_\theta + \alpha h_\phi,\end{aligned} \tag{43}$$

with

$$\begin{aligned}h_\theta &= -\left(1 + Q + h_x^{\mathrm{a}}\right)\xi - \chi\phi,\\ h_\phi &= +\chi\xi - \left(Q + h_x^{\mathrm{a}}\right)\phi.\end{aligned} \tag{44}$$

If D is the differential operator $D = \mathrm{d}/\mathrm{d}\tau$ and $u = Q + h_{\mathrm{x}}^{\mathrm{a}}$, the characteristic equation for the set of linear differential equations (43), (44) becomes:

$$D^2 + \left[\alpha + 2\left(\alpha u - \chi\right)\right] D + u\left(1+u\right) = 0. \tag{45}$$

Its determinant proves to be strictly negative to the lowest order in α and χ implying complex conjugate eigenvalues, μ. The focus $m_x = 1$ is stable when $\mathrm{Re}\ \mu > 0$, and unstable in the opposite case. Converting to real time units, (43) and (44) lead to elliptical precession proportional to

$$\exp(-t/t_0)\cos(\omega t + \Phi_0), \tag{46}$$

where

$$\begin{aligned}t_0 &= \frac{1}{\gamma_0 M_{\mathrm{S}}\left(\chi_{\mathrm{crit}} - \chi\right)},\\ \omega &= \gamma_0 M_{\mathrm{S}}\sqrt{u\left(1+u\right) - \left(\chi_{\mathrm{crit}} - \chi\right)^2},\end{aligned} \tag{47}$$

with,

$$\chi_{\mathrm{crit}} = \alpha\left(\frac{1}{2} + Q + h_x^{\mathrm{a}}\right) \cong \frac{\alpha}{2} \quad \text{if} \quad Q, h_x^{\mathrm{a}} \ll 1. \tag{48}$$

Therefore, magnetization motion is expected to be exponentially damped out for χ less χ_{crit} (low currents) and exponentially amplified as the current increases to the point where $\chi > \chi_{\mathrm{crit}}$. As is usual, the characteristic time t_0 goes to infinity when $\chi = \chi_{\mathrm{crit}}$. On the other hand, the angular precession frequency ω goes through a maximum equal to Kittel's resonance frequency $\omega_{\mathrm{K}} = \gamma_0 M_{\mathrm{S}}\sqrt{Q + h_x^{\mathrm{a}}}\,\sqrt{1 + Q + h_x^{\mathrm{a}}}$ at the transition.

In first approximation (Q, $h_x^{\mathrm{a}} \ll 1$), the critical current is proportional to the damping parameter, the soft-layer thickness and to the square of the saturation magnetization. It is represented by the line labeled χ_1 in Fig. 22 that displays the phase diagram of the magnetization states in the $h_x^{\mathrm{a}} > 0$ half-space[2]. As anticipated, direct numerical estimates of the critical current for the onset of precessional motion as a function of applied field are in perfect agreement with (48). Stability calculations such as these can be used to determine the stability ranges of the different configurations for different anisotropies and applied fields. *Bazaliy* et al. [146–148] and *Xi* et al. [149,150] have carried out a variety of such calculations for various magnetic configurations.

As an illustration, using typical parameters for $\mathrm{Ni_{80}Fe_{20}}$ ($\mu_0 M_{\mathrm{S}} = 1$ T, $\alpha = 0.01$, $d = 2.5$ nm), one arrives at a critical current density $J_{\mathrm{c1}} \approx 3 \times 10^{10}$ A/m^2 or 30 mA/µm^2 for a 100 % polarization in zero field. For a typical elliptical element with lateral dimensions 130×70 nm^2, this would correspond to a current ≈ 0.2 mA. A three-times higher current would be required for a more realistic electron polarization of 1/3.

4.3.2 Precessional States: Stability Range

In the previous section, we have shown that for current greater than the critical current, the static state becomes unstable to precession. However, this calculation does not address the issue of how large the precession becomes because the precession amplitude is determined by the nonlinearities of the system. For currents just above the critical current, we compute numerically the finite precession amplitude that occurs. Concentrating first on the case with no applied field, $h_x^{\mathrm{a}} = 0$, simulations do show that stable magnetization trajectories may be found over a finite current range. Trajectories open up with increasing current beyond J_{c1}, as shown in Fig. 20 and Fig. 21. This opening up can be monitored by the minimal and maximal ϕ values. In the immediate vicinity of J_{c1}, ϕ_{Max} increases with an infinite slope, a characteristic signature of a simple Hopf bifurcation (Fig. 21). Note that we use a double x-scale in Fig. 21: the top scale refers to the current density (A/µm^2), the lower to the corresponding χ/α value. As shown in Appendix B, χ/α is the true reduced current density and damping-related variable in this nonlinear magnetization dynamics problem.

Soon after the onset of precessional states, the precession frequency decreases almost linearly with increasing current density (Fig. 21). Close to a second threshold, J_{c2}, trajectories pile up against a typical eye-shaped curve (Fig. 20), that is later going to be identified with a homoclinic cycle (see

[2] Unless otherwise stated, all simulations in Sect. 4.3.3 and Sect. 4.4 use parameters mimicking a $130 \times 65 \times 2.5\,\mathrm{nm}^3$ $\mathrm{Ni_{80}Fe_{20}}$ elliptical platelet, namely $M_{\mathrm{S}} = 80\,\mathrm{kilo} \cdot \mathrm{A/m}$, $H_{\mathrm{K}} = 29.05\,\mathrm{kilo} \cdot \mathrm{A/m}$ ($Q = 0.0365$), $\gamma_0 = 2.21\,10^5\,(\mathrm{s} \cdot \mathrm{A/m})^{-1}$, $\alpha = 0.01$.

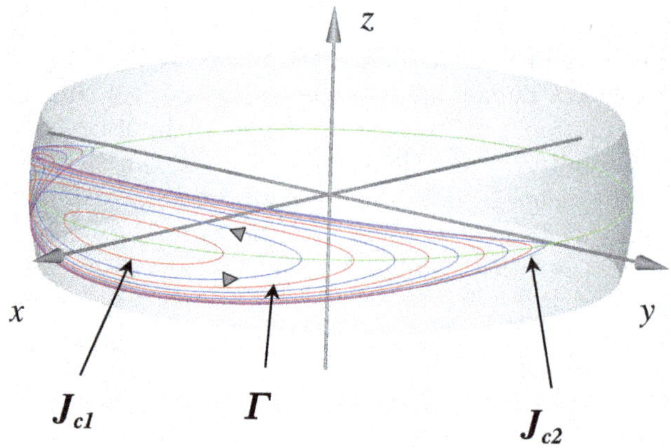

Fig. 20. Closed-orbits characteristic of precessional states for current densities $J = 0.11 - 0.13$ in steps of $0.0025\,\mathrm{A}/\mu\mathrm{m}^2$ and $J = 0.13175\,\mathrm{A}/\mu\mathrm{m}^2$ in the single-spin approximation. $h_x = 0$. The locus of the extrema of ϕ indicated by the *light line* is slightly inclined with respect to the equatorial plane. See text for details

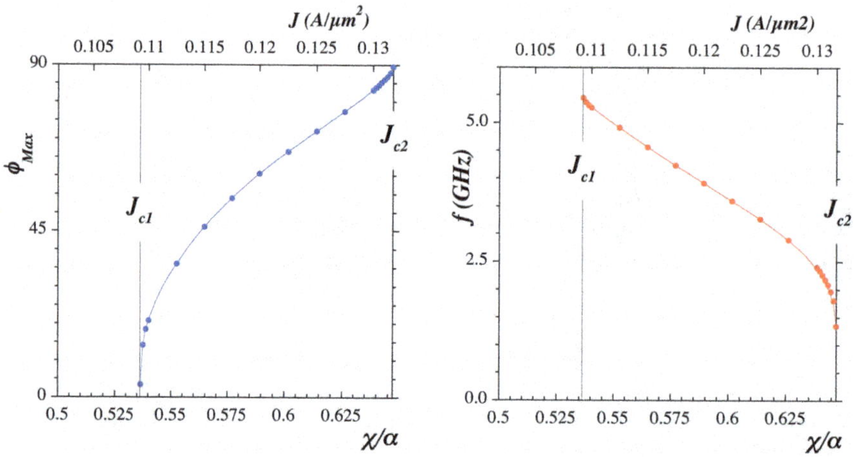

Fig. 21. Orbit opening as a function of current density (*left*) and associated precession frequency (*right*)

Appendix B). Whilst doing so, the velocity of the magnetization direction along a given trajectory, Γ, on the surface of the unit sphere, namely

$$\frac{\mathrm{d}s}{\mathrm{d}\tau} = \sqrt{\left(\frac{\mathrm{d}\theta}{\mathrm{d}\tau}\right)^2 + \sin^2\theta\left(\frac{\mathrm{d}\phi}{\mathrm{d}\tau}\right)^2} \tag{49}$$

is seen to drop to almost zero at the apices of the trajectory (Fig. 25a). The spin torque is, however, largest there. From a static point of view, it has to be

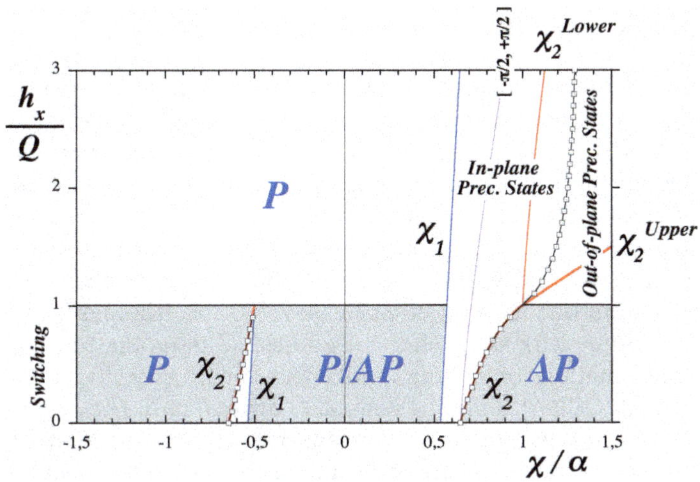

Fig. 22. Phase diagram in the $h_x > 0$ half-space as a function of χ/α, a dimensionless quantity proportional to the current density. The *line* $h_x/Q \cong 1$ schematically separates regions where switching may take place (i.e., between neighboring *shaded areas*) from regions where only a positive current may upset the parallel alignment enforced by the applied field. Curves labeled χ_1 correspond to transitions between stable (P or AP) states and precessional states of the type shown in Fig. 20 (In-plane Prec. States). *Open square symbols* delineate the transition between an in-plane prec. state and a reversed state ($h_x/Q < 1$) or an out-of-plane prec. state ($h_x/Q > 1$) according to numerical micromagnetics in the single-spin approximation. *Curves* labeled χ_2 correspond to the critical switching current vs. field according to Melnikov's method. When $h_x > Q$, the critical current χ_2/α for transition between an in-plane and an out-of-plane prec. state may not be estimated directly, although bounds to χ_2/α may be defined (*curves* labeled χ_2^{Lower} and χ_2^{Upper}: see Appendix B for details). The *faint line* between χ_1 and χ_2–χ_2^{Lower} is the locus for in-plane prec. states with π span in the equatorial (sample) plane

compensated by another torque: the latter derives from the demagnetization field, hence a slight slanting of the trajectory that, in first approximation, obeys

$$m_z^{\text{Apex}} + \chi m_y^{\text{Apex}} = 0$$

or,

$$\frac{\pi}{2} - \theta^{\text{Apex}} = \chi \sin\left(\phi^{\text{Apex}}\right) . \tag{50}$$

It also ought to be noted that the velocity distribution is not symmetrical (see Fig. 25a), another result of magnetization trajectories slanting. Lastly, close to J_{c2}, the precession frequency collapses to zero (Fig. 21).

4.3.3 Switching and Out-of-Plane Precessional States

Increasing the current density beyond J_{c2} leads to one of two possible events. As long as $h_x^{a} < Q$, the system switches to a configuration with the magnetization opposite to the ground state, in the $-\hat{x}$-direction. As shown in Fig. 23 (top), switching takes place when magnetization trajectories bifurcate close to one of the saddle points S_1 and S_2 [151]. Then, depending on the exact current density, switching takes place via either a clockwise or a counterclockwise motion. As first noticed by *Sun* [145], as soon as the magnetization has moved from one attraction basin (say F_1) into the next (say F_2) the spin-transfer torque acts as an effective additional damping term as evidenced by the small number of ringing oscillations before reaching equilibrium despite the small Gilbert damping constant. Switching in the single-spin approximation at 0 K is therefore to be viewed as a three-stage event (Fig. 23 (top)): during the initial phase, the magnetization creeps out of its initial equilibrium position through precession around the demagnetizing field that is parametrically amplified under the action of the spin-transfer torque. It is indeed a parametric excitation because the spin-transfer torque is a function of the magnetization orientation. Following bifurcation (second phase), a more highly damped precession-type motion (third stage) drives the magnetization into its new equilibrium state. Spin-transfer-induced switching therefore appears to qualitatively mirror a field-induced precessional switching event [50, 152–154] where the initial phase proves ringing free whereas relaxation within the new potential well may imply a large number of ringing oscillations. Critical currents vs. applied field for bifurcation into a reversed state ($h_x < Q$) are shown as square symbols in Fig. 22.

In the previous section, the onset of precessional states could be derived from elementary perturbation theory. However, the switching current is distinctly different from the critical current for the onset of precessional states, as seen in numerical simulations. However, no closed form could be established for the switching current starting from simple perturbation theory. As shown in Appendix B, such a result may be derived from *Melnikov's* method [155, 156], a method applying to weakly perturbed time-periodic Hamiltonian systems for which unperturbed trajectories in space may be derived from the energy landscape. Within the relevant field interval $-Q \le h_x \le Q$ and within the framework of our simple energetics, the critical current for switching is found to be equal to:

$$\frac{\chi_2}{\alpha} = r^2\sqrt{1+Q}\left(\frac{h\xi_{\mathrm{S}} + \sqrt{Q}\sqrt{1-\frac{h_x^2}{Q^2}}}{h\sqrt{1-\frac{h_x^2}{Q^2}} + r^2\sqrt{Q}\xi_{\mathrm{S}}}\right), \tag{51}$$

with $h = \frac{h_x}{Q}\sqrt{\frac{Q}{1+Q}}$, $r^2 = 1-h^2$, $\xi_{\mathrm{S}} = \cos^{-1}\left(-\frac{h}{\sqrt{Q}\sqrt{1-h^2}}\right)$ (see Appendix B for details). Notable switching-current values include $\frac{\chi_2}{\alpha} = \frac{1}{2}, \frac{2}{\pi}\sqrt{1+Q}$ and

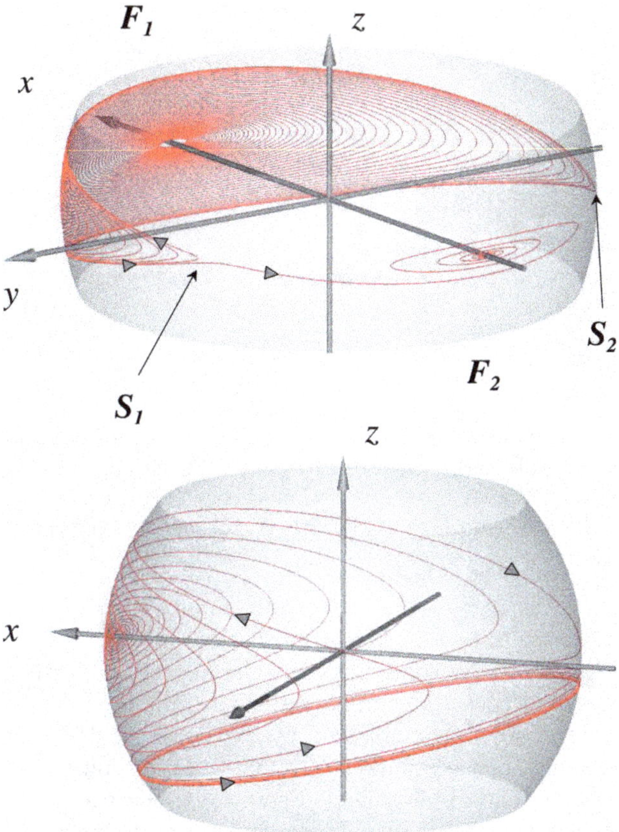

Fig. 23. *Top:* example of switching trajectory ($h_x^{\mathrm{a}} = Q/2$, $\chi/\alpha = 0.77$); F_1 and F_2 are fixed points due to sample energetics, whereas S_1 and S_2 are saddle points that correspond to a velocity minimum along the switching trajectory. *Bottom:* bifurcation to an out-of-plane precessional state ($h_x^{\mathrm{a}} = 2Q$, $\chi/\alpha = 1.75$)

1 for $h_x = -Q, 0, +Q$, respectively. The reduced critical current density for switching in the interval $-Q \leq h_x \leq Q$ ($-H_{\mathrm{K}} \leq H_x \leq H_{\mathrm{K}}$) is displayed as the line labeled χ_2 in Fig. 22. Clearly, the agreement with numerical simulations in the single-spin approximation proves excellent. Lastly, the time necessary for switching is very close to the time necessary to reach the bifurcation point and depends sharply on the current density as shown in Fig. 24a.

When $h_x^{\mathrm{a}} > Q$, switching in the conventional sense does not occur because the field exceeds the effective anisotropy field. Instead, a bifurcation takes place between a closed orbit that is essentially symmetrical with respect to the equatorial plane ($z = 0$) and another closed orbit either above or below the equatorial plane as shown in Fig. 23 (bottom). Similarly to "in-plane" precessional trajectories, the velocity along "out-of-plane" trajectories also proves to be highly nonuniform, as shown in Fig. 25b.

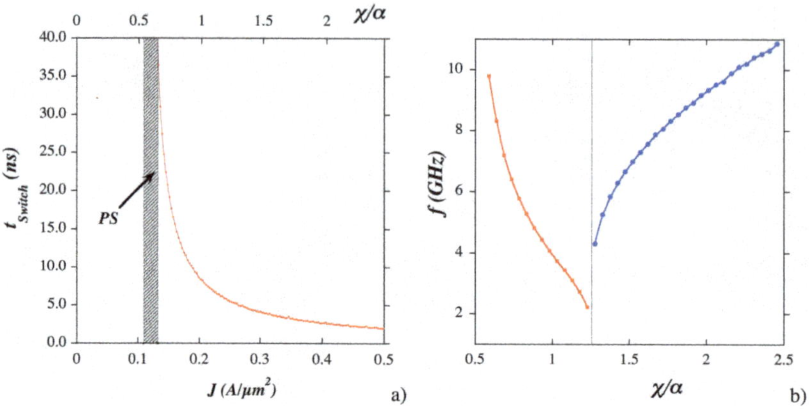

Fig. 24. (**a**) Switching time vs. current density at zero temperature and field. In this example, the switching time behaves roughly as $(J - J_{c1})^{\eta}$ with $\eta \approx -0.9$, i.e., a value close to expectation based on perturbation theory ($\eta = -1$ [145]); (**b**) Frequency of precessional states (in-plane to the left, out-of-plane to the right) for $h_x = 2Q$ at zero temperature

Here also, critical currents vs. applied field for bifurcation into an "out-of-plane" precessional state are shown as square symbols in Fig. 22. Note that precessional states are expected to be precursor states to both the "P" to "AP" and "AP" to "P" transitions, as confirmed by experiments [157]. At zero temperature, the frequency of precessional states follows trends indicated in Fig. 24b for a given value of the applied field. Essentially, the frequency decreases with increasing current for "in-plane" precessional states whereas the opposite behavior characterizes "out-of-plane" precessional states. Lastly, the case $h_x^{\mathrm{a}} = Q$ ($H_{\mathrm{x}}^{\mathrm{a}} = H_{\mathrm{K}}$) proves singular. Mathematically, the switching time diverges to ∞. However, such a situation is not robust against thermal fluctuations and leads to intermittency, as shown in Fig. 4.

4.4 Langevin Dynamics

The results in the previous sections were computed at zero temperature. They provide a reasonably accurate description of measurements made at low (liquid He) temperatures. However, most experiments are done at room temperature where thermal effects become much more important. Finite temperatures cause the system to fluctuate between states. For the most part, these states are close to the zero-temperature state, but large fluctuations can cause transitions over barriers into other stable or metastable states. Thus, finite temperatures change the transition between states from deterministic to statistical. Thermal effects can be modeled through statistical descriptions in Fokker–Planck approaches [158–160] or barrier-crossing models [33,41]. Alternatively, thermal effects can be included in dynamical simulations through

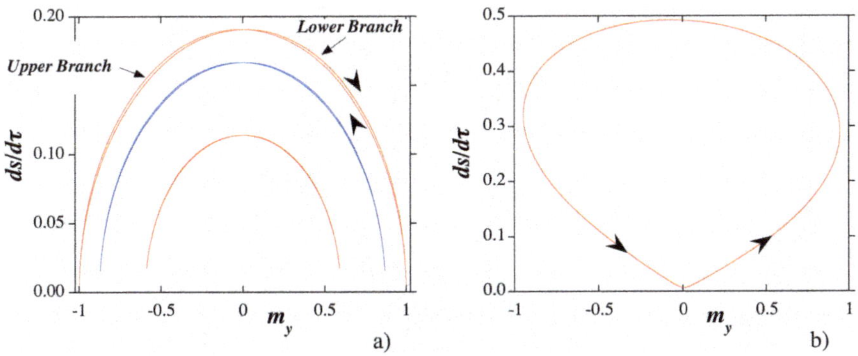

Fig. 25. Velocity along closed orbits. (**a**) $h_x^a = 0$, $J = 0.1150$, 0.1225, 0.13175 A/μm^2, i.e., $\chi/\alpha = 0.5647, 0.60155, 0.64697$; (**b**) $h_x^a = 2Q$, $m_z > 0$ (*upper branch*), $J = 0.256$ A/μm^2, i.e., $\chi/\alpha = 1.2571$. Motion is counterclockwise and clockwise for the *upper* and *lower branches*, respectively

the addition of a random field, H_{Rd}, to the effective field. Each component of H_{Rd} is uncorrelated both in space and time and obeys a purely Gaussian distribution with zero average value and variance μ:

$$\begin{aligned} \langle H_{\mathrm{Rd}} \rangle &= 0 \\ \langle H_{\mathrm{Rd}}^{\mathrm{i}}(t)\, H_{\mathrm{Rd}}^{\mathrm{j}}(t') \rangle &= \mu \delta_{ij} \delta(t - t') \\ \mu &= \frac{2k_{\mathrm{B}}T}{\mu_0 \gamma_0 M_{\mathrm{S}} V} \alpha\,, \end{aligned} \tag{52}$$

as derived from the application of the fluctuation dissipation theorem [161]. Here, V is the volume of the sample for macrospin simulations and the volume of the simulation cells for micromagnetic simulations (see next section).

Several authors have studied the effect of thermal fluctuations in macrospin models and simulations [33, 41, 158–160, 162, 163]. The main effect is that thermal fluctuations promote transitions between metastable (or stable) zero-temperature configurations. These fluctuations move the phase boundaries shown in Fig. 22 and give them a degree of indeterminacy. Figure 26 illustrates the stochastic nature of finite-temperature switching. It shows the distribution of switching times for two pairs of fields and currents. For one pair, the system is stable (and would never switch) at zero temperature, and for the other it is unstable and would switch with a unique switching time at zero temperature.

Generally speaking, switching-time distributions appear to be pseudo-log-normal, with a rapidly increasing distribution width as soon as the current density falls below the switching threshold at 0 K. Experiments do not yield switching-time distributions for preset current densities but rather switching probabilities vs. current density for preset waiting times or current ramp rates [33, 42, 157]. Measured distributions of switching currents are substan-

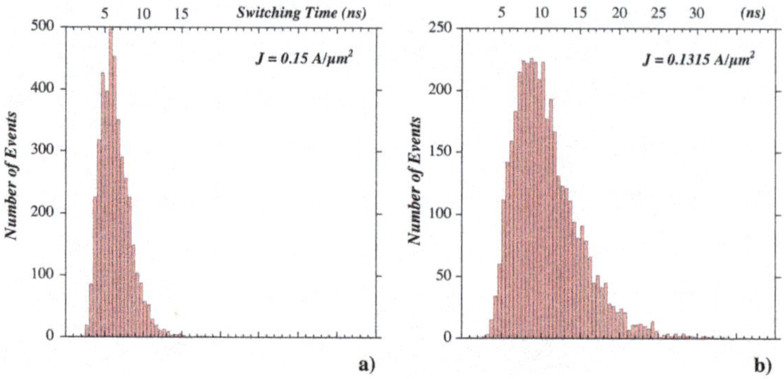

Fig. 26. Switching-time distribution at 300 K for $H_x = 0$. Average over 4096 trajectories. (**a**) $J = 0.15\,\mathrm{A}/\mu\mathrm{m}^2$: current density slightly above the switching threshold at 0 K; (**b**) $J = 0.1315\,\mathrm{A}/\mu\mathrm{m}^2$: current density just below the switching threshold at 0 K

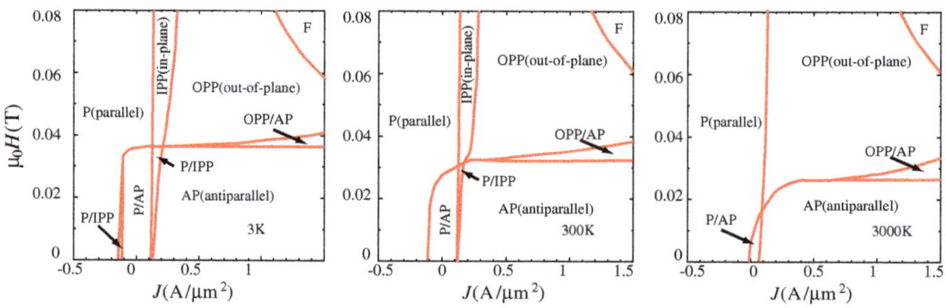

Fig. 27. Dynamic phase diagrams for a 130 nm by 70 nm elliptical Co free layer of thickness 3 nm. *Left panel*: $T = 3\,\mathrm{K}$; *middle panel*: $T = 300\,\mathrm{K}$; *Right panel*: $T = 3000\,\mathrm{K}$. For fixed H, a bistable region labeled A/B exhibits the A state when J is scanned from left to right and the B state when J is scanned from right to left. The correspondence needed to compare with Fig. 6 is $10^8\,\mathrm{A/cm}^2 \leftrightarrow 10\,\mathrm{mA}$. "F" refers to a configuration with a fixed magnetization direction that is not along the easy axis

tially skewed on the side of the small current densities (absolute value). Unfortunately, typical experimental waiting times (ramp rates) prove far too long (too slow) for direct comparisons to be made between experimental data and Langevin dynamics predictions, even in the single-spin regime. An important conclusion of [42] is that the barrier height appears mostly independent of the free layer thickness, a disproof of any thermal activation model over a single barrier. Indeed, as shown below, full-scale micromagnetic simulations lead to genuinely complex representations of spin-transfer-induced switching.

Russek et al. [162] and *Xiao* et al. [163] have computed the effect of temperature on phase diagrams like that shown in Fig. 22. The evolution of the phase boundaries with temperature in illustrated in Fig. 27. These phase diagrams were computed by sweeping the current at fixed field as was done in the experiment in Fig. 6. However, as noted above, the field sweep rates in the calculation are orders of magnitude faster than those in the experiment. The effect of the slower sweep rate can be qualitatively captured by considering higher temperatures, hence the calculation at the unphysical temperature of 3000 K. The 3000 K simulation gives some indication of how the 300 K simulation might look if the sweep rate was comparable to that of the experiment. This indication is only approximate because the tradeoff between sweep rate and temperature depends on the energy barrier between the competing configurations and this energy barrier varies throughout the phase diagram. These finite-temperature simulations show trends toward better agreement with the results in Fig. 6. Two differences are apparent. The simulation does not show the "W" phase, in which there is no apparent precessional peak, but that has a resistance distinctly different from the antiparallel state. On the other hand, there is no indication of the high-current, high-field fixed phase "F" in the experimental data.

Temperature also affects the details of the precessional dynamics [162]. In the single-spin and zero-temperature limit, precessional states are characterized by a frequency that is just a function of current density, in-plane anisotropy and damping parameter. In the frequency domain precessional states are thus defined by a pure Dirac delta function. Raising the temperature introduces stochasticity in the magnetization trajectories. As a result of the action of the stochastic field, magnetization motion also displays fluctuations when compared to the 0 K limit. Of major interest here is a quantity that measures the influence of a value of the function $\boldsymbol{m}$ at time t on the value at time $t+\xi$, a quantity called the autocorrelation function (see, e.g., [164]). The power-spectrum density (PSD) of a scalar quantity $x(t)$ is the Fourier transform of the autocorrelation function (Wiener–Khinchin theorem). It is the "power" emitted within the interval $[f, f+df]$ (natural units W/Hz). Physically, a long correlation time corresponds to nicely peaked power spectra whereas short correlation times yield flat spectra. Lastly, it ought to be stressed that care is necessary when estimating PSDs from data $x(t)$ extending over finite times (see, e.g., [165] for an overview on windowing and averaging methods).

Figures 28 and 29 ($H_x = 2H_\mathrm{K}$) display the main results to be expected from the single-spin model at low temperature in a current regime allowing for precessional states of the first kind (see Fig. 20). Due to symmetry of the closed magnetization trajectories with respect to $\boldsymbol{x}$, the easy magnetization axis, the fundamental frequency for a given current density is given by the power spectrum density of the m_y magnetization component, whereas the PSD deduced from m_x corresponds to twice the fundamental frequency. Thus, depending on the observable, PSDs may reveal either the fundamental frequency, its first harmonics or a mixture of both, as well as higher harmonics. Hopefully, however, ambiguity in the spectral response should in most

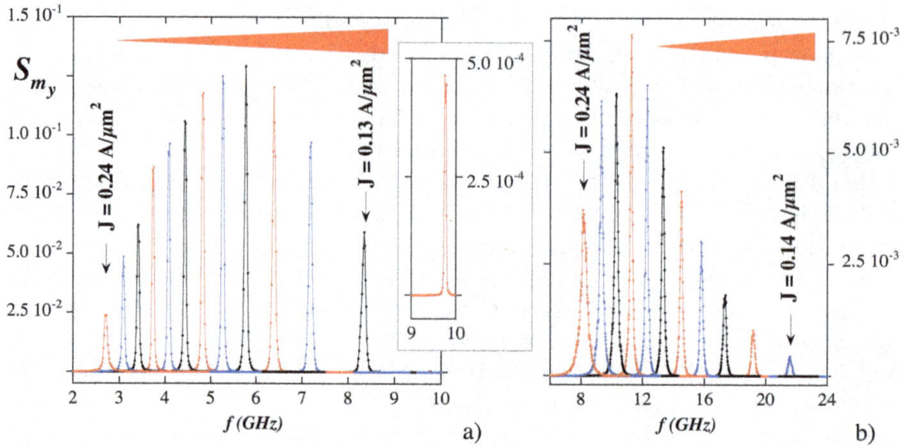

Fig. 28. m_y power spectrum density (PSD) vs. current density at 4.2 K for $H_x = 2H_K$. Current-density step: 0.01 A/μm². "Low" current-density regime: precessional states essentially symmetrical with respect to the sample plane. (**a**) fundamental frequency $f(J)$. The frequency decreases with increasing current (redshift), as symbolized by the *elongated triangle*. *Inset*: PSD very close to Kittel's resonance frequency; (**b**) 3rd harmonics. Pollution from higher harmonics has been removed

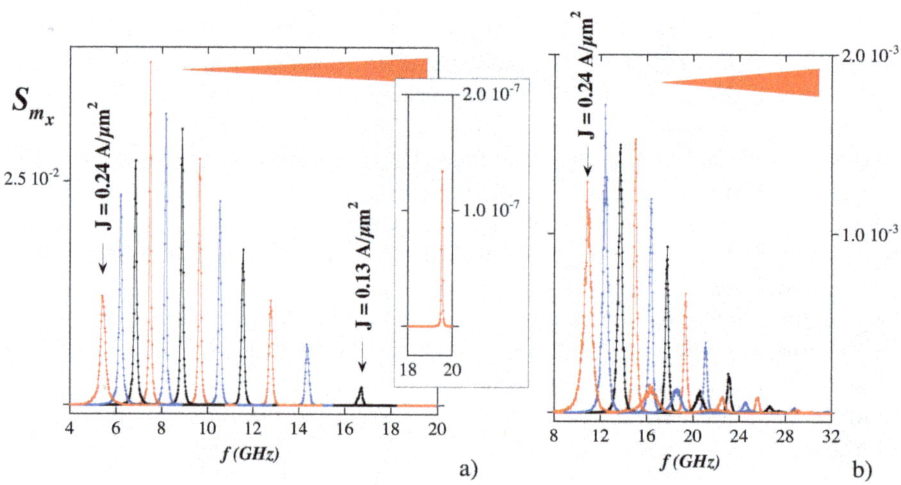

Fig. 29. m_x Power-spectrum density vs. current density at 4.2 K for $H_x = 2H_K$. Current-density step: 0.01 A/μm². "Low" current-density regime: precessional states essentially symmetrical with respect to the sample plane. (**a**) lowest frequency. *Inset*: PSD in the immediate vicinity of twice Kittel's resonance frequency; (**b**) harmonics

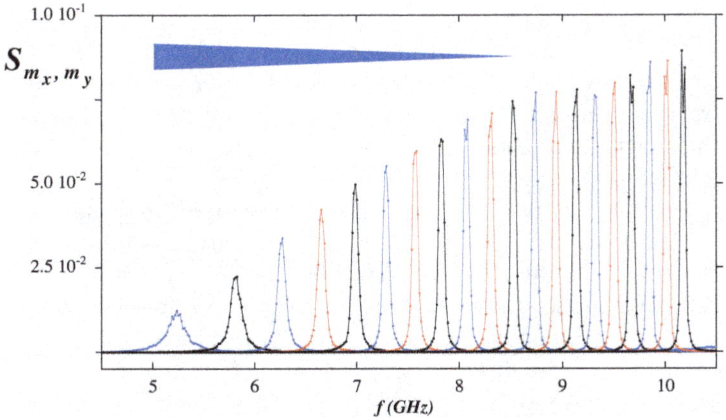

Fig. 30. Power-spectrum density vs. current density at 4.2 K for $H_x = 2H_K$. "High" current-density regime: out-of-plane precessional states. The frequency increases with increasing current (blueshift), as symbolized by the *elongated blue triangle.* The m_x and m_y spectra are indistinguishable in this regime. Current density ranging from 0.27 A/μm² to 0.45 A/μm² in steps of 0.1 A/μm²

cases be lifted due to the fact that, at the onset of sustained precession, the fundamental frequency is equal to Kittel's resonance frequency, which only depends on materials parameters and applied field (see Sect. 4.3.3). Figures 28 and 29 show that the peak in the PSD initially grows with rising current density as the frequency decreases (the so-called "redshift" regime owing to [38, 39]). Further increases of the current density lead to a decrease of the peak in the PSD and power is gradually transferred into higher harmonics, namely $3, 5, \ldots, (2n+1)f$ for the m_y PSD, $2, 4, \ldots, (2n)f$ for the PSD stemming from m_x. In this precessional regime, in agreement with the phase diagram in Fig. 22 and general perturbation theory (Appendix B) closed magnetization orbits gradually open up to extremely wide excursion angles (from typically $\pm\pi/3$ for $J = 0.13\,\mathrm{A}/\mu\mathrm{m}^2$ to $\pm\pi$ for $J = 0.24\,\mathrm{A}/\mu\mathrm{m}^2$). A plot of the log of the power-spectrum density vs. frequency (not shown) indicates the presence of higher harmonics to orders up to about 20 with a close to exponential decay of the PSD maximum with increasing frequency (typically 30 dB per octave). Note also that at these very low temperatures and a usual damping parameter ($\alpha = 0.01$ in these simulations), the linewidth remains extremely narrow with full width at half-maximum in the 70 MHz to 80 MHz range.

As explained above, a further increase of the current density leads to the onset of a new precessional regime characterized by out-of-plane precessional states (see Fig. 23 (bottom)). The power-spectrum density in this regime is displayed in Fig. 30. As expected from 0 K simulations, out-of-plane precessional states are characterized by a blueshift in frequency with increasing current density with, roughly speaking, a $\frac{1}{2}$ power-law dependence of the frequency vs. current density. It is noteworthy that in this regime the PSDs associated with m_x or m_y are indistinguishable, a property that might help in establishing a clear separation between precessional states in the "low-" and "high-" current regimes (in other words, the PSD should remain essentially independent of the exact orientation of the applied field in the vicinity of the easy-magnetization axis).

Remarkably enough, many of the predictions of the simple single-spin approximation have clear transcriptions in the room-temperature point-contact experiments of *Rippard* et al. [38]. The existence of a redshift and a blueshift regime has been clearly identified, although the latter has only been observed under high perpendicular applied field ($> 0.6\,\mathrm{T}$). Similarly, precessional states in the redshift regime have been observed only for easy-axis fields exceeding a threshold (50 mT) well above any estimated effective anisotropy field in these systems. At large applied fields, frequency-domain measurements of the magnetoresistance (reported in units of $\mathrm{V/Hz^{1/2}}$) show extremely narrow magnetoresistance peaks that behave at least qualitatively similarly to the PSDs shown in Figs. 28 and 29. MR spectra do imply some mixing between f and $2f$ contributions, the source of which has been attributed to misalignment between the magnetization directions at equilibrium under field within the reference and free layers. It is generally estimated that a misalignment of less than 10° proves sufficient to restore PSDs of comparable magnitude for the fundamental frequency and its first harmonics. Higher harmonics have not been observed but, clearly, theoretical expectations ought to be convoluted with the experimental spectral bandwidth and take into account detailed features of the magnetoresistance response such as shown in Fig. 14. On the other hand, a major discrepancy between experimental data and theoretical expectations appears as soon as the temperature is raised in the simulations: the linewidth grows with roughly the square root of the temperature, reaching typical values around 300 MHz at 77 K and some 700 MHz at 300 K for a damping parameter still equal to its nominal value as extracted from, e.g., an FMR experiment. Now, raising the anisotropy will mainly affect the resonance frequency at the onset of precessional states, but not so much the linewidth that remains essentially controlled by the damping parameter. It thus turns out that a few tens of MHz full width at half-maximum at room temperature requires an effective damping parameter in the 10^{-4} range, the origin of which remains unknown at this stage.

The experiments of *Rippard* et al. [38] use a lithographically defined point contact with continuous magnetic films (see Fig. 1). The exchange coupling of the dynamic part of the system to an extended magnetic film raises doubts on the applicability of the macrospin approximation. The experiments of *Kiselev* et al. [36, 166], on the other hand, use a nanopillar geometry with a finite free layer. We expect the macrospin model to be a better approximation for these samples. Unfortunately, the results' complexity extends far beyond macrospin predictions, calling for full micromagnetic simulations.

4.5 The Micromagnetic Regime

In contrast to the single-spin approximation, micromagnetics allows for nonuniform magnetization distributions across the area (and thickness) of both the free- and hard-layer elements. As stressed in Sect. 4.3.1, the magnetization now becomes a function of position within a given element and is subject to the spin-transfer torque as well as to the action of an effective field, itself a function of the magnetization distribution. Both the spin-transfer torque and the effective field are thus position dependent within the considered element [(29), (32)]. Excitations such as spinwaves or magnetization waves are now allowed within the elements of the stack. Moreover, the effect of the Oersted field may now be taken into account.

There have been a number of calculations using this approach [167–176]. These have clarified some of the discrepancies between measurements and macrospin simulations. In particular, *Lee* et al. [173] have shown that in the region where experiments show the "W" phase, the magnetization is far from uniform. Vortices enter and leave the system reducing the resistance from the full antiparallel value without exhibiting well-defined precession. *Berkov* and *Gorn* [176], have shown that a postulated distribution of local anisotropies qualitatively reproduces the current dependence of the precession frequency and linewidth seen in experiments of *Kiselev* et al. [36]. While full micromagnetic simulations do a better job of describing the behavior of the system, they do it with a greater degree of uncertainty because many unmeasured details of the system are important, included among these are the exchange stiffness and saturation magnetization, the distribution of anisotropies, the details of the shape, particularly the edges of the sample, etc.

Even for a perfectly characterized sample, however, micromagnetics alone does not provide a fully consistent model of spin-transfer-induced excitations and switching since, up to this date, it has not been twinned with a truly 3D transport model [140,141]. So far, micromagnetics calculations have relied on 1D transport theory for the evaluation of the spin-transfer torque. Additional approximations are often made. In most instances, micromagnetic simulations rely on the assumption of a uniform and static magnetization distribution within the hard (pinned) layer as well as a uniform current density across the element. It is often further assumed that the polarization $\mathcal{P}(\boldsymbol{r})$ is constant, irrespective of the relative orientations of the magnetization within the hard and soft elements. According to the results of Sect. 3.7, lateral diffusion and rescattering are anticipated to have a stabilizing effect for a net electron flow from the free layer into the adjacent nonmagnet and a destabilizing effect in the reverse geometry. Owing to prevailing conventions, lateral diffusion and rescattering would thus be anticipated to improve and degrade spatial coherence for the *P*–*AP* and *AP*–*P* transitions, respectively.

We start this section by considering the fate of precessional states ($H_x > H_{\mathrm{K}}$) within small elements when shifting from the single-spin approximation to the full micromagnetic regime [175]. These states are substantially the

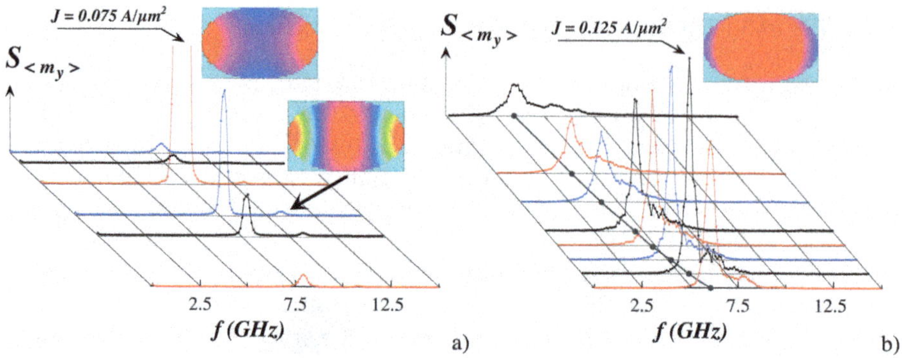

Fig. 31. Micromagnetic regime: $\langle m_y \rangle$ power-spectrum density vs. current density at 300 K for $H_x = 3H_K$. (**a**) Low current-density regime ($J = 0.025, 0.05, 0.06, 0.075, 0.085, 0.09\,\mathrm{A/\mu m^2}$); Insets: eigenmodes at 8.2 and 11.2 GHz. (**b**) Redshift regime at higher current densities ($J = 0.1, 0.125, 0.150, 0.175, 0.2, 0.25, 0.3, 0.4\,\mathrm{A/\mu m^2}$); *Inset*: spatial PSD distribution at 5.4 GHz ($J = 0.125\,\mathrm{A/\mu m^2}$). $Ni_{80}Fe_{20}$-like materials parameters (exchange constant: $A = 1.0 \times 10^{-11}$ J/m, saturation magnetization $M_S = 80$ kilo · A/m, damping parameter $\alpha = 0.01$). Elliptical free element: thickness $d = 2.5$ nm, long and short axes 115 nm and 70 nm, respectively. For this set of material parameters and dimensions, the shape-induced anisotropy field amounts to ≈ 19.5 kilo · A/m (≈ 245 Oe). After [175]

same but there are important differences between the two descriptions. At low current densities and room temperature, power is emitted at basically constant frequency within a well-defined current-density range. Two peaks in the power spectrum density extracted from the time variation of $\langle m_y \rangle$, i.e., the m_y magnetization component averaged over the volume of the element, are observed. They correspond to two vibration eigenmodes of elliptical elements [177] as depicted in Fig. 31a. The largest emitted power corresponds to a mode characterized by enhanced in-phase magnetization precession in the vicinity of the apices of the elliptical element. For this mode, the emitted power first grows with increasing current density, then decreases. It is worth noticing that the characteristic frequencies of these eigenmodes differ from Kittel's frequency (8.2 GHz and 11.2 GHz for the former vs. 9.18 GHz for the latter for the dimensions and material parameters considered). Over a large current-density range beyond a second current threshold (Fig. 31b) the PSD vs. current density exhibits the redshift behavior also characteristic of the single-spin approximation for $\chi > \chi_{\mathrm{crit}}$ (48). In the redshift regime, precessional states correspond to an essentially coherent precession of the magnetization within the central zone of the element as illustrated in the inset of Fig. 31b. Results contained in Fig. 31 are, qualitatively at least, consistent with the experimental results of *Kiselev* et al. (notably Figs. 1c and 1d in [36]). The results above differ from the simulation data of *Lee* et al. [173] who found that

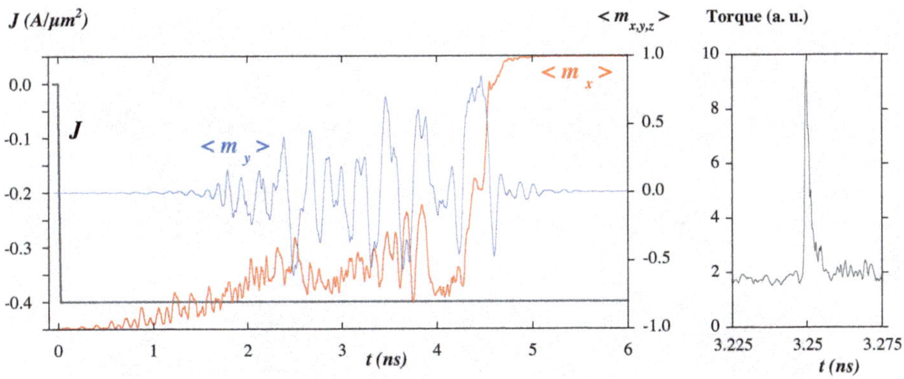

Fig. 32. Switching event in a $170 \times 80 \times 2.5\,\mathrm{nm}^3$ elliptical element at 0 K owing to numerical micromagnetics. *Left*: average magnetization components $\langle m_x \rangle$ and $\langle m_y \rangle$ vs. time. *Right*: Torque spike linked to vortex-pair annihilation. Current density: $J = -0.4\,\mathrm{A}/\mu\mathrm{m}^2$, polarization $\mathcal{P} = 0.3$. Co/CoFe-like material parameters (exchange constant: $A = 1.3 \times 10^{-11}\,\mathrm{J/m}$, saturation magnetization $M_\mathrm{S} = 150\,\mathrm{kilo} \cdot \mathrm{A/m}$, damping parameter $\alpha = 0.006$)

precessional states could only be maintained over a very narrow current density range for fields typically equal to $3H_\mathrm{K}$ in Co elements. Actually, in their respective attempts to fit *Kiselev's* et al. experiments, *Lee* et al. [173] and *Berkov* and *Gorn* [176] rely on markedly different assumptions with respect to material parameters (exchange stiffness, existence or not of a depressed saturation magnetization) and microstructure. These analyses also differ in the conversion of the time-dependent micromagnetic magnetization distributions into a measurable GMR signal (*Berkov* and *Gorn* apply the field along a rather skewed direction with respect to the ellipse long axis, *Lee* et al. do not). Altogether, assuming that all of the simulations referred to above are technically sound, it seems clear that simulation results prove extremely sensitive to a wide range of parameters that remain only poorly known in these tiny magnetic elements.

We turn now to a micromagnetic switching event. For simplicity, we keep the fixed layer frozen and fix the spin current coming from it to the free layer, thus, the free layer is treated as a soft element flooded with a uniform density of spin-polarized electrons. The Oersted field is assumed to originate from an elliptical crosssection cylinder with height 50 nm meant to represent the pillar height with the soft layer located at half-height across the pillar. We further assume that the hard layer is left unpatterned so that the soft element is only subject to its self-demagnetizing field, exchange interactions and the spin-transfer torque with $\mathcal{P}(\boldsymbol{r})$ constant. Symmetry in the spin torque is lifted through the application of a minute constant field along $\boldsymbol{y}$. The current is assumed to start flowing at time $t = 0$ and follow a simple step function.

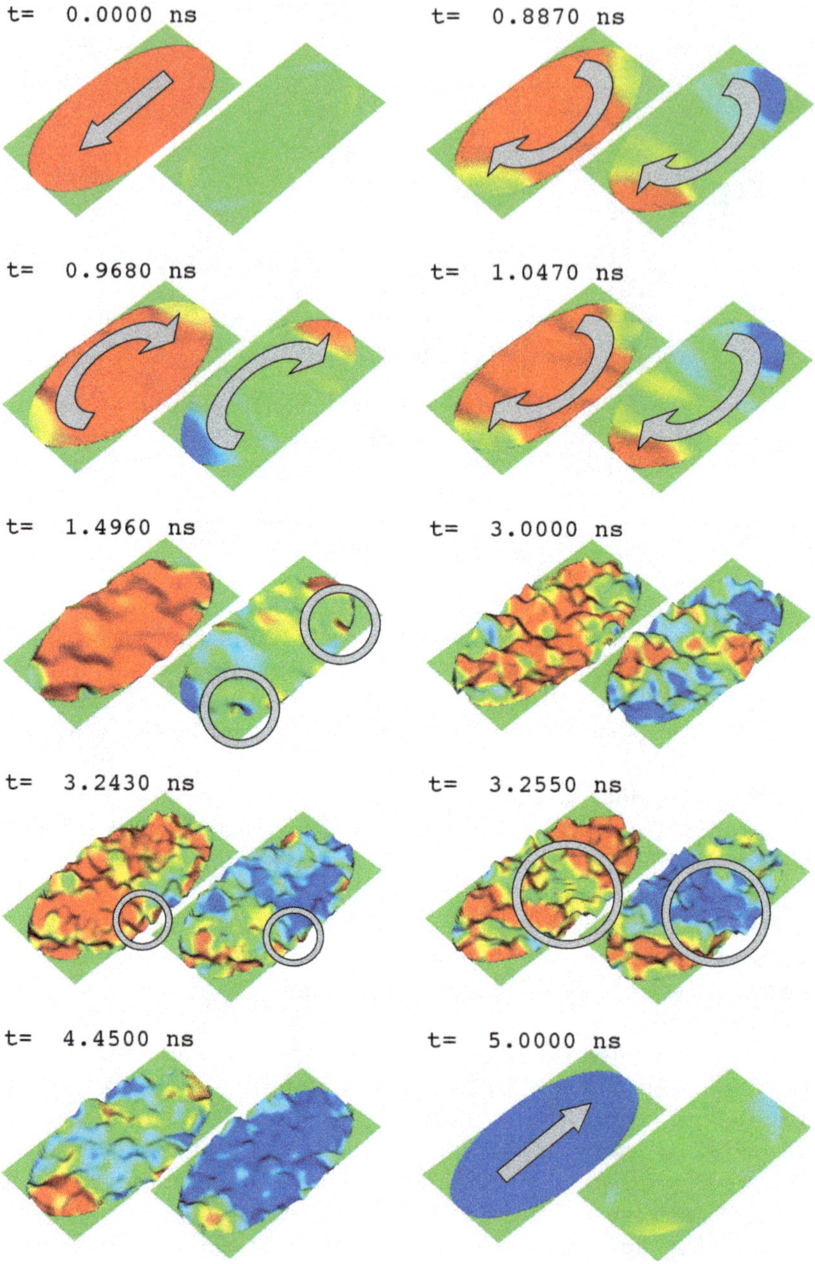

Fig. 33. Magnetization distribution vs. time corresponding to Fig. 32 ($T = 0\,\mathrm{K}$). Color map extending from red (-1) to blue ($+1$) for the m_x and m_y magnetization components (the *left* and *right image* of each images pair, respectively). The surface of the element distorts according to the sign and amplitude of the m_z magnetization component. See text for details

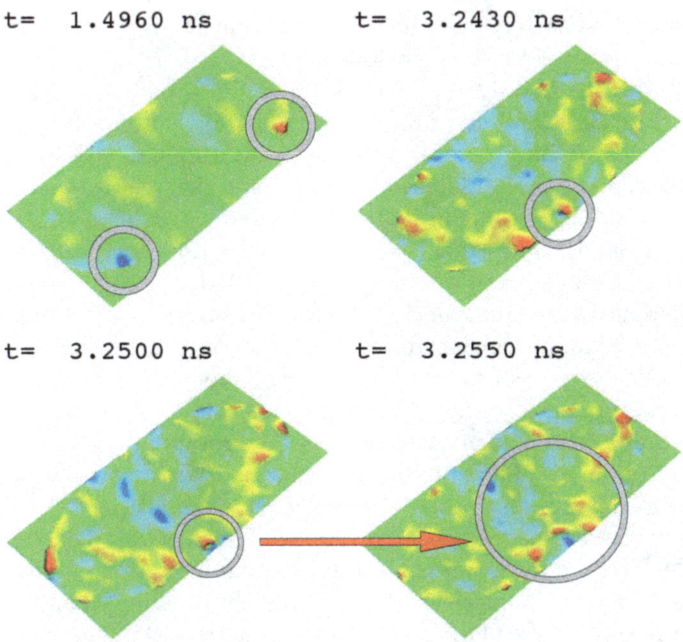

Fig. 34. Vortex nucleation ($t = 1.496\,\text{ns}$) and pair annihilation (starting at $t = 3.243\,\text{ns}$) during the switching event described in Fig. 33. Color representation and surface distortion according to the sign and amplitude of the m_z magnetization component. Note the extremely sharp transition between the blue and red colors within the gray circle for $t = 3.25\,\text{ns}$, followed by the emission of a magnetization wave for $t = 3.255\,\text{ns}$

The current density is chosen to stand slightly above the switching threshold current density at zero temperature.

Three phases characterize the switching event. During the first phase, lasting about 1.5 ns, the average value of the transverse magnetization component $\langle m_y \rangle$ remains almost zero, as shown in Fig. 32. So does the $\langle m_z \rangle$ component. Due to the inhomogeneous magnetization and effective field distribution, including the Oersted field, the spin torque proves most efficient close to the apices of the elliptical element, leading to a beating between two mirror-symmetric C-type states (Fig. 33) with a period of $\approx 160\,\text{ps}$. Magnetization waves develop smoothly ($t = 1.0\,\text{ns}$ to $1.5\,\text{ns}$). When t reaches 1.496 ns, the first vortex pair is nucleated along the lower rim of the element (see Fig. 33 and Fig. 34).

During the second phase, the magnetization motion becomes complicated with extreme excursions of both the m_y and m_z components between -1 and $+1$, as displayed in Fig. 33 ($t = 2.5\,\text{ns}$ to $4.4\,\text{ns}$). However, the average transverse magnetization component $\langle m_y \rangle$ undergoes globally increasing oscillations away from its ground state. This phase is equivalent to the parametric pumping phase characteristic of the single-spin model.

Switching, the third phase, only occurs if partial resynchronization takes place ($t = 4.45\,\text{ns}$, Fig. 33). This feature is quite general. With increasing current density, sub-ns switching speeds may be achieved in simulations, in agreement with experiments [178, 179]. However, prior to switching, the element still contains winding vortex–antivortex pairs indicated, e.g., by the gray circles in Fig. 33 and Fig. 34 for $t = 3.243\,\text{ns}$ that are seen to decay through the emission of magnetization waves ($t = 3.255\,\text{ns}$ in Fig. 33 and Fig. 34). In most instances, the annihilation process manifests itself through a sharp peak in the maximum torque recorded across the simulation area (see Fig. 32, right). Vortex pairs are termed winding if both the circulation of the magnetization and the core magnetization [180] change sign between the elements of the pair so that the magnetization would rotate by 360° along a line joining the vortex cores. The annihilation of such a vortex pair would normally imply the nucleation and propagation of a Bloch point.

Although numerical micromagnetics can be done in a sophisticated enough manner to treat Bloch points rather satisfactorily [181], annihilation of winding vortex pairs is usually a numerical artifact [182], especially when the simulation mesh implies a single plane of nodes in these very thin elements. If such structures in the magnetization do occur, the fine details of the simulation become unreliable as the results become strongly mesh and time-step dependent, even at zero temperature. Readers are, in this respect, cautioned to examine all simulations of spin-transfer-induced dynamics critically.

Fortunately, a markedly different picture emerges from exchange-biased stacks whereby the biasing direction is set at an angle with respect to the shape-induced easy-magnetization axis. We therefore now consider a fully patterned stack composed of an exchange-biased "hard" layer, a normal metal spacer and a free layer, all with an identical elliptical cross section. In our model calculations, the biasing direction is set at an angle of +45° with respect to the long axis of the elliptical hard-layer element, the magnetization of which we assume to be uniform. The free element is now subject to a rather large demagnetizing field that enforces two noticeably asymmetrical magnetization distributions at equilibrium: for the dimensions considered, the magnetization distribution within the free element is oriented at a mean angle close to −155° (almost opposite) for the "*AP*" configuration, but at −45° (almost perpendicular rather than parallel) for the "*P*" configuration (see Fig. 35). Due to biasing at a skewed angle, the potential wells for the "*P*" and "*AP*" states are also strongly asymmetrical giving a much easier "*P*" to "*AP*" transition as compared to the "*AP*" to "*P*" transition. Also due to the shallow potential well of the "*P*" state, spontaneous "*P*" to "*AP*" transitions are observed in numerical simulations for temperatures as low as 40 K. At 20 K on the other hand, $\langle m_y \rangle$ undergoes $\approx 6°$ to 7° peak-to-peak thermal fluctuations that do not prove large enough to jeopardize the long-

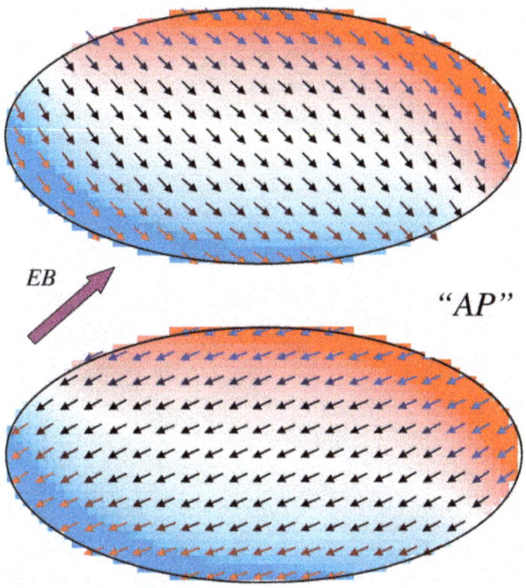

Fig. 35. "P" and "AP" states of a fully patterned elliptical pillar exchange biased at a skewed angle. Soft and hard elements dimensions: $130 \times 65 \times 2.5\,\mathrm{nm}^3$; spacer thickness: 5 nm. Biasing direction EB at 45°. $Ni_{80}Fe_{20}$-type material parameters. The background color coding follows the m_z magnetization component and is determined solely by the demagnetization field from the biased fixed layer, and hence does not differ much between the two states. Note, however, that in the "AP" state the magnetization is largely antiparallel to the fixed layer, while in the "P" state it is almost perpendicular

term stability of the "P" state. A typical "P" to "AP" switching event is depicted in Fig. 36.

The skewed biasing causes the greater simplicity of the switching event in Fig. 36 as compared to that in Fig. 00. The large angle between the fixed-layer and free-layer magnetizations provides a substantial spin-transfer torque the instant the current is switched on. This shortens or eliminates the growing oscillation phase illustrated in Fig. 32. The major features of the reversal process are, 1. the existence of weak thermal fluctuations at the temperature considered, an essentially coherent magnetization rotation that takes place within less than 1 ns, 2. the existence of postswitching oscillations that decay over times also less than 1 ns.

Similar features may be observed in Fig. 37 depicting $\langle m_{x,y} \rangle$ vs. time at zero temperature with the current density as a parameter.

Coherence in the time domain is emphasized in Fig. 38a that displays an average over 400 trajectories of $\langle m_{\mathrm{x,y}} \rangle$ vs. time for two values of the current density. Remarkably, the essential characters of a switching trajectory at zero temperature are preserved. The corresponding switching-time histogram is shown in Figs. 38b and c; unsurprisingly, the histograms prove rather narrow. It may also be shown that coherence in the time domain is similarly

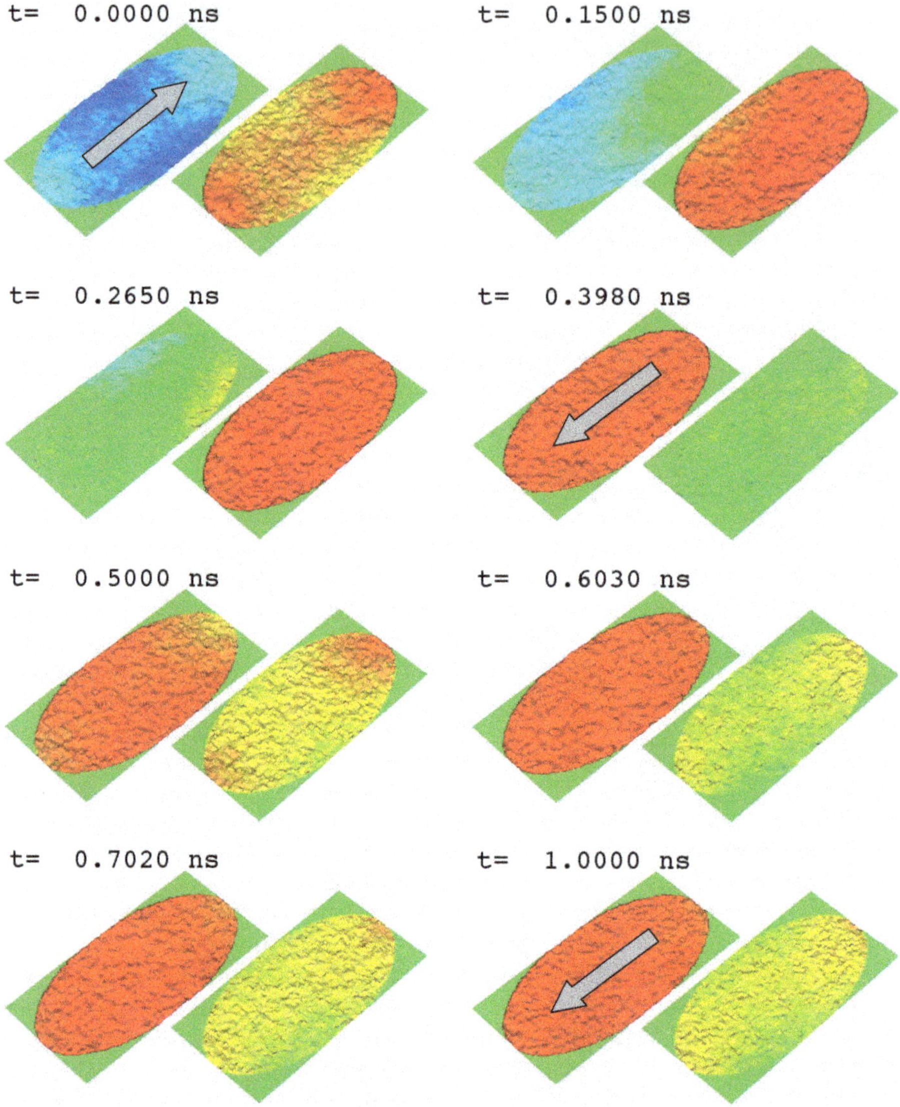

Fig. 36. Typical switching event in a $130 \times 65 \times 2.5\,\mathrm{nm}^3$ elliptical element with the hard layer fully patterned and biased at a skewed angle. Micromagnetic simulations at 20 K. $Ni_{80}Fe_{20}$-like material parameters (damping parameter $\alpha = 0.01$). Same representation conventions as in Fig. 33. Switching at $t = 0.265\,\mathrm{ns}$. Subsequent $\langle m_y \rangle$ maxima and minima at $t = 0.398, 0.603$ and $t = 0.5, 0.702\,\mathrm{ns}$, respectively. Note that thermal fluctuations at time $t = 0$ and $t = 1\,\mathrm{ns}$ appear dissimilar: the current still flows at $t = 1\,\mathrm{ns}$ and enhances damping

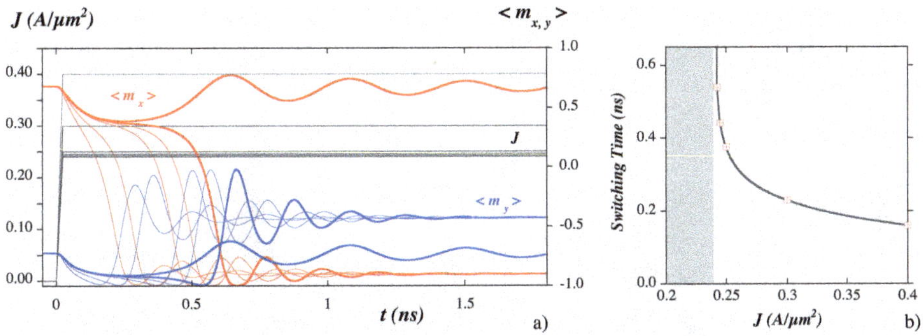

Fig. 37. (**a**) Switching trajectories as a function of current density at 0 K owing to numerical micromagnetics. Dimensions and material parameters identical to those in Fig. 36. The *thick lines* show trajectories across the current-density bifurcation threshold $0.2425 > J > 0.24\,\mathrm{A/\mu m^2}$. (**b**) Switching time vs. current density. The switching time is defined as the time for which $\langle m_x \rangle = 0$. Current densities: $J = 0.4, 0.3, 0.25, 0.245, 0.2425, 0.24\,\mathrm{A/\mu m^2}$

preserved for low current-density excitations (Figs. 38d and e), in particular in the subcritical regime, i.e., for current densities below the onset of precessional states. Although no attempt was made to reproduce the experimental data of *Krivorotov* et al [183], micromagnetic simulations permit a qualitative understanding of these results. Still, a proper description of correlated data in the time and frequency domains owing to conventional micromagnetics appears almost out of reach because of the extremely narrow linewidths extracted from the experiments.

Two additional remarks deserve to be made, namely 1. fully patterned pillars with skewed exchange biasing do, when viewed from a micromagnetic perspective, allow for extremely coherent magnetization rotation, at least for a "P" to "AP" transition, in a way that is conceptually very similar to spin-transfer-induced switching in systems with a fourfold shape anisotropy [167], 2. precessional states may, in such systems, be excited for fast-rising currents with both positive or negative current densities whether starting from the "P" or the "AP" state, a fact directly linked to the non-collinearity between the mean magnetization direction and the electrons' spin polarization.

Altogether, this section shows that the behaviors observed in experiments can be qualitatively explained by the torque originally proposed by Slonczewski. However, all of the calculations described here have ignored many of the features of the spin-transport calculations, that is, the polarization $\mathcal{P}$ in (28) has been treated as a constant, and the angular dependence described in (26) has been ignored. So far, the comparison between experiments and simulations appears to be somewhat insensitive to the details of the polarization, at least given the other uncertainties about the systems. Only the size of the polarization has been important, and that can only be determined from experiment if the size of the damping is accurately known. It appears that the

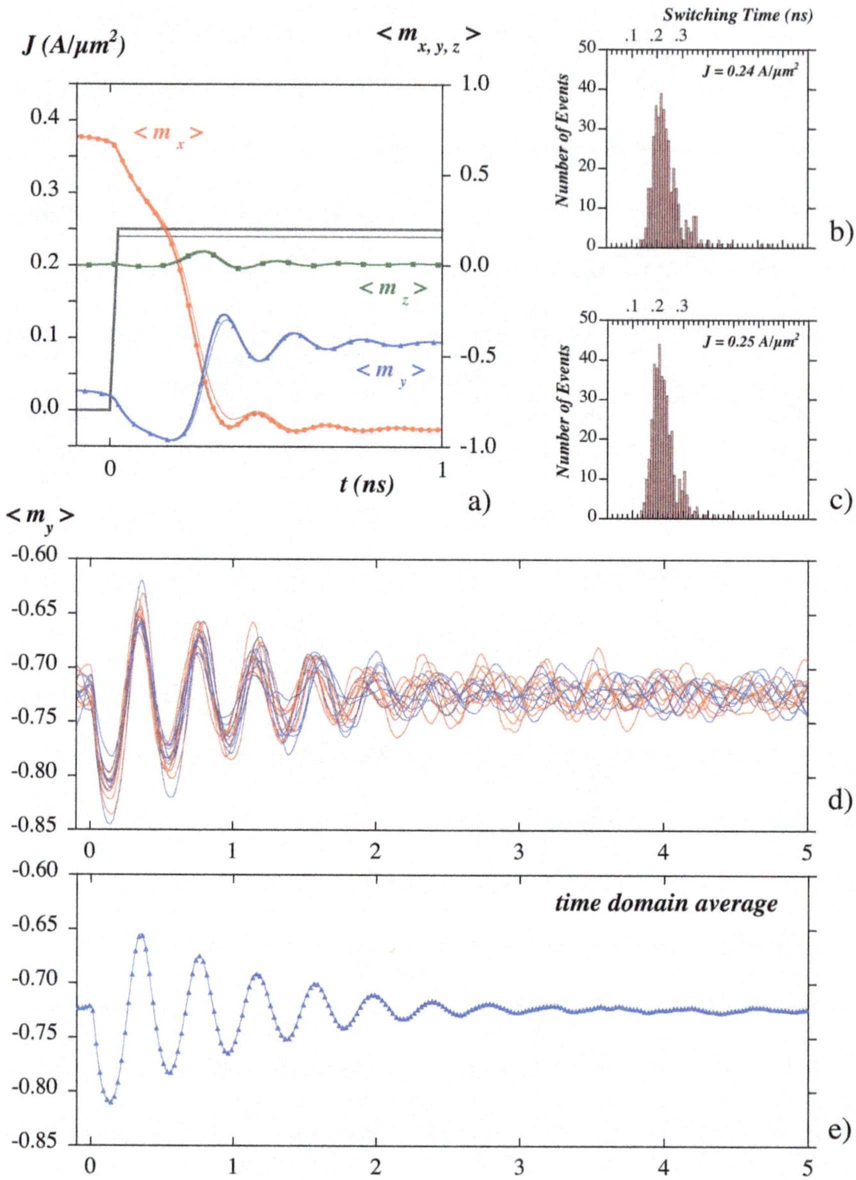

Fig. 38. Full-scale micromagnetics for a hard layer biased at a skewed angle: (**a**) time-domain averages of "P" to "AP" switching trajectories at 20 K for current density $J = 0.24$ (*thin lines*) and $0.25\,\mathrm{A}/\mu\mathrm{m}^2$ (*full symbols*). Over 400 simulated trajectories, 19 (14) failed to display switching within the maximum allowed time for switching in the simulations, namely 10 ns, for $J = 0.24$ (0.25) $\mathrm{A}/\mu\mathrm{m}^2$; (**b**) and (**c**) corresponding switching-time histograms; (**d**) selected $\langle m_y \rangle$ vs. time trajectories in the subcritical regime at 4.2 K; (**e**) time-domain average over 100 trajectories such as shown in (**d**)

best possibility for testing the results of calculations like those in Sect. 3 is in measurements of biased samples with $Ni_{80}Fe_{20}$ free layers. $Ni_{80}Fe_{20}$ layers have smaller fluctuations in local anisotropies and biased samples appear to behave like macrospins in micromagnetic simulations. The recent measurements of *Smith* et al. [137] are significant progress in this direction.

5 Summary

Since the concept of spin-transfer torques was first proposed by Slonczewski and Berger, there has been remarkable experimental and theoretical progress. Experiments have observed and characterized a variety of different behaviors including hysteretic switching, precession, and two-level fluctuations. In this Chapter we have tried to review the physical picture that has been developed to explain this variety of behaviors.

A generally accepted description of these systems has evolved from a hierarchy of models based on a separation of time and length scales. At the most basic level, quantum-mechanical calculations describe the behavior of electrons spins close to the interface between ferromagnetic and nonmagnetic layers. The exchange interaction in the ferromagnet rotates noncollinear spins in complicated ways. Calculations of this effect show that the component of the spin current along the magnetization is conserved as electrons transmit through and reflect from the interface. However, the transverse component of the incident spin current is effectively absorbed close to the interface and transferred to the magnetization. This absorption of the transverse-spin current is the origin of the torque that drives the magnetization dynamics measured in these systems.

In such systems, any quantum-mechanical coherence due to multiple scattering from interfaces tends to be washed out. The lack of coherence allows the results of the quantum-mechanical calculations to enter as boundary conditions into semiclassical transport calculations, which is the next level in the hierarchy of models. The semiclassical calculations give the spin currents flowing through the different layers and the torques on the magnetizations. A variety of semiclassical transport equations, the Boltzmann equation, the drift-diffusion equation, and circuit theory all give qualitatively the same result. The magnitude of the torque depends on the polarization of the current flowing from the fixed layer to the free layer, and the polarization of the current depends on everything in the sample within a few spin diffusion lengths, both upstream and downstream. This dependence leads to the result that the torque for small deviations away from antiparallel alignment is generally much stronger than the torque for equivalent deviations away from parallel alignment. These two consequences are the important observable predictions of the transport calculations.

The electron velocities are high enough and spin-flip scattering lifetimes short enough that from the point of view of transport, the magnetization can be thought of as fixed. This separation of time scales allows the torques to be computed for fixed magnetic configurations and the resulting torques then used to describe the magnetization dynamics. Macrospin or micromagnetic simulations of the dynamics are the final level of model in the hierarchy. Macrospin models accurately describe the system when the magnetization remains uniform throughout the sample. The resulting behavior separates naturally into two regimes depending on the size of the applied field.

For applied fields less than the coercive field of the free layer, the system is stable in either a parallel or antiparallel alignment. For large electron flow from the fixed layer to the free layer, the stable alignment is parallel and for electron flow in the opposite direction it is antiparallel. There is a region of bistability for small currents. The transitions from one stable region to the other are preceded by a small region of in-plane precession. In-plane precession describes a state in which the magnetization precesses around the easy axis, which lies in-plane. As the transition to the opposite configuration is approached, the amplitude of the precession gets large, with a substantial out-of-plane component. For applied fields larger than the coercive field, the applied field inhibits switching because the antiparallel state is never stable. Instead, with increasing electron flow from the free into the fixed layer, the system goes from parallel through the in-plane precession region into an out-of-plane precession configuration.

At finite temperature, the transitions between the different configurations become statistical rather than deterministic. The transitions can be modeled either through direct integration of the Landau–Lifshitz–Gilbert equation including a term to generate thermal fluctuations or through modified barrier-crossing models. In this Chapter we briefly described transition-time distributions for cases in which the switching time is not impractically long. In some regimes, particularly for high fields, the effective barrier in both directions is sufficiently small that the systems can make thermally driven transitions in both directions. Then the magnetization exhibits two-level fluctuations with the associated low-frequency noise in the resistivity.

Full micromagnetic simulations, in which the magnetization is allowed to become nonuniform, show that the magnetization does indeed become nonuniform. These simulations show that nonuniform dynamics may explain some of the discrepancies between the predictions of the macrospin model and experiment. Unfortunately, the results of the simulations are sensitive to many unmeasured details of the sample, so that such simulations are simply suggestive rather than rigorously conclusive.

Taken together, the theoretical work described in this review suggests that the form of the torque originally envisioned by Slonczewski is capable of explaining the experimentally observed phenomena. However, the confirmation is not yet quantitative. There are only limited experimental observations that confirm the detailed angular dependence of the torque as predicted by a series of transport calculations. In fact, all of the micromagnetic simulations to date ignore these details. One reason it is difficult to quantitatively test the model is the variability from sample to sample. Samples are still difficult enough to fabricate that the results from nominally identical samples differ substantially in detail. Detailed testing of models will probably depend on experimental developments that allow sufficient characterization of the samples to constrain the theoretical models.

Additional theoretical developments may be necessary. For example, it may be necessary to integrate a fully three-dimensional transport calculation with the micromagnetics in order to describe the nonuniform dynamics. The standard form used to describe damping in all of the calculations is phenomenological. A more accurate form may be necessary to describe the precessing states in these systems. Finally, when the models are quantitatively tested, they may not be consistent with the measured results, necessitating and driving a deeper understanding of the behavior of spins in magnetic multilayers.[3]

Acknowledgements

J. M. gratefully acknowledges enlightening discussions with André Thiaville, Gonçalo Albuquerque and Benoit Montigny. He also thanks G. Albuquerque and Yoshinobu Nakatani for assistance with advanced graphics and acknowledges partial support from the European Community's Human Potential Programme contract HPRN-CT-2002-00318, ULTRASWITCH. M. S. thanks his collaborators Andrew Zangwill and Jiang Xiao with whom most of his understanding has been developed. He also thanks Gerrit Bauer, Arne Brataas, and Yaroslav Tserkovnyak for informative discussions on spin pumping. Both thank Robert McMichael, Stephen Russek, André Thiaville and Andrew Zangwill for critical readings of the manuscript.

A Appendix Drift-Diffusion Solution for a Single Interface

In the nonmagnet, the spin accumulation decays away from the interface with a length given by the spin accumulation length

$$s_x(z) = s(0)\exp\left(z/l_{\text{sf}}^{\text{NM}}\right). \tag{53}$$

According to (16), the gradient of this spin accumulation determines the spin current in the nonmagnet. In the ferromagnet, the spin accumulation also decays away from the interface

$$\delta s(z) = \delta s(0)\exp\left(-z/l_{\text{sf}}^{\text{FM}}\right). \tag{54}$$

However, the spin current also depends on the total current because the conductivity is spin dependent. Using (18) and adjusting the internal electric

[3] Completed on June 15, 2005.

field so that the total current is constant and there is no charge accumulation, the spin accumulation and the spin current satisfy

$$Q_{xz}(z) = P_\sigma \frac{\hbar}{2} j - D_z \nabla_z \delta s(z), \tag{55}$$

where $D_z = (D_\uparrow \sigma_\downarrow + D_\downarrow \sigma_\uparrow)/(\sigma_\uparrow + \sigma_\downarrow)$. See [92] for more details. The bulk equations give relations between the spin accumulation and the spin current at the interface

$$\begin{aligned} 0 &= -Q_{xz}(0) - \frac{D}{l_{\mathrm{sf}}^{\mathrm{NM}}} s_x(0), \\ 0 &= P_\sigma \frac{\hbar}{2} j - Q_{xz}(0) + \frac{D_z}{l_{\mathrm{sf}}^{\mathrm{FM}}} \delta s(0). \end{aligned} \tag{56}$$

Here, D is the diffusion constant in the non-magnetic material. In the absence of spin-flip scattering, the longitudinal spin current is conserved across the interface.

Equation (19) gives the boundary conditions for the spin currents and spin electrochemical potentials. The latter are related to the spin accumulation through the densities of states

$$\delta s = \frac{\hbar}{2} (\mathcal{N}_\uparrow \mu_\uparrow - \mathcal{N}_\downarrow \mu_\downarrow), \tag{57}$$

where $\mathcal{N}_\uparrow = (\partial n / \partial \mu)_\uparrow$. In the bulk, the internal electric field adjusts itself so there is no charge accumulation. However, we must include the dipole layer that develops across the interface when a current is flowing. We do this by allowing for charge accumulation in the ferromagnet just at the interface, but then ignore this charge accumulation for the behavior away from the interface. The interface dipole is specified through

$$\delta n(0) = \mathcal{N}_\uparrow \mu_\uparrow + \mathcal{N}_\downarrow \mu_\downarrow . \tag{58}$$

The boundary conditions in terms of densities and currents are then

$$\begin{aligned} \frac{\hbar}{2} j_z &= C_1 \frac{\hbar}{2} \delta n(0) + C_2 \delta s(0) - C_3 s_x(0), \\ Q_{xz}(0) &= C_2 \frac{\hbar}{2} \delta n(0) + C_1 \delta s(0) - C_4 s_x(0). \end{aligned} \tag{59}$$

The coefficients are

$$\begin{aligned} C_1 &= \frac{1}{2} \left(\frac{1}{R_\uparrow \mathcal{N}_\uparrow} + \frac{1}{R_\downarrow \mathcal{N}_\downarrow} \right), \quad C_3 = \frac{1}{2} \left(\frac{1}{R_\uparrow \mathcal{N}} - \frac{1}{R_\downarrow \mathcal{N}} \right), \\ C_2 &= \frac{1}{2} \left(\frac{1}{R_\uparrow \mathcal{N}_\uparrow} - \frac{1}{R_\downarrow \mathcal{N}_\downarrow} \right), \quad C_4 = \frac{1}{2} \left(\frac{1}{R_\uparrow \mathcal{N}} + \frac{1}{R_\downarrow \mathcal{N}} \right). \end{aligned} \tag{60}$$

$\mathcal{N}$ is the per-spin density of states in the nonmagnet. The two boundary conditions, (59) together with the boundary values of the bulk solutions, (56) give four equations in four unknowns that can be determined in terms of the specified current j_z. Then using the bulk solutions, (53) and (54) gives curves like those in Fig. 10.

The same procedure is used for collinear transport in more complicated structures. However, in each finite-thickness layer the spin accumulation is a superposition of exponentials decaying in opposite directions. The same procedure is used for noncollinear transport as well, but needs to be augmented by the noncollinear boundary conditions, (20). Values for the parameters that enter such calculations are given in [92].

B Appendix Precession and Spin Transfer in Phase Space: Melnikov's Method

Melnikov's method [155, 156] applies to weakly perturbed time-periodic Hamiltonian systems for which unperturbed trajectories may be derived from the energy landscape. The material below owes much to the work of *Valet*, *Serpico* et al. and *Bertotti* et al. [184–187] although the following results have been derived independently. We first establish the nature of unperturbed trajectories with the energy as a parameter in the simple case of a zero applied field and show that the switching current is well described as the current that satisfies (35) for a specific trajectory of the unperturbed system. We then extend results obtained for $h_x^{\mathrm{a}} = 0$ to the $h_x^{\mathrm{a}} \neq 0$ case.

Equation (38) together with the conservation of the modulus of $\boldsymbol{m}$ define stable magnetization orbits in the absence of damping, namely,

$$\begin{aligned} Q\left(1-m_x^2\right) - 2m_x h_x^{\mathrm{a}} + m_z^2 &= 2w, \\ m_x^2 + m_y^2 + m_z^2 &= 1. \end{aligned} \tag{61}$$

The velocity field reads:

$$\frac{\mathrm{d}\,\mathbf{m}}{\mathrm{d}\tau} = -\left[\begin{pmatrix} m_x \\ m_y \\ m_z \end{pmatrix} \times \begin{pmatrix} Qm_x + h_x^{\mathrm{a}} \\ 0 \\ -m_z \end{pmatrix}\right] = \begin{pmatrix} m_y m_z \\ -\left[(1+Q)\,m_x + h_x^{\mathrm{a}}\right] m_z \\ \left(Qm_x + h_x^{\mathrm{a}}\right) m_y \end{pmatrix}. \tag{62}$$

Magnetization directions where the torque vanishes define the fixed points for this problem. These are the parallel state, $m_x = 1$, labeled F_1; the antiparallel state, $m_x = -1$, labeled F_2; two points with $m_y = 0$ and $(1+Q)m_x + h_x^{\mathrm{a}} = 0$ (m_z can be positive or negative); and two points S_1 and S_2 with $m_z = 0$ and $Qm_x + h_x^{\mathrm{a}} = 0$ (m_y can be positive or negative). Depending on the applied field h_x^{a}, these points can be stable, unstable or saddle points. For zero applied field, F_1 and F_2 are both stable, S_1 and S_2 are saddle points, and the other two points are unstable.

Let us first look at unperturbed trajectories in the $h_x^{\mathrm{a}} = 0$ case. The energy relationship then simply reads

$$m_z^2 - Qm_x^2 = 2w - Q, \tag{63}$$

indicating that unperturbed trajectories are defined through the intersection of hyperbolic cylinders (63) with the unit sphere. The saddle points are now located at $m_y = \pm 1$ and all possible trajectories within the attraction basin of F_1 span the (0, $Q/2$) energy range. Variables can easily be separated, and, e.g., one finds that $\mathrm{d}m_y/\mathrm{d}\tau$ may be expressed as:

$$\begin{aligned} \frac{\mathrm{d}m_y}{\mathrm{d}\tau} &= \sqrt{2w\left(1+Q-2w\right)}\sqrt{1-\frac{1}{1+Q-2w}m_y^2}\sqrt{1-\frac{Q}{2w}m_y^2}\,, \\ m_z &> 0\,. \end{aligned} \tag{64}$$

Letting $u = \sqrt{Q/2w}\ m_y$, one obtains readily:

$$\frac{\mathrm{d}u}{\sqrt{1-u^2}\sqrt{1-k^2u^2}} = \sqrt{Q\left(1+Q-2w\right)}\mathrm{d}\tau, \tag{65}$$

with $k^2(w) = 2w/Q\,(1+Q-2w)$. Thus,

$$\begin{aligned} \tau - \tau_0 &= \frac{1}{\sqrt{Q\left(1+Q-2w\right)}}\int_0^{\mathrm{u}} \frac{\mathrm{d}u}{\sqrt{1-u^2}\sqrt{1-k^2u^2}}, \\ &= \frac{1}{\sqrt{Q\left(1+Q-2w\right)}}F\left[\psi, k\left(w\right)\right], \qquad , \end{aligned} \tag{66}$$

where $u = \sin(\psi)$; $k \in [0,1]$ and $F(\psi,\, k)$ is the elliptic integral of the first kind. The period for any unperturbed trajectory within the specified energy span amounts to four times the time necessary for m_y to move from 0 to $m_y^{\mathrm{Max}} = \sqrt{2w/Q}$, namely,

$$\begin{aligned} T\left(w\right) &= \frac{1}{\gamma_0 M_{\mathrm{S}}}\frac{4}{\sqrt{Q\left(1+Q-2w\right)}}\int_0^1 \frac{\mathrm{d}u}{\sqrt{1-u^2}\sqrt{1-k^2u^2}} \\ &= \frac{1}{\gamma_0 M_{\mathrm{S}}}\frac{4}{\sqrt{Q\left(1+Q-2w\right)}}K\left(k\right), \qquad , \end{aligned} \tag{67}$$

where $K(k)$ is the complete elliptic integral of the first kind. Periodic orbits obeying (65) and (66) (unperturbed system) in the vicinity of a fixed point F_1 are shown in Fig. 39 (top). The velocity along such trajectories as deduced from (66) proves highly nonuniform in agreement with the results of numerical simulations (see Fig. 25a). When w increases towards $Q/2$, orbits pile up against the heteroclinic orbits (they link two separate saddle points) forming a homoclinic cycle Γ^0 displayed as a red line in Fig. 39 (top) similar to the numerical results of Fig. 20. In addition, the period increases with w and diverges (the frequency drops to 0) when $w = Q/2$ as shown in Fig. 40.

Conditions for the application of Melnikov's theory, a general perturbation theory, are now met, namely, the existence of a homoclinic orbit to a saddle point for $\alpha = \chi = 0$, the existence of a continuous family of periodic orbits within Γ^0 and the fact that the period is a differentiable function of the

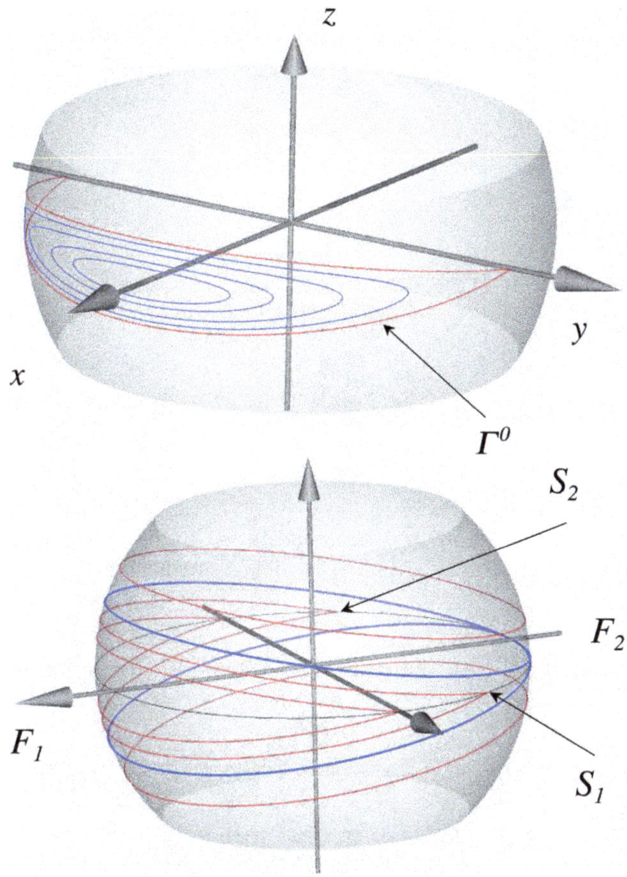

Fig. 39. *Top*: unperturbed trajectories for $w \in [0, Q/2]$; *Bottom*: homoclinic cycles ($h_x^{\mathrm{a}} \in [0, Q]$) and cycles ($h_x^{\mathrm{a}} > Q$) vs. reduced field $h_x^{\mathrm{a}} = 0, \frac{Q}{2}, Q, 3Q$. Equation (73) defines the corresponding reduced energies. S_1 and S_2 are the saddle points pertaining to the $h_x^{\mathrm{a}} = \frac{Q}{2}$ homoclinic cycle

energy of the Hamiltonian system. Then, to order $\mathscr{O}(\alpha)$, there exist closed orbits of the perturbed system, the proximity of which to unperturbed orbits is governed by Melnikov's function G. In order to estimate the latter, it proves convenient first to rewrite (39) according to

$$\begin{aligned}\frac{\mathrm{d}\theta}{\mathrm{d}\tau} &= -\frac{1}{\sin\theta}\frac{\mathrm{d}w}{\mathrm{d}\phi} - \alpha\sin\theta\frac{\mathrm{d}\phi}{\mathrm{d}\tau} + \chi p_\theta, \\ \sin\theta\frac{\mathrm{d}\phi}{\mathrm{d}\tau} &= \frac{\mathrm{d}w}{\mathrm{d}\theta} + \alpha\frac{\mathrm{d}\theta}{\mathrm{d}\tau} + \chi p_\phi.\end{aligned} \tag{68}$$

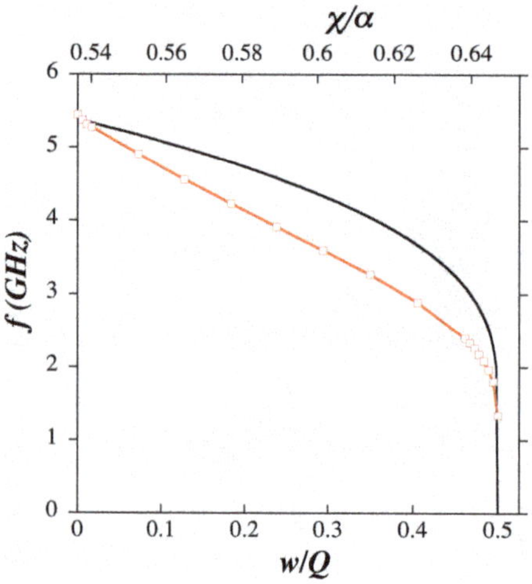

Fig. 40. Frequency for unperturbed orbits in the $w \in [0, Q/2]$ range in zero applied field. Also shown is the frequency of precessional states in the single-spin approximation as a function of χ/α (open symbols) as determined by numerical simulation

Since we are interested in Hamiltonian trajectories, (68) readily transforms into

$$\begin{aligned} \frac{\mathrm{d}\theta}{\mathrm{d}\tau} &= -\frac{1}{\sin\theta}\frac{\mathrm{d}w}{\mathrm{d}\phi} + \alpha\left(-\frac{\mathrm{d}w}{\mathrm{d}\theta} + \frac{\chi}{\alpha}p_\theta\right) &&= f_1 + \alpha g_1, \\ \sin\theta\frac{\mathrm{d}\phi}{\mathrm{d}\tau} &= \frac{\mathrm{d}w}{\mathrm{d}\theta} + \alpha(-\frac{1}{\sin\theta}\frac{\mathrm{d}w}{\mathrm{d}\phi} + \frac{\chi}{\alpha}p_\phi) &&= f_2 + \alpha g_2. \end{aligned} \tag{69}$$

Melnikov's function would, in the (θ, ϕ) plane, classically be built from (69) owing to $G\left(w, \frac{\chi}{\alpha}\right) = \int_0^{\mathrm{T}} (f_1 g_2 - f_2 g_1)\, \mathrm{d}\tau$, or

$$G\left(w, \frac{\chi}{\alpha}\right) = \int_0^{T(w)} \left\{ \left(\frac{\mathrm{d}\boldsymbol{m}}{\mathrm{d}\tau}\right)^2 - \frac{\chi}{\alpha}\left[\frac{\mathrm{d}\boldsymbol{m}}{\mathrm{d}\tau}\cdot(\boldsymbol{m}\times\boldsymbol{p})\right]\right\} \mathrm{d}\tau, \tag{70}$$

if a static variable could be associated with $\sin\theta\frac{\mathrm{d}\phi}{\mathrm{d}\tau}$. In the limit $\sin\theta \to 1$, a condition easily fulfilled due to the strong ellipticity of the Hamiltonian trajectories, $G\left(w, \frac{\chi}{\alpha}\right)$ is cleanly approximated by (70), a relation equivalent to (35), but now evaluated along an unperturbed trajectory with period $T(w)$.

It follows directly from the definitions above that the onset of precessional states should coincide with the current density that zeroes Melnikov's function for an infinitesimal unperturbed trajectory around F_1, whereas the critical current density for switching in zero applied field should coincide with the current that zeroes Melnikov's function for the unperturbed homoclinic cycle. Note that, for $w = Q/2$, one gets $m_z = \pm\sqrt{Q}m_x$, an equation defining planes in space, indicating that the homoclinic cycle is the locus of the intersection of the unit sphere with the above-mentioned planes.

Although we have completely characterized the Hamiltonian trajectories in the case $h_x^{\mathrm{a}} = 0$, it actually proves unnecessary. Indeed, as noted earlier [156, 185], the time integral in (70) may be written as a line integral owing to:

$$G\left(w, \frac{\chi}{\alpha}\right) = \oint_{\Gamma} \left[\left(\frac{\mathrm{d}\boldsymbol{m}}{\mathrm{d}\tau}\right) - \frac{\chi}{\alpha}\left(\boldsymbol{m} \times \boldsymbol{p}\right)\right] \cdot \mathrm{d}\boldsymbol{m}. \tag{71}$$

Extension to the $h_x^{\mathrm{a}} \neq 0$ case is straightforward. With increasing h_x^{a} the saddle points S_1 and S_2 gradually move (Fig. 39, bottom) in the equatorial plane in the direction of the fixed point F_2, according to:

$$\cos\left(\phi^{\max}\right) = -\frac{h_x^{\mathrm{a}}}{Q}. \tag{72}$$

Rewriting the energy expression yields $2w = Q\left(1 + m_x^2\right) + m_z^2$, or:

$$2w = Q + \frac{h_x^2}{Q}, \tag{73}$$

where $h_x \equiv h_x^{\mathrm{a}}$. Equation (73) defines the relation between the applied field and the energy level for the homoclinic bifurcation. When w reaches Q (or, equivalently, h_x reaches Q), $\phi^{\max}$ reaches π, so that the two saddle points S_1 and S_2 merge at $m_x = -1$ (Fig. 39, bottom). Similar arguments hold true for $h_x < 0$ ($-Q \leq h_x \leq 0$) with the two saddle points S_1 and S_2 merging at $m_x = +1$ for $h_x = -Q$.

In summary, in the whole interval $-Q \leq h_x \leq Q$, the homoclinic cycle as a function of parameter $w\left(h_x\right)$ is defined as the intersection of the unit sphere with planes defined by

$$m_z = \pm\left(\frac{h_x}{\sqrt{Q}} + \sqrt{Q}m_x\right), \tag{74}$$

and the critical current density for switching vs. h_x is defined as the current density that zeroes Melnikov's function (71) for the corresponding homoclinic trajectory. Proper parameterization of the homoclinic cycles allows for an analytical expression for the switching threshold (51) in which ξ_{S} is a measure of the opening of the homoclinic cycle in the equatorial plane.

Altogether, Melnikov's method both establishes a clear distinction between the critical currents for the onset of precessional states and switching, and expresses the critical current for switching vs. applied field as the ratio of line integrals that may be evaluated analytically for the present simple energy ansatz. It also emphasizes the fact that $\frac{\chi}{\alpha}$ is the fundamental parameter of this particular nonlinear dynamical system. Ultimately, within the framework of our simple energetics, Q stands as the only remaining free parameter in the reduced units of Fig. 22.

As h_x keeps growing beyond Q, the unique saddle point forces Hamiltonian trajectories to split into two subsets still satisfying relation (74), one above, the second under the equatorial plane (Fig. 39, bottom). For $h_x > Q$, however, Melnikov's theory may no longer be directly applied, although it is still possible to look for solutions satisfying (71). The critical current density χ/α deduced from this approach is proportional to $\cong \frac{h_x}{Q}$ in the limit $Q \ll 1$, a result easily reached through application of Stoke's theorem to unperturbed out-of-plane orbits. It turns out that, for $h_x > Q$, the critical current density necessary to the onset of out-of-plane precessional states appears to be bound, on the one hand, by the current density that zeroes the Melnikov function G for the last homocline cycle $w = Q$ (curve labeled χ_2^{Lower} in Fig. 22) and on the other hand by the current density that zeroes G for the unperturbed out-of-plane orbit (curve labeled χ_2^{Upper}).

References

[1] J. C. Slonczewski: J. Magn. Magn. Mater. **159**, L1 (1996)
[2] J. C. Slonczewski: J. Magn. Magn. Mater. **195**, L261 (1999)
[3] L. Berger: Phys. Rev. B **54**, 9353 (1996)
[4] L. Berger: J. Appl. Phys. **90**, 4632 (2001)
[5] J. Slonczewski: United States Patent #5 695 864, Dec. (1997)
[6] M. Tsoi, A. G. M. Jansen, J. Bass, W. C. Chiang, M. Seck, V. Tsoi, P. Wyder: Phys. Rev. Lett. **80**, 4281 (1998)
[7] E. B. Myers, D. C. Ralph, J. A. Katine, R. N. Louie, R. A. Buhrman: Science **285**, 867 (1999)
[8] J. E. Wegrowe, D. Kelly, Y. Jaccard, P. Guittienne, J. P. Ansermet: Europhys. Lett. **45**, 626 (1999)
[9] J. E. Wegrowe, D. Kelly, X. Hoffer, P. Guittienne, J. P. Ansermet: J. Appl. Phys. **89**, 7127 (2001)
[10] D. Kelly, J. E. Wegrowe, T. K. Truong, X. Hoffer, J. P. Ansermet: Phys. Rev. B **68**, 7947 (2003)
[11] J. Z. Sun: J. Magn. Magn. Mater. **202**, 157 (1999)
[12] J. A. Katine, F. J. Albert, R. A. Buhrman, E. B. Myers, D. C. Ralph: Phys. Rev. Lett. **84**, 3149 (2000)
[13] J. Grollier, V. Cros, A. Hamzić, J. M. George, H. Jaffrès, A. Fert, G. Faini, J. B. Youssef, H. L. Gall: Appl. Phys. Lett. **78**, 3663 (2001)
[14] J. Grollier, V. Cros, H. Jaffrès, A. Hamzic, J. M. George, G. Faini, J. B. Youssef, H. L. Gall, A. Fert: Phys. Rev. B **67**, 174402 (2003)
[15] A. Fert, V. Cros, J.-M. George, J. Grollier, H. Jaffrès, A. Hamzić, A. Vaurès, G. Faini, J. B. Youssef, H. L. Gall: J. Magn. Magn. Mater. **272–276**, 1706 (2004)
[16] F. B. Mancoff, S. E. Russek: IEEE Trans. Magn. **38**, 2853 (2002)
[17] W. H. Rippard, M. R. Pufall, T. J. Silva: Appl. Phys. Lett. **82**, 1260 (2003)
[18] M. R. Pufall, W. H. Rippard, T. J. Silva: Appl. Phys. Lett. **83**, 323 (2003)
[19] S. Urazhdin, N. O. Birge, W. P. Pratt, J. Bass: Phys. Rev. Lett. **91**, 146803 (2003)
[20] B. Özyilmaz, A. D. Kent, D. Monsma, J. Z. Sun, M. J. Rooks, R. H. Koch: Phys. Rev. Lett. **91**, 67203 (2003)

[21] Y. Ji, C. L. Chien, M. D. Stiles: Phys. Rev. Lett. **90**, 106601 (2003)
[22] K. J. Lee, Y. Liu, A. Deac, M. Li, J. W. Chang, S. Liao, K. Ju, O. Redon, J. P. Nozières, B. Dieny: J. Appl. Phys. **95**, 7423 (2004)
[23] J. Hayakawa, K. Ito, M. Fujimori, S. Heike, T. Hashizume, J. Steen, J. Brugger, H. Ohno: J. Appl. Phys. **96**, 3440 (2004)
[24] M. Covington, M. AlHajDarwish, Y. Ding, N. J. Gokemeijer, M. A. Seigler: Phys. Rev. B **69**, 184406 (2004)
[25] M. Covington, A. Rebei, G. J. Parker, M. A. Seigler: Appl. Phys. Lett. **84**, 3103 (2004)
[26] Y. Jiang, S. Abe, T. Ochiai, T. Nozaki, A. Hirohata, N. Tezuka, K. Inomata: Phys. Rev. Lett. **92**, 167204 (2004)
[27] Y. M. Huai, F. Albert, P. Nguyen, M. Pakala, T. Valet: Appl. Phys. Lett. **84**, 3118 (2004)
[28] R. Moriya, K. Hamaya, A. Oiwa, H. Munekata: Jpn. J. Appl. Phys. Part 2 – Lett. & Express Lett. **43**, L825 (2004)
[29] Y. W. Liu, Z. Z. Zhang, J. G. Wang, P. P. Freitas, J. L. Martins: J. Appl. Phys. **93**, 8385 (2003)
[30] G. D. Fuchs, N. C. Emley, I. N. Krivorotov, P. M. Braganca, E. M. Ryan, S. I. Kiselev, J. C. Sankey, D. C. Ralph, R. A. Buhrman, J. A. Katine: Appl. Phys. Lett. **85**, 1205 (2004)
[31] A. Deac, O. Redon, R. C. Sousa, B. Dieny, J. P. Nozières, Z. Zhang, Y. Liu, P. P. Freitas: J. Appl. Phys. **95**, 6792 (2004)
[32] M. Tsoi, A. G. M. Jansen, J. Bass, W. C. Chiang, V. Tsoi, P. Wyder: Nature **406**, 46 (2000)
[33] E. B. Myers, F. J. Albert, J. C. Sankey, E. Bonet, R. A. Buhrman, D. C. Ralph: Phys. Rev. Lett. **89**, 196801 (2002)
[34] A. Fabian, C. Terrier, S. S. Guisan, X. Hoffer, M. Dubey, L. Gravier, J. P. Ansermet, J. E. Wegrowe: Phys. Rev. Lett. **91**, 257209 (2003)
[35] M. R. Pufall, W. H. Rippard, S. Kaka, S. E. Russek, T. J. Silva, J. Katine, M. Carey: Phys. Rev. B **69**, 214409 (2004)
[36] S. I. Kiselev, J. C. Sankey, I. N. Krivorotov, N. C. Emley, R. J. Schoelkopf, R. A. Buhrman, D. C. Ralph: Nature **425**, 380 (2003)
[37] I. N. Krivorotov, N. C. Emley, A. G. F. Garcia, J. C. Sankey, S. I. Kiselev, D. C. Ralph, R. A. Buhrman: Phys. Rev. Lett. **93**, 166603 (2004)
[38] W. H. Rippard, M. R. Pufall, S. Kaka, S. E. Russek, T. J. Silva: Phys. Rev. Lett. **92**, 27201 (2004)
[39] W. H. Rippard, M. R. Pufall, S. Kaka, T. J. Silva, S. E. Russek: Phys. Rev. B **70**, 100406(R) (2004)
[40] S. I. Kiselev, J. C. Sankey, I. N. Krivorotov, N. C. Emley, M. Rinkoski, C. Perez, D. C. R. R. A. Buhrman: Phys. Rev. Lett. **93**, 36601 (2004)
[41] R. H. Koch, J. A. Katine, J. Z. Sun: Phys. Rev. Lett. **92**, 88302 (2004)
[42] F. J. Albert, N. C. Emley, E. B. Myers, D. C. Ralph, R. A. Buhrman: Phys. Rev. Lett. **89**, 226802 (2002)
[43] F. B. Mancoff, R. W. Dave, N. D. Rizzo, T. C. Eschrich, B. N. Engel, S. Tehrani: Appl. Phys. Lett. **83**, 1596 (2003)
[44] M. Tsoi, J. Z. Sun, S. S. P. Parkin: Phys. Rev. Lett. **93**, 36602 (2004)
[45] M. AlHajDarwish, H. Kurt, S. Urazhdin, A. Fert, R. Loloee, W. P. Pratt, Jr., J. Bass: Phys. Rev. Lett. **93**, 157203 (2004)

[46] M. AlHajDarwish, A. Fert, W. P. Pratt, J. Bass: J. Appl. Phys. **95**, 6771 (2004)
[47] T. Y. Chen, Y. Ji, C. L. Chien, M. D. Stiles: Phys. Rev. Lett. **93**, 26601 (2004)
[48] B. Özyilmaz, A. D. Kent, J. Z. Sun, M. J. Rooks, R. H. Koch: Phys. Rev. Lett. **93**, 176604 (2004)
[49] B. T. Jonker, A. T. Hanbicki, D. T. Pierce, M. D. Stiles: J. Magn. Magn. Mater. **277**, 24 (2004)
[50] J. Miltat, G. Albuquerque, A. Thiaville: in B. Hillebrands, K. Ounadjela (Eds.): *Spin Dynamics in Confined Magnetic Structures I* (Springer Verlag Berlin Heidelberg 2002) pp. 1–33
[51] X. Waintal, E. B. Myers, P. W. Brouwer, D. C. Ralph: Phys. Rev. B **62**, 12317 (2000)
[52] K. Xia, P. J. Kelly, G. E. W. Bauer, A. Brataas, I. Turek: Phys. Rev. B **65**, 220401 (2002)
[53] M. D. Stiles, A. Zangwill: Phys. Rev. **66**, 14407 (2002)
[54] P. Fulde: *Electron Correlations in Molecules and Solids* (Springer-Verlag, New York 1995)
[55] K. Held, D. Vollhardt: Eur. Phys. J. B **5**, 473 (1998)
[56] I. Yang, S. Y. Savrasov, G. Kotliar: Phys. Rev. Lett. **87**, 216405 (2001)
[57] W. Kohn, L. J. Sham: Phys. Rev. A **140**, 1133 (1965)
[58] U. von Barth, L. Hedin: Phys. Rev. B **4**, 1629 (1972)
[59] O. Gunnarsson, B. I. Lundqvist: Phys. Rev. B **13**, 4274 (1976)
[60] R. O. Jones, O. Gunnarsson: Rev. Mod. Phys. **61**, 689 (1989)
[61] V. L. Moruzzi, J. F. Janak, A. R. Williams: *Calculated Electronic Properties of Metals* (Pergamon, New York 1978)
[62] K. M. Schep, J. B. A. N. van Hoof, P. J. Kelly, G. E. W. Bauer, J. E. Inglesfield: J. Magn. Magn. Mater. **177**, 1166 (1998)
[63] M. D. Stiles, D. R. Penn: Phys. Rev. B **61**, 3200 (2000)
[64] K. Xia, P. J. Kelly, G. E. W. Bauer, I. Turek, J. Kudrnovsky, V. Drchal: Phys. Rev. B **63**, 64407 (2001)
[65] G. E. W. Bauer, K. M. Schep, K. Xia, P. J. Kelly: J. Phys. D **35**, 2410 (2002)
[66] M. D. Stiles: J. Appl. Phys. **79**, 5805 (1996)
[67] L. Berger: IEEE Trans. Magn. **31**, 3871 (1995)
[68] C. Heide, P. E. Zilberman, R. J. Elliott: Phys. Rev. B **63**, 64424 (2001)
[69] C. Heide: Phys. Rev. Lett. **87**, 197201 (2001)
[70] C. Heide: Phys. Rev. B **65**, 54401 (2001)
[71] S. Zhang, P. M. Levy, A. Fert: Phys. Rev. Lett. **88**, 236601 (2002)
[72] A. Shpiro, P. Levy, S. Zhang: Phys. Rev. B **67**, 104430 (2003)
[73] A. Rebei, W. N. G. Hitchon, R. W. Chantrell: Phys. Lett. A **346**, 371 (2005)
[74] S. Urazhdin: Phys. Rev. B **69**, 134430 (2004)
[75] L. Berger: J. Appl. Phys. **81**, 4880 (1997)
[76] L. Berger: Phys. Rev. B **59**, 11465 (1999)
[77] L. Berger: J. Appl. Phys. **91**, 6795 (2002)
[78] M. V. Tsoi, V. S. Tsoi: JETP Lett. **73**, 98 (2001)
[79] M. Tsoi, J. Z. Sun, M. J. Rooks, R. H. Koch, S. S. P. Parkin: Phys. Rev. B **69**, 100406 (2004)
[80] Y. Tserkovnyak, A. Brataas, G. E. W. Bauer: Phys. Rev. B **67**, 140404 (2003)
[81] M. D. Stiles: J. Magn. Magn. Mater. **200**, 322 (1999)

[82] P. Bruno: J. Phys. – Condens. Matter **11**, 9403 (1999)
[83] D. E. Bürgler, P. Grünberg, S. O. Demokritov, M. T. Johnson: in K. H. J. Buschow (Ed.): *Handbook of Magnetic Materials*, vol. 13 (2001)
[84] M. D. Stiles: in D. Mills and J. A. C. Bland (Eds.): *Contemporary Concepts of Condensed Matter Science I*, (Elsevier New York 2006) pp. 51–77
[85] S. Yuasa, T. Nagahama, Y. Suzuki: Science **297**, 234 (2002)
[86] H. Itoh, J. Inoue, A. Umerski, J. Mathon: Phys. Rev. B **68**, 174421 (2003)
[87] G. E. W. Bauer, Y. V. Nazarov, D. Huertas-Hernando, A. Brataas, K. Xia, P. J. Kelly: Mater. Sci. Eng. B-Solid State Mater. Adv. Technol. **84**, 31 (2001)
[88] D. M. Edwards, F. Federici, J. Mathon, A. Umerski: Phys. Rev. B **71**, 54407 (2005)
[89] M. D. Stiles, A. Zangwill: J. Appl. Phys. **91**, 6812 (2002)
[90] L. Berger: IEEE Trans. Magn. **34**, 3837 (1998)
[91] L. Berger: J. Appl. Phys. **89**, 5521 (2001)
[92] M. D. Stiles, J. Xiao, A. Zangwill: Phys. Rev. B **69**, 54408 (2004)
[93] T. Valet, A. Fert: Phys. Rev. B **48**, 7099 (1993)
[94] A. Brataas, Y. V. Nazarov, G. E. W. Bauer: Phys. Rev. Lett. **84**, 2481 (2000)
[95] A. Brataas, Y. V. Nazarov, G. E. W. Bauer: Europhys. Phys. J. B **22**, 99 (2001)
[96] J. C. Slonczewski: J. Magn. Magn. Mater. **247**, 324 (2002)
[97] J. Xiao, A. Zangwill, M. D. Stiles: Phys. Rev. B **70**, 172405 (2004)
[98] A. Brataas, G. E. W. Bauer, P. J. Kelly: Physics Reports **427**, p. 157–255 (2006)
[99] J. Zhang, P. M. Levy, S. Zhang, V. Antropov: Phys. Rev. Lett. **93**, 256602 (2004)
[100] P. M. Levy, J. Zhang: Phys. Rev. B **70**, 132406 (2004)
[101] J. Zhang, P. M. Levy: Phys. Rev. B **71**, 184417 (2005)
[102] J. Zhang, P. M. Levy: Phys. Rev. B **71**, 184426 (2005)
[103] Y. B. Bazaliy, B. A. Jones, S. C. Zhang: Phys. Rev. B **57**, R3213 (1998)
[104] J. P. Ansermet: IEEE Trans. Magn. **40**, 358 (2004)
[105] J. Grollier, P. Boulenc, V. Cros, A. Hamzić, A. Vaurès, A. Fert, G. Faini: J. Appl. Phys. **95**, 6777 (2004)
[106] A. Thiaville, Y. Nakatani, J. Miltat, N. Vernier: J. Appl. Phys. **95**, 7049 (2004)
[107] Z. Li, S. Zhang: Phys. Rev. Lett. **92**, 7049 (2004)
[108] A. Yamaguchi, T. Ono, S. Nasu, K. Miyake, K. Mibu, T. Shinjo: Phys. Rev. Lett. **92**, 7049 (2004)
[109] G. Tatara, H. Kohno: Phys. Rev. Lett. **92**, 7049 (2004)
[110] X. Waintal, M. Viret: Europhys. Lett. **65**, 427 (2004)
[111] M. Tsoi, R. E. Fontana, S. S. P. Parkin: Phys. Rev. Lett. **83**, 2617 (2003)
[112] L. Berger: J. Appl. Phys. **49**, 2156 (1978)
[113] L. Berger: J. Appl. Phys. **50**, 2137 (1979)
[114] Y. Tserkovnyak, A. Brataas, G. E. W. Bauer: Phys. Rev. Lett. **88**, 117601 (2002)
[115] A. Brataas, Y. Tserkovnyak, G. E. W. Bauer, B. I. Halperin: Phys. Rev. B **66**, 60404 (2002)
[116] Y. Tserkovnyak, A. Brataas, G. E. W. Bauer, B. I. Halperin: Rev. Mod. Phys. **77**, 1375 (2005)
[117] D. L. Mills: Phys. Rev. B **68**, 14419 (2003)

[118] E. Šimánek, B. Heinrich: Phys. Rev. B **67**, 144418 (2003)
[119] E. Šimánek: Phys. Rev. B **68**, 224403 (2003)
[120] R. Urban, G. Woltersdorf, B. Heinrich: Phys. Rev. Lett. **87**, 217204 (2001)
[121] S. Mizukami, Y. Ando, T. Miyazaki: J. Magn. Magn. Mater. **239**, 42 (2002)
[122] S. Ingvarsson, L. Ritchie, X. Y. Liu, G. Xiao, J. C. Slonczewski, P. L. Trouilloud, R. H. Koch: Phys. Rev. B **66**, 214416 (2002)
[123] P. Lubitz, S. F. Cheng, F. J. Rachford: J. Appl. Phys. **93**, 8283 (2003)
[124] B. Heinrich, R. Urban, G. Woltersdorf, E. Šimánek: J. Appl. Phys. **93**, 7545 (2003)
[125] K. Lenz, T. Tolinski, J. Lindner, E. Kosubek, K. Baberschke: Phys. Rev. B **69**, 144422 (2004)
[126] A. G. Aronov: JETP Lett. **24**, 32 (1977)
[127] M. Johnson, R. H. Silsbee: Phys. Rev. B **35**, 4959 (1987)
[128] P. C. van Son, H. van Kempen, P. Wyder: Phys. Rev. Lett. **58**, 2271 (1987)
[129] I. Zutic, J. Fabian, S. D. Sarma: Rev. Mod. Phys. **76**, 323 (2004)
[130] A. Vedyayev, N. Ryzhanova, B. Dieny, P. Dauguet, P. Gandit, J. Chaussy: Phys. Rev. B **55**, 3728 (1997)
[131] D. Huertas-Hernando, G. E. W. Bauer, Y. V. Nazarov: J. Magn. Magn. Mater. **240**, 174 (2002)
[132] G. E. W. Bauer, Y. Tserkovnyak, D. Huertas-Hernando, A. Brataas: Phys. Rev. B **67**, 944 (2003)
[133] J. Manschot, A. Brataas, G. E. W. Bauer: Phys. Rev. B **69**, 92407 (2004)
[134] J. Manschot, A. Brataas, G. E. W. Bauer: Appl. Phys. Lett. **85**, 3250 (2004)
[135] A. A. Kovalev, A. Brataas, G. E. W. Bauer: Phys. Rev. B **66**, 224424 (2002)
[136] A. Fert, V. Cros, J.-M. George, J. Grollier, H. Jaffrès, A. Hamzić, A. Vaurès, G. Faini, J. B. Youseff, H. L. Gall: J. Magn. Magn. Mater. **69**, 184406 (2004)
[137] N. Smith, J. A. Katine, J. R. Childress, M. J. Carey: IEEE Trans. Magn. **42**, 144 (2006)
[138] J. W. P. Pratt: (private communication)
[139] S. Urazhdin, R. Loloee, W. P. Pratt: Phys. Rev. B **71**, 100401 (2005)
[140] L. Berger: J. Magn. Magn. Mater. **278**, 185 (2004)
[141] J. Hamrle, T. Kimura, T. Yang, Y. Otani: Phys. Rev. B **71**, 94434 (2005)
[142] M. L. Polianski, P. W. Brouwer: Phys. Rev. Lett. **92**, 26602 (2004)
[143] F. J. Albert, J. A. Katine, R. A. Buhrman, D. C. Ralph: Appl. Phys. Lett. **77**, 3809 (2000)
[144] A. Hubert, R. Schäfer: *Magnetic Domains* (Springer-Verlag, Berlin 1998)
[145] J. Z. Sun: Phys. Rev. B **62**, 570 (2000)
[146] Y. B. Bazaliy, B. A. Jones, S. C. Zhang: J. Appl. Phys. **89**, 6793 (2001)
[147] Y. B. Bazaliy, B. A. Jones: Physica B **329**, 1290 (2003)
[148] Y. B. Bazaliy, B. A. Jones, S. C. Zhang: Phys. Rev. B **69**, 94421 (2004)
[149] H. W. Xi, Y. M. Shi: J. Appl. Phys. **96**, 1585 (2004)
[150] H. W. Xi, K. Z. Gao, Y. M. Shi: Appl. Phys. Lett. **84**, 4977 (2004)
[151] Y. Acremann, C. H. Back, M. Buess, D. Pescia, V. Pokrovsky: Appl. Phys. Lett. **79**, 2228 (2001)
[152] H. W. Schumacher, C. Chappert, P. Crozat, R. C. Souza, P. P. Freitas, J. Miltat, J. Fassbender, B. Hillebrands: Phys. Rev. Lett. **90**, 17201 (2003)
[153] H. W. Schumacher, C. Chappert, R. C. Souza, P. P. Freitas, J. Miltat: Phys. Rev. Lett. **90**, 17204 (2003)
[154] T. Devolder: this volume

[155] V. K. Melnikov: On the stability of the center for time periodic perturbations, Trans. Moscow Math. Soc. **12**, 1–57 (1963)
[156] J. Guckenheimer, P. Holmes: *Nonlinear Oscillations, Dynamical Systems and Bifurcations of Vector Fields* (Springer-Verlag, New York 1983)
[157] T. Devolder, P. Crozat, C. Chapppert, J. Miltat, A. A. Tulapurkar, Y. Suzuki, T. K. Yagami: Phys. Rev. B **71**, 184401 (2005)
[158] Z. Li, S. Zhang: Phys. Rev. B **69**, 134416 (2004)
[159] D. M. Apalkov, P. B. Visscher: Phys. Rev. B **72** 180405(R) (2005)
[160] D. M. Apalkov, P. B. Visscher: J. Magn. Magn. Mater. **286**, 370 (2005)
[161] J. W. F. Brown: Phys. Rev. **130**, 1677 (1963)
[162] S. E. Russek, S. Kaka, W. H. Rippard, M. R. Pufall, T. J. Silva: Phys. Rev. B **71**, 104425 (2005)
[163] J. Xiao, A. Zangwill, M. D. Stiles: Phys. Rev. B **72**, 14446 (2005)
[164] C. W. Gardiner: *Handbook of Stochastic Methods* (Springer-Verlag, Berlin Heidelberg New York 2002)
[165] W. H. Press, S. A. Teukolsky, W. T. Wetterling, B. P. Flannery: *Numerical Recipes in Fortran 77: The Art of Scientific Computing* (Cambridge University Press 2001)
[166] S. I. Kiselev, J. C. Sankey, I. N. Krivorotov, N. C. Emley, A. G. F. Garcia, R. A. Buhrman, D. C. Ralph: Phys. Rev. B **72**, 64430 (2005)
[167] J. Miltat, G. Albuquerque, A. Thiaville, C. Vouille: J. Appl. Phys. **89**, 6982 (2001)
[168] Z. Li, S. Zhang: Phys. Rev. B **68**, 24404 (2003)
[169] X. C. Zhu, J. G. Zhu, R. M. White: J. Appl. Phys. **95**, 6630 (2004)
[170] X. C. Zhu, J. G. Zhu: J. Appl. Phys. **95**, 7318 (2004)
[171] J. G. Zhu, X. C. Zhu: IEEE Trans. Magn. **40**, 182 (2004)
[172] J. G. Zhu, N. Kim, Y. C. Zhou, Y. F. Zheng, J. W. Chang, K. C. Ju, X. C. Zhu, R. M. White: IEEE Trans. Magn. **40**, 2323 (2004)
[173] K.-J. Lee, A. Deac, O. Redon, J.-P. Nozières, B. Dieny: Nature Mater. **3**, 877 (2004)
[174] D. Berkov, N. Gorn: Phys. Rev. B **71**, 52403 (2005)
[175] B. Montigny, J. Miltat: J. Appl. Phys. **97**, 10C708 (2005)
[176] D. Berkov, N. Gorn: Phys. Rev. B **72**, 94401 (2005)
[177] R. D. McMichael, M. D. Stiles: J. Appl. Phys. **97**, 10J901 (2005) and references cited therein
[178] A. A. Tulapurkar, T. Devolder, K. Yagami, P. Crozat, C. Chappert, A. Fukushima, Y. Suzuki: Appl. Phys. Lett. **85**, 5358 (2004)
[179] T. Devolder, A. A. Tulapurkar, K. Yagami, P. Crozat, C. Chappert, A. Fukushima, Y. Suzuki: J. Magn. Magn. Mater. **286**, 77 (2005)
[180] J. Miltat, A. Thiaville: Science **298**, 555 (2002)
[181] A. Thiaville, J. M. Garciá, R. Dittrich, J. Miltat, T. Schrefl: Phys. Rev. B **67**, 94410 (2003)
[182] M. Labrune, J. Miltat: J. Appl. Phys. **75**, 2156 (1994)
[183] I. N. Krivorotov, N. C. Emley, J. C. Sankey, S. I. Kiselev, D. C. Ralph, R. A. Buhrman: Science **307**, 228 (2005)
[184] T. Valet: (private communication – 2004)
[185] C. Serpico, M. d'Aquino, G. Bertotti, I. D. Mayergoyz: J. Appl. Phys. **95**, 7052 (2004)

[186] C. Serpico, M. d'Aquino, G. Bertotti, I. D. Mayergoyz: J. Magn. Magn. Mater. **290–291**, 502 (2004)
[187] G. Bertotti, C. Serpico, I. D. Mayergoyz, A. Magni, M. d'Aquino, R. Bonin: Phys. Rev. Lett. **94**, 127206 (2005)

Index

Spin- and Energy Relaxation of Hot Electrons at GaAs Surfaces

Torsten Ohms, Kevin Hiebbner, Hans Christian Schneider, and Martin Aeschlimann

Department of Physics, Kaiserslautern University of Technology
67663 Kaiserslautern, Germany
ma@physik.uni-kl.de

Abstract. The mechanisms for spin relaxation in semiconductors are reviewed, and the mechanism prevalent in p-doped semiconductors, namely spin relaxation due to the electron–hole exchange interaction, or Bir–Aronov–Pikus mechanism, is presented in some depth. It is shown that the solution of Boltzmann-type kinetic equations allows one to obtain quantitative results for spin relaxation in semiconductors that go beyond the original Bir–Aronov–Pikus relaxation-rate approximation. Experimental results using surface-sensitive two-photon photoemission techniques show that the relaxation time of the electron spin polarization in p-doped GaAs at a semiconductor/metal surface is several times longer than the corresponding bulk spin-relaxation times. A theoretical explanation of these results in terms of the reduced density of holes in the band-bending region at the surface is presented.

1 Introduction

Semiconductor technology has relied on the manipulation of electronic charges since the invention of the transistor. The spin degree of freedom of the carriers has been mostly ignored in traditional electronics. In the last few years there has been a push towards the control of the spin dynamics of charged carriers *independently* of their charge. The exploitation of the spin degree of freedom in electronic [1] and optoelectronic devices [2] and the basic physics associated with the control of spins have been dubbed "spintronics."

Outside of semiconductor electronics, the control of spins has gained tremendous importance in basic and applied physics in connection with magnetic recording techniques [3]. Metallic materials, which are routinely used for these purposes, have some disadvantages compared to semiconductors. For instance, their carrier densities are not freely controllable and they do not possess bandgaps, which makes their use in optoelectronic devices difficult. It would therefore be advantageous to combine some features of magnetic materials with the versatility of semiconductors. However, the decay of the spin polarization in semiconductors limits all information processing and storage capabilities in semiconductors, which is a big difference from conventional electronics that is based on the conserved electronic charge. Spin relaxation and methods to control spin relaxation thus constitute central problems of spintronics. A promising result on the physical limit of spin manipulation by

B. Hillebrands, A. Thiaville (Eds.): Spin Dynamics in Confined Magnetic Structures III, Topics Appl. Physics **101**, 309–340 (2006)

relaxation phenomena is the measurement of a spin lifetime of about 1 ns in n-doped GaAs at low temperatures [4], but the reason for this remarkable result remains an area of active research.

One important requirement for spin-electronic devices, such as the spin transistor, is the efficient injection of spin-polarized carriers into semiconductors. The straightforward implementation by using ferromagnetic metal contacts has proved to be difficult mainly due to the different conductivities of ferromagnetic metals and semiconductors [5], but tunnel injection into nonmagnetic semiconductors was realized by using incoherent electrical tunneling into the target semiconductor from ferromagnetic metals [6]: In tunneling contacts the injected current is proportional to the density of states of the respective material. Different densities of states for spin-up and spin-down electrons in the ferromagnet therefore give rise to a spin-polarized tunneling current leading to injected spin polarizations of up to 30 %. An alternative to the spin injection via interfaces between ferromagnetic metals and semiconductors is the use of magnetic semiconductors instead of ferromagnets. For instance, doping II–VI and III–V semiconductors with Mn can lead to paramagnetic, e.g., BeMnZnSe, and even ferromagnetic compounds such as GaMnAs [7]. Aside from the intrinsic importance of these structures, they can be used for spin injection into nonmagnetic semiconductors [3, 8]. However, these techniques suffer from characteristic drawbacks. For instance, paramagnetic semiconductors based on II–VI compounds, require high magnetic fields. The ferromagnetic GaMnAs, on the other hand, has a relatively low Curie temperature of 150 K and works with spin-aligned holes, for which momentum-space scattering in bulk (and quantum-well) semiconductors severely limits the lifetimes. Since these problems with electrical spin injection persist, optical techniques are important not only for probing the spin alignment of the carrier system, but also for the creation of a well-controlled spin alignment at arbitrary temperatures. The experimental techniques described in the following rely on the creation of spin-polarized electrons by optical fields and the emission of carriers from the material using laser pulses, and have the advantage that the spin polarization can be obtained directly from the carriers, which are freed from the material.

In this contribution, we briefly review some of the mechanisms that lead to the relaxation of spin polarization in semiconductors. We then focus on p-doped GaAs and its prevalent spin-relaxation mechanism, the electron–hole exchange interaction, to show that the relaxation of the spin polarization is determined by the full dynamics of the electronic distribution functions and cannot be described using a simple relaxation rate. Instead, we present theoretical results for the spin-polarization dynamics in p-doped semiconductors using a Boltzmann-equation approach. In the second part of this contribution we show how the experimental technique of two-photon photoemission (2PPE) can be used to extract information about spin dynamics at semiconductor surfaces and how it can be applied to semiconductor/metal interfaces like Schottky contacts. Experimental results are presented and theoretically explained.

2 Review of Spin-Flip Processes in GaAs

In this section we first give an overview of the electron and hole states accessible by optical excitation of GaAs, and the possibility to create spin-polarized electrons by optical excitation. We then describe the origin of the three most important processes that lead to spin relaxation. Since they are all connected to the band structure of semiconductors, we first discuss some generalities on the semiconductor band structure, and how it can be approximately calculated close to the center of the Brillouin zone.

2.1 Optical Orientation of Photoexcited Carriers

Optical orientation refers to the creation of a nonequilibrium spin alignment, or preferential spin orientation, by excitation with polarized electromagnetic fields. As mentioned above, in semiconductors this is the most important and versatile process to create spin alignment, or as it is commonly called, spin polarization. In the following, we investigate the microscopic *spin polarization*

$$P = \frac{n_\uparrow - n_\downarrow}{n_\uparrow + n_\downarrow} \tag{1}$$

defined in terms of the microscopic, time- and momentum- (or kinetic energy) dependent carrier densities in the spin-up and spin-down electron bands. The dynamics of the microscopic spin polarization P determines the relaxation of the macroscopic spin polarization, which is often described by a phenomenological T_1 time. We therefore refer to the *decay of the electron spin polarization simply as spin relaxation.* The experimental and theoretical results on spin relaxation presented in the following are obtained without external magnetic fields, and should therefore not be confused with the dephasing of *coherent* spin dynamics under the influence of magnetic fields [9, 10], whose macroscopic counterpart is often described by a T_2 time, see also Sect. 3.1. Spin-polarized carriers can be detected in experiments and exploited in electro-optical devices due to the polarization of the emitted light when spin-polarized electrons and holes recombine and emit photons. The band structure of GaAs near the fundamental band edge [11] that will be the basis of the following discussion consists of:

- electrons with total spin $S = \frac{1}{2}$ and spin projection quantum number $s = +\frac{1}{2} \equiv \uparrow$ and $s = -\frac{1}{2} \equiv \downarrow$,
- heavy holes with total angular momentum $J = \frac{3}{2}$ and projection quantum number $j = \pm\frac{3}{2}$,

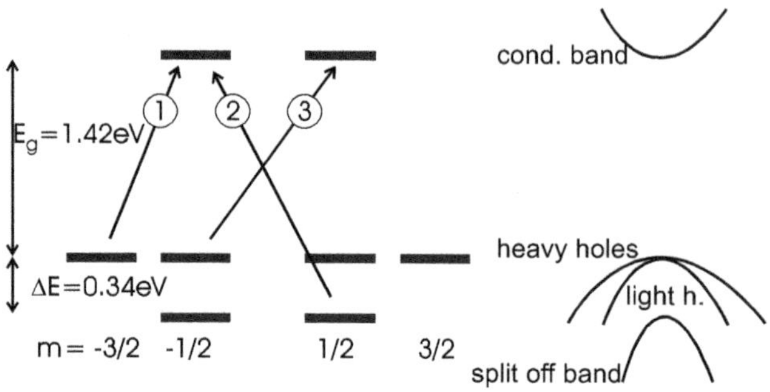

Fig. 1. Transitions between electron and hole states induced by circularly polarized light in GaAs at the zone center (*left*) and schematic band structure (*right*). The angular momentum projection quantum numbers m shown in the figure apply to all bands above the respective ms. The relative strengths of the transitions are indicated. For excitation photon energies of less than 1.76 eV, only heavy- and light-hole transitions can be driven

- light holes with total angular momentum $J = \frac{3}{2}$ and projection quantum number $j = \pm\frac{1}{2}$,
- holes in the "split-off" band with total angular momentum $J = \frac{1}{2}$ and projection quantum number $j = \pm\frac{1}{2}$,

as shown schematically in Fig. 1. More specifically, one has for the spherically symmetric conduction-band wavefunctions at $k = 0$ in real-space representation

$$\langle \boldsymbol{x}, \sigma | S, s \rangle = u_c(r)\chi_s(\sigma)\,, \qquad \sigma = \uparrow, \downarrow\,, \tag{2}$$

where $r = |\boldsymbol{x}|$ is the modulus of $\boldsymbol{x}$. The light- and heavy-hole valence band wavefunctions at $k = 0$ are given by

$$\langle \boldsymbol{x}, \sigma | J = \frac{3}{2}, j = \frac{3}{2} \rangle = Y^1_{+1}(\hat{\boldsymbol{x}})\chi_\uparrow(\sigma)u_\mathrm{v}(r)\,, \tag{3}$$

$$\langle \boldsymbol{x}, \sigma | J = \frac{3}{2}, j = \frac{1}{2} \rangle = \frac{1}{\sqrt{3}}\left[\sqrt{2}Y^1_0(\hat{\boldsymbol{x}})\chi_\uparrow(\sigma) + Y^1_{+1}(\hat{\boldsymbol{x}})\chi_\downarrow(\sigma)\right] u_\mathrm{v}(r)\,, \tag{4}$$

$$\langle \boldsymbol{x}, \sigma | J = \frac{3}{2}, j = -\frac{1}{2} \rangle = \frac{1}{\sqrt{3}}\left[\sqrt{2}Y^1_0(\hat{\boldsymbol{x}})\chi_\downarrow(\sigma) + Y^1_{-1}(\hat{\boldsymbol{x}})\chi_\uparrow(\sigma)\right] u_\mathrm{v}(r)\,, \tag{5}$$

$$\langle \boldsymbol{x}, \sigma | J = \frac{3}{2}, j = -\frac{3}{2} \rangle = Y^1_{-1}(\hat{\boldsymbol{x}})\chi_\downarrow(\sigma)u_\mathrm{v}(r)\,, \tag{6}$$

and for the split-off hole wavefunctions one has

$$\langle \boldsymbol{x}, \sigma | J = \frac{1}{2}, j = \frac{1}{2} \rangle = \frac{1}{\sqrt{3}}\left[\sqrt{2}Y^1_1(\hat{\boldsymbol{x}})\chi_\downarrow(\sigma) - Y^1_0(\hat{\boldsymbol{x}})\chi_\uparrow(\sigma)\right] u_\mathrm{v}(r)\,, \tag{7}$$

$$\langle \boldsymbol{x}, \sigma | J = \frac{1}{2}, j = -\frac{1}{2} \rangle = \frac{1}{\sqrt{3}}\left[Y^1_0(\hat{\boldsymbol{x}})\chi_\downarrow(\sigma) - \sqrt{2}Y^1_{-1}(\hat{\boldsymbol{x}})\chi_\uparrow(\sigma)\right] u_\mathrm{v}(r)\,. \tag{8}$$

Here, the Y^l_m are the spherical harmonics that depend on the direction $\hat{\boldsymbol{x}}$ of the vector $\boldsymbol{x}$, $\chi_{\rm s}(\sigma) = \delta_{\sigma,s}$ are spinors, and $u_{\rm c}(r)$, $u_{\rm v}(r)$ contain the radial dependences, which can be obtained from a band-structure calculation. From (2) and (3)–(8) the vectorial dipole matrix elements $\boldsymbol{d} = e\boldsymbol{x}$ for electron–hole transitions can be computed. For instance, choosing the angular momentum quantization axis $\hat{z}$ perpendicular to the crystal surface and assuming excitation with polarized light propagating in the z-direction, one has to evaluate the matrix elements

$$\langle S, s|\hat{\sigma}_\pm|J, j\rangle \tag{9}$$

between electron and hole wavefunctions and the polarization vector $\hat{\boldsymbol{\sigma}}_\pm = (\hat{\boldsymbol{x}} \pm \mathrm{i}\hat{\boldsymbol{y}})/\sqrt{2}$. Because $\hat{\sigma}_\pm \propto Y^1_{\pm 1}$ this is relatively easy to do. For example, one obtains for the ratio between electron to heavy-hole and electron to light-hole transitions

$$\left|\frac{\langle S=\frac{1}{2},\downarrow|\hat{\sigma}_+|J=\frac{3}{2}, j=-\frac{3}{2}\rangle}{\langle S=\frac{1}{2},\downarrow|\hat{\sigma}_+|J=\frac{3}{2}, j=-\frac{1}{2}\rangle}\right|^2 = 3\,. \tag{10}$$

These relative magnitudes are indicated in Fig. 1. It is apparent that selective excitation of light- and heavy-hole transitions will result in an electronic spin polarization of $P = 0.5$. In the following, we always assume that an electronic spin polarization is created in this way, and that no split-off holes are populated in the optical excitation process. We will therefore use the abbreviations

$$|j\rangle \equiv |J = \frac{3}{2}, j\rangle, |s\rangle \equiv |S = \frac{1}{2}, s\rangle\,. \tag{11}$$

Once a spin polarization is created by optical fields, several mechanisms are known to destroy the spin polarization. These spin-relaxation mechanisms have been investigated in detail for bulk and, more recently, for quantum-well semiconductors. In the following, we will briefly review the Elliott–Yafet (EY), Dyakonov–Perel (DP), and Bir–Aronov–Pikus (BAP) mechanisms, after some general properties of band-structure calculations near the fundamental band edge have been discussed.

2.2 Band-Structure Properties

The band structure of III–V compounds is generally simple at the zone center ($k = 0$). For GaAs, the zone-center wavefunctions of the bands at the fundamental bandgap are given by (2)–(6). For finite k the energy eigenstates and dipole matrix elements for the bands participating in optical transitions can be determined by perturbation theory starting from the $k = 0$ states. This is usually done by dividing the bands into those energetically close to the transitions of interest and remote bands. In the case of the interaction with optical fields, the important bands are the heavy hole, light hole, and electron states. Following [12–14], our starting point is the general form of single-particle carrier states in a periodic potential

$$\psi_{n,\boldsymbol{k}}(\boldsymbol{x},\sigma) = \frac{1}{\sqrt{L^3}} \mathrm{e}^{\mathrm{i}\boldsymbol{k}\cdot\boldsymbol{x}} u_{n,\boldsymbol{k}}(\boldsymbol{x},\sigma)\,, \tag{12}$$

with the crystal volume L^3 and the lattice-periodic Bloch function $u_{n,\boldsymbol{k}} = |n,\boldsymbol{k}\rangle$. The indices n run over all bands in the semiconductor. A band-structure calculation in general attempts an approximate solution of

$$H_{\text{carrier}}\psi_{n,\boldsymbol{k}} = \left(\frac{1}{2m_0}p^2 + V_{\text{lattice}} + H_{\text{SO}}\right)\psi_{n,\boldsymbol{k}} = \epsilon_{n,\boldsymbol{k}}\psi_{n,\boldsymbol{k}}\,, \tag{13}$$

where $H_{\text{SO}} \propto \boldsymbol{L}\cdot\boldsymbol{S}$ is the spin-orbit interaction, $\boldsymbol{p} = -\mathrm{i}\hbar\nabla$ the linear momentum operator and V_{lattice} the periodic potential of the crystal lattice. Equations (12) and (13) lead directly to

$$\left(H_{\text{carrier}} + H_{kp} + \frac{\hbar^2}{2m_0}k^2\right)|n,\boldsymbol{k}\rangle = \epsilon_{n,\boldsymbol{k}}|n,\boldsymbol{k}\rangle\,, \tag{14}$$

with

$$H_{kp} = \frac{\hbar}{m_0}\boldsymbol{k}\cdot\boldsymbol{p}\,. \tag{15}$$

At $k = 0$, the electron and hole states of interest are eigenstates of the total angular momentum operator and are given by (2) and (3)–(8). This set of states will be denoted by $\mathscr{D}$ in the following. Together with the remote bands at $k = 0$, the states $\mathscr{D}$ satisfy

$$H_{\text{carrier}}|n\rangle = \epsilon_{\mathrm{n}}|n\rangle\,, \tag{16}$$

where $|n\rangle = |n, k = 0\rangle$ and $\epsilon_n = \epsilon_{n,\boldsymbol{k}=0}$. In the vicinity of $k = 0$ one can now determine the carrier states $|n,k\rangle$ by including the coupling of the $\mathscr{D}$ states with the remote bands via (14) as a 2nd-order perturbation term with interaction matrix element

$$\langle n'|H_{kp}(k)|n\rangle = \frac{\hbar}{m_0}\boldsymbol{k}\cdot\boldsymbol{p}_{n'n}\,, \tag{17}$$

where

$$\boldsymbol{p}_{n'n} = \langle n'|\boldsymbol{p}|n'\rangle\,. \tag{18}$$

Expanding the states at $\boldsymbol{k} \neq 0$ in terms of the states at $\boldsymbol{k} = 0$

$$|n,\boldsymbol{k}\rangle = \sum_m c_{n,m}(\boldsymbol{k})\,|n\rangle \tag{19}$$

leads to the Hamiltonian matrix

$$\begin{aligned}\mathscr{H}_{mn}(\boldsymbol{k}) &= \left(\epsilon_n + \frac{\hbar^2}{2m_0}k^2\right)\delta_{m,n} \\ &\quad + \frac{1}{2}\frac{\hbar^2}{m_0^2}\sum_{r\neq m,n}\sum_{\alpha,\beta=x,y,z} p^{\alpha}_{n,r}\, p^{\beta}_{r,m}\left(\frac{1}{\epsilon_m-\epsilon_r}+\frac{1}{\epsilon_n-\epsilon_r}\right), \qquad (20)\end{aligned}$$

determining the single-particle energies $\epsilon_{n,\boldsymbol{k}}$ and expansion coefficients in (19) via the eigenvalue problem

$$\sum_{m\in\mathscr{D}} \mathscr{H}_{n'm}(\boldsymbol{k})\, c_{m,n}(\boldsymbol{k}) = \epsilon_{n,\boldsymbol{k}}\, c_{n'n}(\boldsymbol{k})\,. \qquad (21)$$

In (20), $p^{\alpha}_{n'n}$ is the Cartesian α component of $\boldsymbol{p}_{n'n}$ in (18). The approach outlined above is known as $k\cdot p$ theory, but the Hamiltonian (20) is usually replaced by an effective Kohn–Luttinger Hamiltonian containing only a few parameters that can be fitted to experimental results instead of the momentum matrix elements with all the remote bands [11, 12, 14]. In the present case, the general band indices m, n represent the hole quantum numbers j or the electron quantum numbers s. By diagonalizing (21) for each $\boldsymbol{k}$, one obtains the band-structure $\epsilon_n(k)$ and the coefficients $c_{n'n}(\boldsymbol{k})$ determining the states $|n,k\rangle$ via (19). The advantages of this method to obtain the band structure close to $k=0$, are that it avoids a full band-structure calculation, and the effective Hamiltonian can be conveniently parametrized in terms of a few parameters.

The procedure outlined in the previous paragraph yields an effective Hamiltonian that describes the energy splitting of the spin-up and spin-down conduction electron bands in certain crystal directions, and therefore directly yields the Dyakonov–Perel process, as will be discussed below. However, it is important to note that this procedure for obtaining effective Hamiltonians is not restricted to the original Hamiltonian (14), but can also be generalized for additional interactions important to the dynamics, such as the Coulomb interaction, the interaction with phonons, or with impurities. Depending on the form of the interaction this leads to additional contributions to the effective Hamiltonian $\mathscr{H}_{nn'}(\boldsymbol{k})$ of the general form $\mathscr{H}^{\mathrm{ph}}_{nn'}(k,k')$ for phonon (or impurity) interaction, which describes scattering with carrier states $(n,k)\to(n',k')$. For the Coulomb interaction, the effective Hamiltonian is an effective two-particle interaction of the form $\mathscr{V}_{n_1n_1'n_2n_2'}(\boldsymbol{k}_1,\boldsymbol{k}_1',\boldsymbol{k}_2,\boldsymbol{k}_2')$, which describes the transition, or scattering process, $(n_1',k_1')(n_2',k_2')\to(n_1,k_1)(n_2,k_2)$. As will be shown below, incorporating the Coulomb interaction leads to the exchange interaction between electrons and holes, i.e., the Bir–Aronov–Pikus process, and carrier–phonon scattering leads to the Elliott–Yafet process.

2.3 Elliott–Yafet Mechanism

The Elliott–Yafet mechanism is a spin-flip process due to the coupling between electrons and holes combined with phonon-scattering processes, which can be described by the following effective Hamiltonian [15] including terms to 3rd order in the subspace of the interesting electron bands ($s, s' =\uparrow$ or $\downarrow$)

$$\mathcal{H}^{\mathrm{ph}}_{s's}(k',k) = \sum_j \frac{\mathcal{H}_{s'j}(\boldsymbol{k})\langle j|H_{e-p}(k',k)|j\rangle\mathcal{H}_{js}(\boldsymbol{k})}{(\epsilon_j - \epsilon_s)^2}, \tag{22}$$

where H_{e-p} is the Fröhlich Hamiltonian [16] that describes the long-range interaction with phonons. The Hamiltonian matrix (22) includes the effect of coupling due to the nonvanishing matrix elements $\mathcal{H}_{sj}(\boldsymbol{k})$, cf. (20), between electron and hole states. The phonon-scattering process changes the hole momentum $k \to k'$, and therefore, in effect, also changes the mixture of electron and hole states described by the overlap $\mathcal{H}_{sj}(\boldsymbol{k})$ since the latter depends on the wavevector k. The spin-dependent electronic lifetime[1]

$$\frac{1}{2\tau^{\mathrm{EY}}(k)} = \frac{2\pi}{\hbar}\sum_{\boldsymbol{k}'} |\mathcal{H}^{\mathrm{ep}}_{\uparrow\downarrow}(k',k)|^2 \delta(\epsilon_{\uparrow,k} - \epsilon_{\downarrow,k'}) \tag{23}$$

is calculated from (22) with Fermi's Golden Rule [15]. As will be discussed in more detail in Sect. 3.1, spin-dependent lifetimes are usually interpreted as spin-relaxation times [15], i.e., the relaxation time of the spin *polarization*, but this interpretation is only valid for a low density of electrons very close to equilibrium [17]. Approximately evaluating (23) and averaging over k using a Maxwell distribution for electrons, results in the spin-relaxation rate [15]

$$\frac{1}{\tau^{\mathrm{EY}}} = C\left(\frac{k_{\mathrm{B}}T}{E_{\mathrm{g}}}\right)^2 \eta^2 \left(\frac{1-\eta/2}{1-\eta/3}\right)\frac{1}{\tau_{\mathrm{p}}}. \tag{24}$$

Here, E_{g} is the bandgap energy, $\eta = \Delta_{\mathrm{SO}}/(E_{\mathrm{g}} + \Delta_{\mathrm{SO}})$, Δ_{SO} the spin-orbit splitting of the holes, τ_{p} the momentum-relaxation time, and C a constant that equals 2 for the polar interaction with optical phonons [18]. The EY spin-relaxation time is therefore proportional to the momentum-relaxation time.

2.4 Dyakonov–Perel Mechanism

The Dyakonov–Perel (DP) mechanism [19] only occurs in crystals without an inversion center because there matrix elements of $\mathcal{H}_{mn}(\boldsymbol{k})$ linear and cubic in k are not forbidden by symmetry [15]. Calculating the effective Hamiltonian for the electronic subsystem taking into account the spin-orbit interaction in the hole subsystem, one finds in these crystals that the spin degen-

[1] With the factor of 2 included on the LHS of (23), τ^{EY} can be identified with the spin-relaxation time under certain assumptions.

eracy is lifted by the Dresselhaus term, usually written in the form [15, 19]

$$\mathscr{H}_{ss'}(\boldsymbol{k}) = \frac{\hbar}{2}\boldsymbol{\Omega}(\boldsymbol{k}) \cdot \boldsymbol{\sigma}_{ss'} \,. \tag{25}$$

In (25), $\boldsymbol{\Omega}(\boldsymbol{k})$ is a *momentum-dependent* vector (cubic in k), and $\boldsymbol{\sigma}_{ss'}$ the vector of Pauli matrices. This effective interaction has the same effect as an external magnetic field with Larmor frequency $\boldsymbol{\Omega}(\boldsymbol{k})$, which can be calculated by taking into account the influence of the spin-orbit interaction with the holes perturbatively. Momentum-scattering processes with phonons or impurities will therefore change the effective magnetic Larmor frequency $\boldsymbol{\Omega}(\boldsymbol{k})$ experienced by an electron by changing its momentum. An electron therefore experiences a fluctuating magnetic field, which contributes to the electron-spin relaxation. Dyakonov and Perel first calculated the spin-relaxation time due to this effect in the collision-dominated limit, i.e., when the momentum scattering time is *shorter* than the time scale of the electron-spin precession. This situation corresponds to spin relaxation by repeated small precessions, which is similar to the "motional narrowing" in nuclear magnetic resonance. The DP result is [15]

$$\frac{1}{\tau^{\mathrm{DP}}} = \tilde{C}\alpha^2 \frac{(k_{\mathrm{B}}T)^3}{\hbar^2 E_{\mathrm{g}}} \tau_{\mathrm{p}} \,, \tag{26}$$

where $\tilde{C}$ is a dimensionless factor of order unity depending on the scattering mechanism and α is a parameter related to the cubic band-structure term, given approximately by

$$\alpha \simeq \frac{4\eta}{\sqrt{3-\eta}} \frac{m_{\mathrm{e}}}{m_0} \,, \tag{27}$$

and τ_{p} the momentum-relaxation time. In (27), m_{e} is the effective electron mass in the crystal at $k = 0$ and m_0 is the free electron mass. The spin-relaxation time (26) due to the DP process is inversely proportional to the momentum-relaxation time. Therefore, the resulting spin-relaxation time is *slower* for faster momentum scattering, leading to the counterintuitive result that a "dirty" material with strong impurity scattering may have a longer spin lifetime.

2.5 Bir–Aronov–Pikus Mechanism

The Bir–Aronov–Pikus (BAP) process is based on the exchange interaction between electrons and holes, which can be calculated from an effective Hamiltonian in 3rd order in the interaction with the remote bands [20]. Since we will treat the BAP process in more depth, and it is a two-particle interaction, we write the Hamiltonian in 2nd quantization

$$H_{\mathrm{exc}} = \sum_{k_1,k_2,q} \sum_{j,s,j',s'} \langle j's'|\mathscr{V}_{\mathrm{exc}}(\boldsymbol{q})|sj\rangle c^{\dagger}_{j',k_2+q} c^{\dagger}_{s',k_1-q} c_{j,k_1} c_{s,k_2} \,. \tag{28}$$

Here, c_s ($c_s^\dagger$) and c_j ($c_j^\dagger$) are electron and hole destruction (creation) operators [notation as in (11)]. The interaction matrix element [15, 21]

$$\langle j's'|\mathscr{V}_{\mathrm{exc}}(\boldsymbol{q})|sj\rangle = \langle j's'|\mathscr{V}_{\mathrm{SR}}(\boldsymbol{q})|sj\rangle + \langle j's'|\mathscr{V}_{\mathrm{LR}}(\boldsymbol{q})|sj\rangle \tag{29}$$

consists of a long-range part

$$\langle j's'|\mathscr{V}_{\mathrm{LR}}(q)|sj\rangle = \frac{\hbar^2}{m_0^2}\frac{(\boldsymbol{q}\cdot\boldsymbol{p}_{j's})(\boldsymbol{q}\cdot\boldsymbol{p}_{s'j})}{(\epsilon_s-\epsilon_{j'})^2}\,, \tag{30}$$

and a short-range part

$$\langle j's'|\mathscr{V}_{\mathrm{SR}}(q)|sj\rangle = \frac{3}{2}\Delta_{\mathrm{exc,SR}}\langle j',s'|\left[\frac{3}{4}+(\boldsymbol{J}\cdot\boldsymbol{S})\right]|j,s\rangle\,. \tag{31}$$

In (30), the $\boldsymbol{p}_{js}$ are the matrix elements of the momentum operator between the states (11) defined in (18). In (31), $\Delta_{\mathrm{exc,SR}}$ is the excitonic exchange splitting, $\boldsymbol{J}$ is the hole total angular momentum operator and $\boldsymbol{S}$ is the electron-spin operator. For an explicit matrix representation of (30) and (31), see [15, 21, 22]. It should be noted that (28), (30), and (31) are written with the customary definition of two-particle interaction matrix elements [23]. The order of the indices shows that (28) describes an electron–hole exchange interaction process, i.e., an *interband* scattering process.

Using these contributions, BAP derived a k-dependent electron-spin lifetime using Fermi's Golden Rule [15, 20][2]

$$\frac{1}{2\tau_s^{\mathrm{BAP}}(k)} = \frac{2\pi}{\hbar}\sum_{\boldsymbol{q},\boldsymbol{p}}\sum_{j,j'}|\langle j's'|\mathscr{V}(\boldsymbol{k}+\boldsymbol{p})|sj\rangle|^2 n_{j,\boldsymbol{p}}(1-n_{j',\boldsymbol{p}+\boldsymbol{q}}) \times\delta(\epsilon_{s,k}+\epsilon_{j,p}-\epsilon_{s',\boldsymbol{k}+\boldsymbol{q}}-\epsilon_{j',\boldsymbol{p}+\boldsymbol{q}})\,, \tag{32}$$

where $s'\neq s$ is the flipped spin. As mentioned in Sect. 2.3, lifetimes derived from Fermi's Golden Rule [15] are usually identified with the spin relaxation-time, but this identification breaks down for higher electronic densities or pronounced nonequilibrium situations [17]. For a low-density *thermalized* electron distribution, one can thermally average (32) with a Maxwell distribution to obtain [15, 24]

$$\frac{1}{\tau^{\mathrm{BAP}}} = \frac{2a_{\mathrm{B}}^3}{\tau_0}\frac{v_{\mathrm{F}}\epsilon_{s,k}}{v_{\mathrm{B}}\epsilon_{\mathrm{F}}}N_{\mathrm{A}}\,. \tag{33}$$

Here, a_{B} and v_{B} are the Bohr radius and velocity of the exciton, ϵ_{F} and v_{F} the hole Fermi energy and velocity, N_{A} the total concentration of holes, and

$$\frac{1}{\tau_0} = \frac{3\pi}{64}\frac{\Delta_{\mathrm{exc,SR}}^2}{E_{\mathrm{b}}\hbar}\,, \tag{34}$$

where E_{b} is the excitonic binding energy and $\Delta_{\mathrm{exc,SR}}$, as defined above, is the excitonic exchange splitting.

[2] The factor 2 in the following is the same as in (23).

3 Spin-Polarization Dynamics

3.1 Theory of Spin-Relaxation Dynamics

Electron-spin dynamics in general can be described by the reduced one-particle density matrix [25, 26]

$$\rho_{s,s'} = \begin{pmatrix} \rho_{\uparrow\uparrow} & \rho_{\uparrow\downarrow} \\ \rho_{\downarrow\uparrow} & \rho_{\downarrow\downarrow} \end{pmatrix} \equiv \begin{pmatrix} n_\uparrow & \psi \\ \psi^* & n_\downarrow \end{pmatrix}, \tag{35}$$

with the real average electronic occupation numbers and the complex spin coherence ψ that are driven by a magnetic field. The spin coherence is only nonzero if the spin-up and spin-down states are no longer eigenstates of the Hamiltonian, i.e., when the electrons are coupled by an external magnetic field or an effective internal magnetic field as is the case for the Dyakonov–Perel mechanism. The dynamics of electrons in a magnetic field therefore follows the equations of motion for the time-dependent electron distributions $n_\uparrow$, $n_\downarrow$ and the spin coherence ψ as defined by (35) [10, 25]. In semiconductors, both these quantities are subject to carrier–carrier and carrier–phonon scattering. The microscopic description of the different interaction mechanisms does not, in general, allow the introduction of macroscopic T_1 and T_2 times [24]. *If* these approximate quantities can be introduced for interacting electrons in semiconductors, T_1 refers to the relaxation of the macroscopic spin polarization toward its equilibrium $P = 0$ value. For this to happen, the angular momentum has to be transferred out of the electron system to the holes and eventually to the lattice. This has to be kept apart from the decay of the spin coherence that is described by T_2 at the macroscopic level. Thus T_2 is a measure of the dephasing of the electron-spin coherence driven by magnetic fields and usually called "homogeneous broadening" analogous to the case of nuclear magnetic resonance. If there are additional contributions to the dephasing related to "inhomogeneous" broadening due to, e.g., spatial fluctuations of the magnetic field, one sometimes describes the combined effect of the inhomogeneous broadening and the homogeneous broadening on the decay of the macroscopic spin coherence by T_2^* [24].

In a magnetic field one has two contributions to the time development of the spin polarization P: one is due to the *incoherent* spin-flip processes related to T_1 and the other is due to the coherent change of the electronic distribution functions as long as there is a spin coherence driven by a magnetic field. For the study of the spin-polarization relaxation it can therefore be advantageous to consider the case without magnetic fields, as will be done here, because this eliminates the influence of *coherent* magnetic-field-induced spin flips. In the following, we will deal only with the *incoherent* spin-polarization relaxation, i.e., the time development of (1) and its momentum- (or energy-) dependent generalization

$$P_k = \frac{n_{\uparrow,k} - n_{\downarrow,k}}{n_{\uparrow,k} + n_{\downarrow,k}}. \tag{36}$$

A preliminary understanding of the spin relaxation can be achieved using lifetimes as they result from a Fermi's Golden Rule treatment as outlined in Sect. 2.5. Consider for definiteness the BAP process, for which the spin lifetime is described by (32). Since Fermi's Golden Rule yields lifetimes, this is equivalent to a relaxation-time approximation of the form [17]

$$\left.\frac{\partial}{\partial t} n_s(k)\right|_{\mathrm{BAP}} = -\frac{1}{2\tau_s^{\mathrm{BAP}}(k)}\left[n_s(k) - f_s(k)\right], \tag{37}$$

where $f_s(k)$ is an equilibrium (Fermi–Dirac) distribution, to which the system relaxes. In addition to this relaxation of the electronic distributions, one has to take into account that the scattered electrons, which flip their spins during the scattering processes, have to be accounted for in the number of carriers in the electronic band with opposite spin. Using this reasoning for a nonzero electronic-spin polarization and an *unpolarized* hole system, i.e., $n_{j=\frac{3}{2},k} = n_{j=-\frac{3}{2},k}$ and $n_{j=\frac{1}{2},k} = n_{j=-\frac{1}{2},k}$, one finds from (32)

$$\tau_\uparrow^{\mathrm{BAP}}(k) = \tau_\downarrow^{\mathrm{BAP}}(k), \tag{38}$$

and further obtains

$$\left.\frac{\partial}{\partial t} P_k\right|_{\mathrm{BAP}} = -\frac{P_k}{\tau_s^{\mathrm{BAP}}(k)}. \tag{39}$$

This result reflects the physical situation that the spin polarization decays, if the different populations relax towards equilibrium according to (37), but the scattered electrons show up in the band with opposite spin with the same momentum after the scattering process. The information about the energy and momentum dependence of spin-flip scattering is therefore lost in this approximation. As will be shown below, the treatment of electron dynamics as scattering processes without the relaxation-time approximation leads to a more general description of spin-polarization dynamics due to the exchange interaction. It is interesting to note that the field of spin relaxation has been dominated by rate-equation approximations for decades [27], but efforts have been made recently to go beyond this approximation for the description of spin-dependent scattering phenomena [26, 28–30].

Since the experimental results discussed below are for heavily p-doped GaAs, the following discussion will focus on the spin dynamics due to the electron–hole exchange interaction (28), i.e., the BAP process. To go beyond Fermi's Golden Rule one derives the dynamical equations for the relevant correlation functions in the electronic ensemble, which are the distribution functions for electrons

$$n_{sk}(t) = \langle c_{sk}^\dagger c_{sk} \rangle, \tag{40}$$

defined with time-dependent creation and destruction operators in the Heisenberg picture. The symbols $\langle \cdots \rangle$ designate a statistical average over the ensemble of electrons. The dynamics of the carrier distributions n_{sk} is then determined by the Heisenberg equation of motion

$$\frac{\hbar}{\mathrm{i}} \frac{\partial}{\partial t} \langle c_{sk}^{\dagger} c_{sk} \rangle = \left\langle [H, c_{sk}^{\dagger} c_{sk}] \right\rangle . \tag{41}$$

With a two-particle interaction Hamiltonian such as H_{exc} in (28), one runs into the well-known hierarchy problem that the equations of motion for the carrier distributions n_{sk} do not close, i.e., they couple to correlation functions containing two creation and two destruction operators. The hierarchy problem can be approximately solved, e.g., using Green's functions [31, 32] or truncation [33, 34] techniques. Technically, the random-phase approximation (RPA) will be employed in the following. One obtains as the dynamical equation for the spin-flip scattering due to electron–hole exchange interaction a Boltzmann equation

$$\left. \frac{\partial}{\partial t} n_{sk} \right|_{\mathrm{BAP}} = -\Gamma_{sk}^{\mathrm{BAP,out}}\, n_{sk} + \Gamma_{sk}^{\mathrm{BAP,in}} \left(1 - n_{sk}\right) , \tag{42}$$

with spin- and momentum-dependent outscattering

$$\begin{aligned} \Gamma_{sk}^{\mathrm{BAP,out}} = \frac{2\pi}{\hbar} \sum_{\boldsymbol{q},\boldsymbol{p}} \sum_{j,j'} |\langle j's'|\mathscr{V}_{\mathrm{exc}}(\boldsymbol{k}+\boldsymbol{p})|sj\rangle|^2\, n_{j,p}(1-n_{j,\boldsymbol{p}+\boldsymbol{q}})(1-n_{s',\boldsymbol{k}-\boldsymbol{q}}) \\ \times\, \delta(\epsilon_{s,k} + \epsilon_{j,p} - \epsilon_{s',\boldsymbol{k}-\boldsymbol{q}} - \epsilon_{j',\boldsymbol{p}+\boldsymbol{q}}) , \end{aligned} \tag{43}$$

and inscattering rates

$$\begin{aligned} \Gamma_{sk}^{\mathrm{BAP,in}} = \frac{2\pi}{\hbar} \sum_{\boldsymbol{q},\boldsymbol{p}} \sum_{j,j'} |\langle j's'|\mathscr{V}_{\mathrm{exc}}(\boldsymbol{k}+\boldsymbol{q})|sj\rangle|^2 (1-n_{j,p}) n_{j',\boldsymbol{p}+\boldsymbol{q}} n_{s',\boldsymbol{k}-\boldsymbol{q}} \\ \times\, \delta(\epsilon_{s,k} + \epsilon_{j,p} - \epsilon_{s',\boldsymbol{k}-\boldsymbol{q}} - \epsilon_{j',\boldsymbol{p}+\boldsymbol{q}}) , \end{aligned} \tag{44}$$

Here, $s' = -s$ is the flipped-spin quantum number, and the interaction matrix element is given by (29). One notices that the outscattering rate is identical to the lifetime obtained by Fermi's Golden Rule (32) if the electronic occupation of the final state, into which the spin-flipped electron is scattered, is small, i.e., if $n_{s',\boldsymbol{k}+\boldsymbol{q}} \ll 1$. Also, a term mimicking the effect of the inscattering term in the Boltzmann equation had to be added to the relaxation-time equation by hand, since Fermi's Golden Rule yields only a lifetime, i.e., a decay time for individual electrons due to the spin-flipping exchange interaction.

Formally, a similar Boltzmann equation describes the time development of the hole-distribution functions under the action of the electron–hole exchange scattering. As will be shown later experimentally and theoretically, the spin-flip scattering (42) due to the electron–hole exchange interaction takes place on a time scale of several tens of picoseconds in moderately to strongly p-doped GaAs. We are here exclusively interested in p-doped GaAs where a large density of holes and a very small density of electrons is present. Thus the hole angular momenta equilibrate due to the rapid momentum scattering and spin-orbit interaction much faster than the electronic spins. We will therefore treat the holes only as a bath in the following calculations.

In addition to the electron–hole exchange interaction, carriers interact also via the *direct* Coulomb interaction. This leads to strongly carrier-density-dependent scattering times that reach several hundred femtoseconds for high carrier densities. We therefore have to include the effect of the fast direct electron–hole Coulomb scattering, which is spin conserving, in addition to the spin-flip electron–hole exchange scattering since the electronic distribution functions in (42) will evolve under the action of the direct scattering quickly. This is described by the well-known Boltzmann equation for the direct Coulomb scattering in the random-phase approximation [35]

$$\begin{aligned} \left.\frac{\partial}{\partial t} n_{sk}\right|_{\mathrm{Coul}} = -\frac{2\pi}{\hbar} \sum_{k'qj} |W_{\mathrm{q}}|^2 \big[& n_{sk}(1-n_{sk+q}) n_{jk'+q}(1-n_{jk'}) \\ & - (1-n_{sk}) n_{sk+q} (1-n_{jk'+q}) n_{jk'} \big] \\ & \times \delta(\epsilon_{jk+q} + \epsilon_{sk} - \epsilon_{jk'} - \epsilon_{sk+q}) \,. \end{aligned} \tag{45}$$

Here, $W_q = V_q/\varepsilon_q$ is the screened Coulomb interaction in momentum space defined in terms of the bare Coulomb interaction

$$V_q = \frac{e^2}{\varepsilon_0 \varepsilon_{\mathrm{bg}} q^2} \,, \tag{46}$$

with the background dielectric function $\varepsilon_{\mathrm{bg}}$. The carrier contribution to the dimensionless dielectric function is calculated in the static approximation $\varepsilon_q = 1/(1+q^2/\kappa^2)$ where κ is the screening wavevector [35].

Equations (42) and (45) define our model that will be used to calculate the dynamics of electronic distribution functions and thus the spin relaxation. We will neglect electron–electron Coulomb scattering because the electronic densities will be kept so low that this process is much slower than the electron–hole Coulomb scattering. We will also use fixed Fermi–Dirac distributions for the high-density hole distributions because the fast hole–hole Coulomb scattering together with the spin-orbit interaction keeps the hole system in thermal equilibrium and prevents the occurrence of hole spin polarization on time scales important for the electronic-spin dynamics.

3.2 Numerical Results

In this section results of the numerical solution of (42) and (45) are presented. The numerical solution is accomplished by transforming the sums into integrals and then using a 4th-order Runge–Kutta algorithm to calculate the time evolution of the electronic distribution functions. The initial conditions in the following are always taken to be quasiequilibrium distributions for electrons

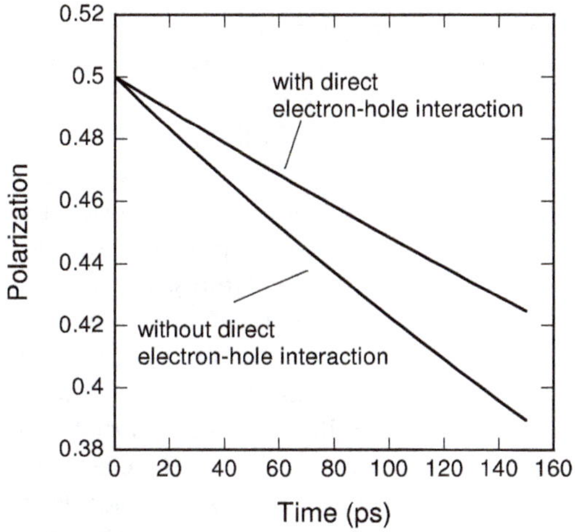

Fig. 2. Computed relaxation of electron-spin polarization $P(t)$ for electron–hole exchange interaction with direct Coulomb interaction (*top curve*) and exchange interaction without Coulomb interaction (*bottom curve*). At $t = 0$, carrier temperatures, densities and spin polarizations are 300 K, $10^{18}\,\mathrm{cm}^{-3}$ and 0 % for holes, and 600 K, $10^{15}\,\mathrm{cm}^{-3}$ and 50 % for electrons, respectively

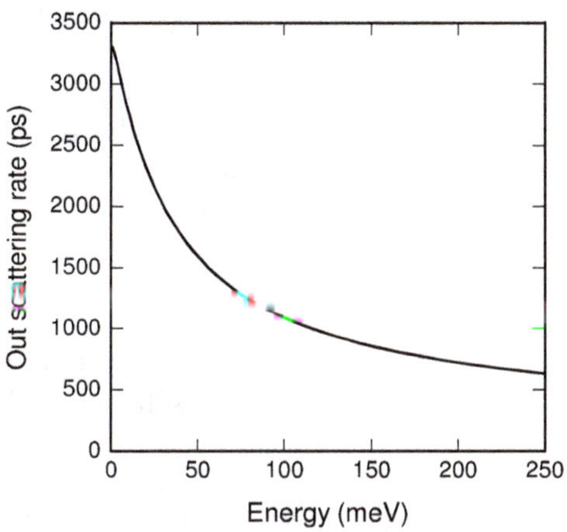

Fig. 3. Calculated outscattering rate (43) vs. electronic energy for the same parameters as in Fig. 2

and holes. After optical excitation of spin-polarized carriers in p-doped GaAs the electrons will quickly equilibrate thermally without losing their spin coherence on this time scale, whereas the holes will lose their spin polarization due to spin-orbit interaction and Coulomb scattering so that one can assume quasiequilibrium conditions with polarized electrons and unpolarized holes.

As a first result of the model we will discuss the interplay between direct Coulomb scattering and exchange scattering. Figure 2 shows the result of a numerical experiment, in which the case with Coulomb scattering is compared to the one where Coulomb scattering is "switched off." For the case of a polarized low electron density at 300 K in the presence of a high density of unpolarized holes at 600 K, one finds that the direct Coulomb interaction has a significant influence on the dynamics of the electron-spin polarization over several tens of picoseconds, even though the characteristic time scale of direct Coulomb scattering is on the order of a few hundred femtoseconds. Thus the direct scattering influences the dynamics of the electron-spin polarization, even though it does not give rise to spin flips. The reason for this behavior is that the exchange scattering alone creates nonequilibrium electron distributions over time, whereas the direct scattering continually drives the electronic distributions into equilibrium with the holes. The resulting competition leads to the faster electron spin-polarization decay if both scattering mechanisms are included. It should also be noted that, even though the time development of the total polarization is plotted here, the time development of the energy-resolved spin polarizations are not much different from the total polarization. In the range of electron and hole temperatures around room temperature, the energy-resolved polarization and the total polarization can be fitted with an exponential decay law, and the time constants are practically equal for all those quantities. In the case of Fig. 2 we find a relaxation time of 950 ps for the electronic polarization.

Figure 3 shows the energy-dependent outscattering rate (43) for electrons due to the exchange interaction with unpolarized holes, with all the parameters being the same as before. It should be noted that the outscattering rate is identical to the Fermi's Golden Rule result in the low electron density limit. This quantity was calculated earlier for $T = 0\,\mathrm{K}$ [22] as a measure of the spin-relaxation time, which was predicted to be strongly energy dependent on the basis of this calculation. However, as mentioned in connection with Fig. 2, no energy dependence of the spin-polarization relaxation is found in our numerical results including the dynamics due to both exchange and direct scattering processes.

Figure 4 shows the computed spin-relaxation time for different hole concentrations. For the range of densities depicted in Fig. 4, the dynamics of electron spins was calculated over 180 ps, and an exponential fit was made to the resulting polarization dynamics. The result is a meaningful measure of the effectiveness of the BAP process for the polarization relaxation that can be compared to experiments. For a high doping concentration of $N_\mathrm{A} = 10^{19}\,\mathrm{cm}^{-3}$ we obtain a spin-relaxation time of 110 ps, which is in very good agreement with recent measurements of the bulk spin relaxation in identical samples using time-resolved Faraday rotation techniques [36]. In principle, one can therefore use the calculated density dependence of the spin-relaxation time to obtain an estimate of the hole density and thus the doping concentration in a GaAs/metal interface (Schottky barrier), where the band

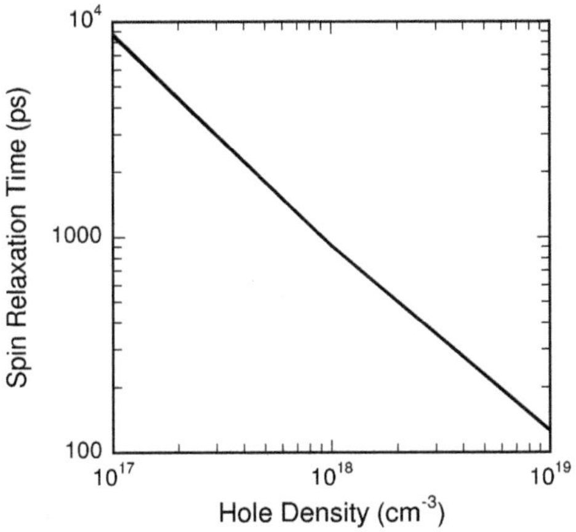

Fig. 4. Calculated spin-relaxation time vs. density of holes for a hole temperature of 300 K, electron density $10^{14}\,cm^{-3}$ and temperature 600 K. The initial polarization is again $P(t = 0) = 50\,\%$

bending near the surface leads to a depletion of holes in the vicinity of the surface. Before we turn to the discussion of the experimental setup and the experimental results, we briefly review the concept of a Schottky contact and the band bending at semiconductor–metal surfaces.

3.3 Spin Decay in a Schottky Barrier

Most metal/semiconductor interfaces act as a diode because the electric current passing through the interface depends exponentially on the forward bias since a depletion layer in the semiconductor builds up [37,38]. These contacts are known as "Schottky contacts," as opposed to ohmic contacts that have a lower resistance. The concept of a depletion layer and the accompanying band bending is most easily described for a semiconductor vacuum surface before moving on to the actual Schottky contact [37].

Consider a uniformly p-doped semi-infinite slab of semiconductor material, where the majority of carriers are holes, and the doping concentration is N_A. The introduction of a surface breaks the crystal symmetry and leads to additional carrier states located near the surface, whose energies lie in the semiconductor bandgap and that are occupied predominantly by holes in a p-doped semiconductor. The holes occupying the surface states lead to a positive surface charge σ and must have come from acceptors in a region of width w beneath the surface, the so-called depletion layer, where there is now a negative space charge $\varrho = -eN_A$ per unit volume, see Fig. 5. (Here, $e > 0$ denotes the magnitude of the electronic charge.) Charge neutrality requires

$$\sigma + \varrho w = 0\,. \tag{47}$$

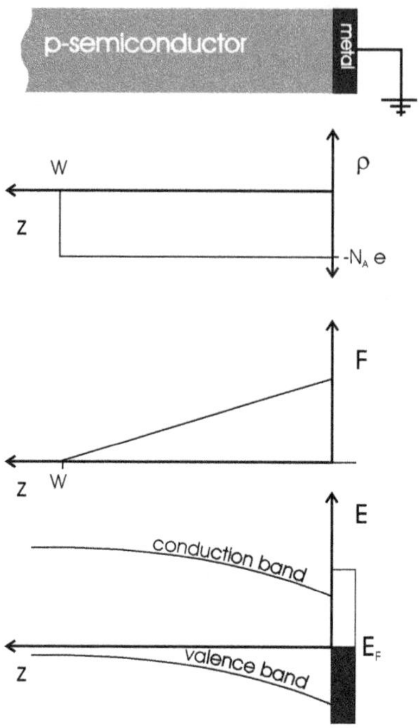

Fig. 5. Schottky contact between p-doped semiconductor and metal. The charge distribution ϱ is approximated as a step function, leading to the electric-field distribution and band lineup shown below

The positive space charge of the depletion layer now causes an electrostatic potential, which is responsible for the characteristic band bending. This, in turn, reduces the surface charge and compensates it almost completely. To see this, one calculates the electrostatic potential $\Phi(z)$ away from the surface induced by the space charge of the depletion layer according to Poisson's law

$$\frac{\mathrm{d}^2}{\mathrm{d}z^2}\Phi(z) = -\frac{\varrho}{\varepsilon_b\varepsilon_0}, \tag{48}$$

where ε_b and ε_0 are the permittivity of the semiconductor and the vacuum, respectively. In solving the equation for $\Phi(z)$ we assume that at the end of the depletion layer ($z = w$) we can set the electrostatic potential Φ and the field $\mathscr{F} = -\mathrm{d}\Phi/\mathrm{d}x$ to zero. We find a simple linear dependence for the electric field

$$\mathscr{F}(z) = -\frac{\varrho}{\varepsilon_b\varepsilon_0}(w - z), \tag{49}$$

and for the potential accordingly

$$\Phi(z) = -\frac{\varrho}{2\varepsilon_b\varepsilon_0}(w - z)^2. \tag{50}$$

Thus the additional potential energy for an electron with charge $q = -e$ in the region $z < w$ is $-e\Phi(z)$, leading to a downward band bending.

With this background discussion one can turn to the metal–semiconductor Schottky contact [37], and consider again the case of a p-type semiconductor. It is assumed that the metal has a chemical potential, which is denoted by the Fermi energy E_{F}, and the semiconductor has a chemical potential μ. Before bringing the two materials into contact the respective chemical potentials are different, with the semiconductor chemical potential lying in the bandgap close to the valence band because of the p-doping, and therefore lower than the metal chemical potential. Bringing the two materials together sets up an electric field due to the different chemical potentials that drives electrons into the semiconductor, thereby creating a negative space charge in the semiconductor and a positive surface charge, just as in the example discussed above. When the materials are joined, these two charge distributions balance each other. Also, the common chemical potential is pinned because the surface (or interface) states are only partially filled. The negative space charge in the semiconductor leads to the band bending described above, as shown in Fig. 5.

When carriers are created in the semiconductor away from the surface by optical excitation near the bandgap, this band configuration is the reason that diffusion of optically excited electrons from the bulk towards the surface yields electrons with a high kinetic energy, i.e., "hot" electrons, at the surface [39].

4 Experiments

4.1 Time- and Spin-Resolved 2PPE

Carrier spin-relaxation measurements in zero applied field have been reported in [9] by means of time-resolved Faraday rotation experiments. In order to isolate the spin decay occurring only in the band-bending region, a surface-sensitive and energy-resolving technique is needed. To explore the spin- and charge dynamics in this region we introduce a novel real-time method. We use a pump-probe approach referred to as time-resolved 2-photon photoemission (TR-2PPE) with the additional option to measure the spin polarization of the emitted electrons [40]. By varying the time delay between the ultrashort pump and probe laser pulses, the spin-dependent population decay of the intermediate (unoccupied) states can be determined. The high surface sensitivity as well as the energy selectivity of the photoemission technique is appropriate to investigate the spin dynamics in the Schottky barrier. From this perspective, TR-2PPE is complementary to Faraday rotation. In TR-2PPE one simply determines the spin polarization P defined in (36) of the transient carrier concentration in a certain energy interval. The energy-resolved spin polarization P is independent of the carrier populations at that energy.

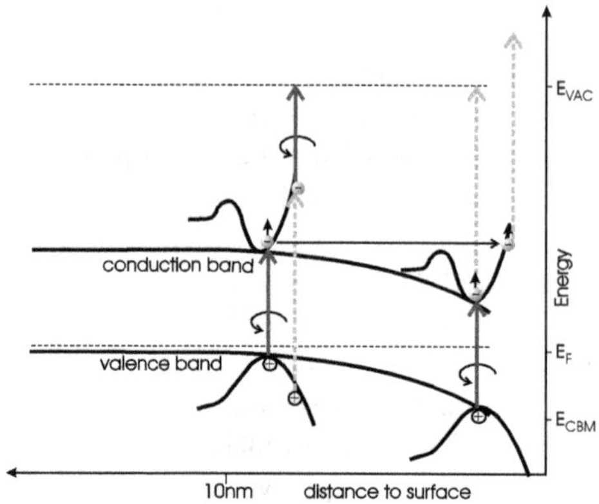

Fig. 6. Bichromatic 2PPE process in a surface region with band bending. By making use of the energy resolution of the photoemisssion technique, we can probe the hot-electron spin dynamics at different energy levels. Parts of the band structure at different distances from the surface are sketched. The *solid arrows* mark the transitions induced by the absorption of 1.55 eV pump photons, and the *dashed arrows* the transitions induced by the 3.1 eV probe photons

A reduced carrier concentration in the investigated energy interval will only lead to an increased statistical error on the measured spin polarization.

In order to study the spin decay in the band-bent region by TR-2PPE, we first excite spin-polarized electrons with a circularly polarized fs-laser pulse at $\hbar\omega = 1.55\,\mathrm{eV}$, slightly larger than the bandgap. Some of the spin-polarized hot electrons will move towards the band-bending area, where the carriers become hot (see Fig. 6) and will undergo different elastic and inelastic relaxation processes. The spin-dependent depletion of the excited-state population in the surface region can be determined by a suitably delayed second laser pulse that photoemits the electrons. The second (probe) laser pulse must be at higher photon energy ($\hbar\omega = 3.1\,\mathrm{eV}$) in order to overcome the vacuum energy.

4.2 Experimental Setup

The laser system used is by now a standard tool for measurements of the type described here. It consists of a mode-locked titanium-doped sapphire ($Ti{:}Al_2O_3$) laser, pumped by a diode laser at about 8 W. This setup generates transform-limited and sech^2 temporally shaped pulses of up to 15 nJ/pulse with a duration of less than 45 fs at a repetition rate of 82 MHz. The wavelengths of the pulses can be tuned from 830 nm to 770 nm, thereby varying the photon energy from 1.49 eV to 1.61 eV. For the time-resolved experiments the pulse train is split by a beam splitter (see Fig. 7). By varying the optical path with a variable delay line, we can shift the two pulse trains by a certain length corresponding to a delay in time.

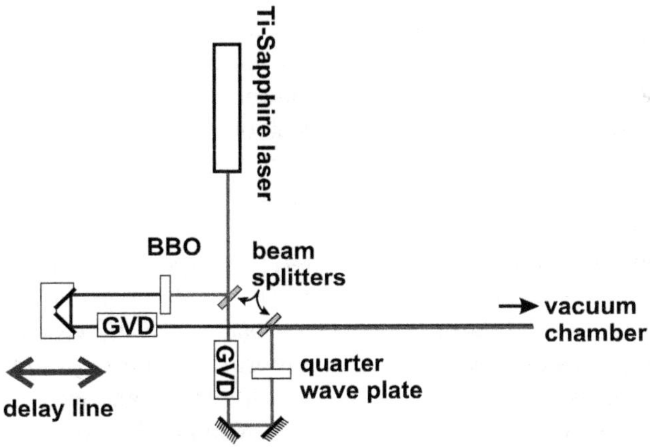

Fig. 7. Schematic setup of the Mach–Zehnder interferometer. The pulses are split up and can be delayed with respect to each other. In addition, we can modify each arm of the interferometer to tailor the beam properties to our specific needs by introducing, e.g., frequency-doubling elements or quarter-wave plates. Each arm then needs individual dispersion compensation

The pulse of one path is frequency doubled by a thin β-barium-borate-(BBO-)crystal to yield photon energies in the range of 3.0 eV to 3.2 eV (in the following designated by *blue*). The pulse of the other path (fundamental, in the following designated by *red*), is transformed from linearly polarized to circularly polarized by introducing a quarter-wave plate. Adjusting the quarter-wave plate, we can change between left-handed and right-handed light. This setup is used for the optical orientation experiments. The frequency-doubled pulses remain linear. After delaying, the pulses are reunited by a second beam splitter and then focused by a lens onto the sample inside the UHV chamber. The polarized laser beam is incident perpendicular on the sample surface, and the electrons are detected at an angle of 45° with respect to the surface normal.

In order to account for the pulse broadening introduced into the system by dispersive elements, a group-velocity dispersion compensation is necessary for both pulse trains. This is accomplished with a prism pair traversed twice by the laser pulse resulting in negative dispersion, which cancels out the usually positive dispersion due to lenses and the vacuum viewport.

The laser beams are adjusted to reach excitation densities as low as $1 \times 10^{16}\,\mathrm{cm}^{-3}$. The excitation densities are of the same order of magnitude for the fundamental laser beam and for the frequency-doubled pulses, although the intensities are approximately 100 times smaller for the 3.1 eV pulses. This is due to the fact that the penetration depth for 3.1 eV photons is 50 times smaller than for the 1.55 eV photons. Considering the laser intensities used and the absorption lengths λ at a wavelength of 800 nm ($\lambda_\mathrm{r} = 730\,\mathrm{nm}$) and 400 nm ($\lambda_\mathrm{b} = 14\,\mathrm{nm}$) in GaAs, we can roughly estimate the average excitation density from the laser power. Assuming a Gaussian profile for our leaser beam, we determine the focus spot size of our lens to 100 μm in diameter. Taking the maximum laser powers available, we can reach excitation densities on the order of 5×10^{18} electron–hole pairs cm^{-3}. By reducing the pulse power (reducing the laser output and/or lowering the frequency-doubling efficiency), more than three orders of magnitude in excitation density are available.

The time-averaged photocurrent at a fixed kinetic energy is measured as a function of the delay between the two beams (two-pulse correlation technique). The nonlinearity of the two-photon process leads to an increase in the 2PPE yield when the pulses are spatially and temporarily superimposed. As long as the two laser pulses overlap in time, it is obvious that an electron can be emitted by absorbing just one photon from each pulse. However, if the pulses are separated in time an excited electron from the first pulse can absorb a photon from the second pulse only as long as the inelastic lifetime of the intermediate state exceeds the delay.

The samples are mounted in a UHV chamber (base pressure 5×10^{-11} mbar) equipped with a cylindrical sector electron-energy analyzer. To investigate the electron dynamics separately for both spin directions, a spin analyzer (SPLEED detector [40, 41]) is mounted on top of the electron-energy analyzer. This makes the measurement of one inplane component of the spin-polarization vector possible. A bias voltage of -14 V is applied to the sample to eliminate the effects of any stray electric and magnetic fields.

As a sample we use a p-doped (100)-oriented GaAs crystal with dopant concentration (zinc) $N_A = 1 \times 10^{19}\,cm^{-3}$. GaAs surfaces are quite reactive when exposed to air and the surface of a sample stored outside a vacuum chamber is oxidized. The samples are cleaned by etching in sulfuric acid and rinsing, followed by heating to approximately 500 °C in the UHV chamber. Prior to the measurements the sample is treated with a small amount of cesium in order to obtain a well-defined Fermi-level pinning and a lowered work function [38].

4.3 Experimental Results for GaAs (100)

By making use of the energy resolution of the photoemisssion technique one can probe the spin polarization of the hot electrons at different energy levels in the band-bending region. Figure 8 shows the spin polarization versus energy at zero delay time when there is temporal overlap of the pump and probe pulses. The reference energy is the conduction-band minimum (CBM) at the surface (E_{CBM}). We observe nearly uniform spin polarization within the band-bending energy range. Taking into account that the electrons close to the conduction band minimum have undergone many elastic and inelastic scattering processes, the result indicates that the spin relaxation in the band-bending region must occur on a much longer time scale than the energy-

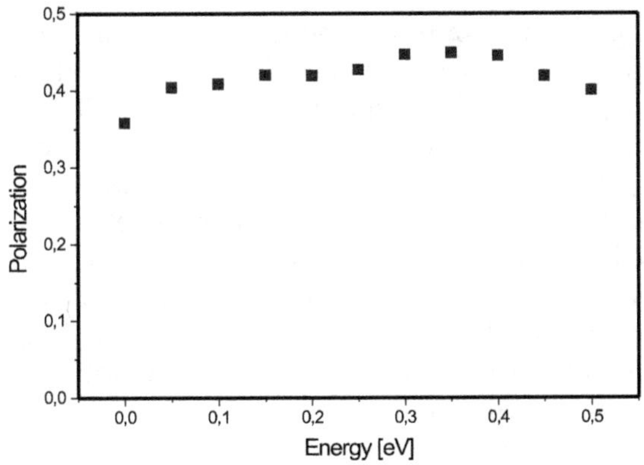

Fig. 8. Spin polarization of photoexcited electrons ($\hbar\omega = 1.55\,\mathrm{eV}$), injected into the (100) surface of a GaAs crystal with band bending. The zero of the energy scale is defined as the conduction-band minimum at the surface (E_{CBM}) and represents the energy of the intermediate state after absorption of the first photon

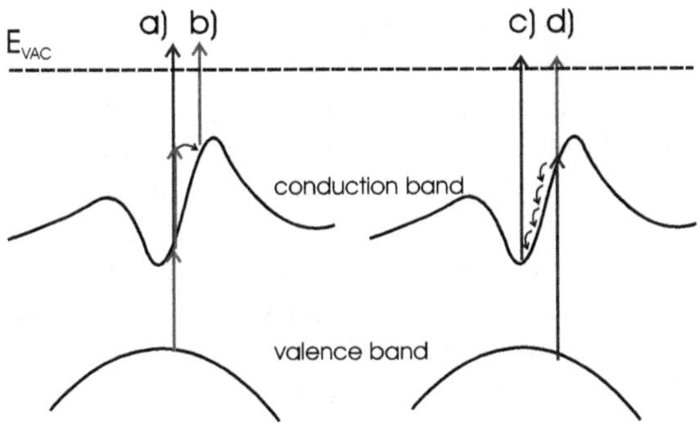

Fig. 9. Multiphoton processes. (**a**) *red–blue* transition yielding spin polarization if the *red* pulse is circularly polarized. (**b**) *triple red* transition yielding also spin polarization. (**c**) *blue–blue* transition involving inelastic scattering events. (**d**) *blue–red* transition. Both (**c**) and (**d**) yield zero spin polarization because the *blue* pulse is linearly polarized. These events occur in overlapping energetic regions, when the experiment is carried out in the band-bending surface area

relaxation time. This has been questioned previously by several authors considering hot electrons [27, 39, 42].

In the following time-dependent measurements we keep the energy of the detected photoelectrons constant and vary the temporal delay between the exciting pulse (which yields the spin polarization) and the probing pulse. As discussed in Sect. 4.2, we use the bichromatic 2PPE method, where pump (*red*) and probe (*blue*) lasers have different photon energies.

In a bichromatic experiment electrons can be excited in two different ways (see Fig. 9). With photons of 1.55 eV average energy and circular polarization (*red*) we create a spin-aligned population with a spin polarization of 50 % in theory or approximately 40 % at room temperature. These electrons are excited with a small excess energy near the Γ point (see Figs. 9a and 9b). The second kind of electrons are excited by linearly polarized photons of 3.1 eV (*blue*) leading to a nonpolarized and highly excited population in the conduction band (see Figs. 9c and 9d). In a second step this population may gain additional energy by other photons. Let us consider electrons absorbing enough energy to reach the vacuum level, which can be properly adjusted by a treatment of the surface with alkali atoms such as Cs. It is obvious that the bichromatic transitions (*red–blue* and *blue–red* transitions) are delay dependent since they involve photons of both pulses. The monochromatic contributions (*blue–blue* or *triple red*) are delay independent since they occur independently of the other laser pulse. Hence, monochromatic transitions will dominate the signal at long delay times ($\Delta t \to \pm\infty$), when the pulses are well separated, whereas the bichromatic events will dominate at delay times within the energetic relaxation time of the electron.

The first event in these multiphoton processes is a direct interband transition and, therefore, not only the intermediate, but also the initial and final state will be energetically distinguishable between the two processes (see Fig. 9). This results in a different kinetic energy of the *blue–red* and the *red–blue* transition photoelectrons and they can be separated by means of an energy analyzer. In the band-bending region, however, we cannot energetically discriminate between the *red* excited populations and the *blue* excited electrons completely. The energy of the initial state (relative to the vacuum energy) varies as a function of the distance to the surface, resulting in partly overlapping energy ranges of *red–blue* and *blue–red* events, see Fig. 6. Next to the intermediate state close to the conduction-band minimum (probed due to a *red–blue* process) one also probes a higher-lying state due to the *blue–red* process. Therefore, working in the bichromatic mode one has to consider that the signal is given by the dynamics of both probed intermediate state. In a semiconductor, however, the dynamics of both probed intermediate states do not interfere with each other. Figure 10a shows the 2PPE photocurrent as a function of the pump-probe delay time. Positive (negative) delay time corresponds to *red–blue* (*blue–red*) transitions. The dotted curve shows the crosscorrelation of the laser pulses (photon fluence) given by the 2PPE yield vs. delay time on a transition-metal surface with very short relaxation times. The result shows that the part of the measured signal due to *blue–red* transitions, which transit via the higher excited intermediate states (see Fig. 9), shows a much faster population decay compared to the *red–blue* part, which probes the dynamics close to the conduction band minimum. Hence, due to the large difference in the population decay between *red–blue* and *blue–red* transitions, we can distinguish very easily between these two (in general interferring) 2PPE processes. The simultaneously measured spin polarization of the photoelectrons is illustrated in Fig. 10b. We can divide our polarization data curves into three different regimes:

Regime I: Excitation with linearly polarized 3.1 eV and emission with circular 1.55 eV photons at a negative delay of 250 fs or more. The comparatively weak signal is dominated by background electrons such as spin-polarized 3-photon processes of 1.55 eV photons (see Fig. 9b) and unpolarized electrons of 2-photon processes of 3.1 eV photons (see Fig. 9c) that lost energy in a scattering event. This regime is characterized by a low count rate and therefore high noise in the spin signal.

Regime II: overlapping regime: the pulses overlap in time and, hence, bichromatic transitions dominate. The left shift of the 2PPE crosscorrelation curve compared to time zero indicates that unpolarized *blue–red* transitions dominate over polarized *red–blue* transitions, resulting in a strongly reduced spin polarization of the detected photoelectrons. In this region we find a strong signal and very accurate spin determination.

Regime III: *red–blue* transition at a delay > 200 fs. The region with the highest spin polarization, since the signal is dominated by optically induced spin-polarized electrons close to the conduction-band edge. These intermediate states are long-lived leading to a high count rate over a wide delay range. Observing the spin polarization with increasing delay time allows the determination of the spin-relaxation time T_1 as long as the energetic relaxation is not much faster than the spin relaxation.

In the following we will discuss the spin-decay processes at the conduction-band minimum and 200 meV above. Figure 11 shows the spin polarization versus positive delay time (*red–blue* transitions) for hot electrons at 200 meV above the conduction-band minimum. The excess energy is a multiple of the longitudinal optical phonon energy $\hbar\omega_{\mathrm{p}} = 36\,\mathrm{meV}$, but it is lower than the split-off valence-band energy $\Delta_{\mathrm{SO}} = 340\,\mathrm{meV}$. The data were corrected for the decaying electron population under the assumption that the background contributions from the two pulses (*blue–blue* and *triple red*) remain constant for all delays. Since the transient population of excited electrons decays quite fast on a time scale of a few 100 fs, the noise level increases considerably over the delay time. The polarization shows no spin relaxation on our time scale of about 20 ps in length. The dash-dotted curve is a fit with $T_1 = 60\,\mathrm{ps}$, which represents the GaAs bulk value expected from our theoretical analysis and experimentally verified for the same sample by means of time-resolved Faraday rotation at $T = 300\,\mathrm{K}$. The fact that the spin polarization remains constant over the investigated delay time indicates that all additional quasielastic scattering processes in the band-bending region, e.g., at steps, defects, and impurities do not cause spin flips. Compared to the bulk, the spin relaxation of hot electrons in the interface is slower.

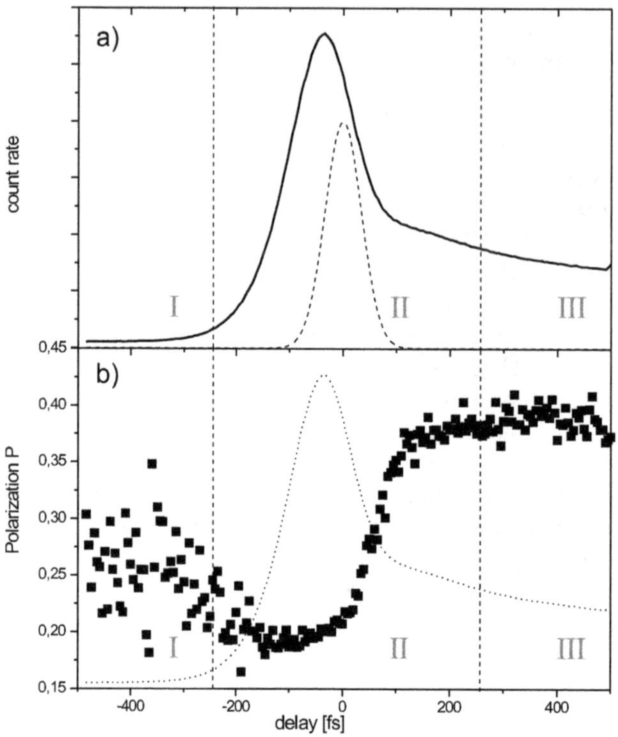

Fig. 10. (**a**) 2PPE photocurrent as a function of the pump-probe delay time. The *dotted curve* shows the crosscorrelation of the laser pulses (photon fluence). (**b**) Spin dynamics around short delays, the dotted curve is a replica from (**a**). We can distinguish 3 regimes. *From left to right*: I. *blue–red* regime: The low count rate is mainly given by background electrons from 3-photon absorption (see Fig. 9). II. Overlapping regime: Spin drops drastically due to highly excited unpolarized electrons, which decay very fast. III. *Red–blue* regime: constant spin polarization on the ps time scale

In Fig. 12 the spin polarization versus positive delay time is shown for electrons at the conduction band minimum of the surface. No direct excitation is possible using photon energies at 1.55 eV (bandgap energy at room temperature: $E_g = 1.42\,\mathrm{eV}$). Time-resolved measurements show that the population increases at zero delay and reaches its maximum at a delay of 100 fs. Hence, the electrons in this energy interval have been generated either in the band-bending region at the surface or in the bulk and have diffused towards the surface. The carriers have undergone many inelastic scattering processes, none of them includes a spin-flip process, as clearly shown in Fig. 12 for longer time delays.

Again, the spin-relaxation rate in the interface is decreased compared to the bulk value (dotted line) as determined by the time-resolved Faraday-

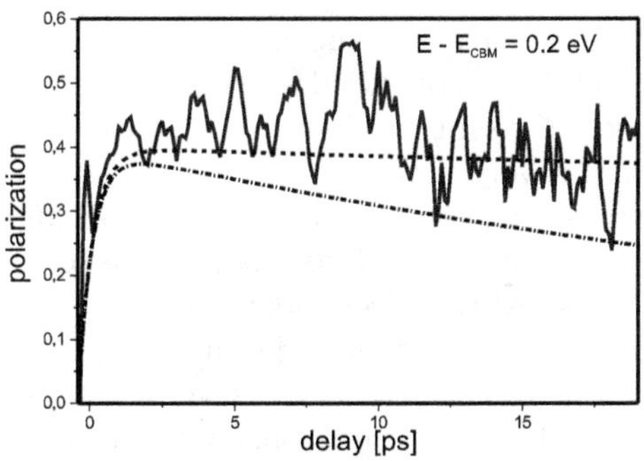

Fig. 11. Spin polarization vs. delay time for hot electrons at 0.2 eV above conduction-band minimum at the surface. The *dashed curve* corresponds to a spin decay on the order of 300 ps. The *dash-dotted curve* represents a spin decay time of 60 ps

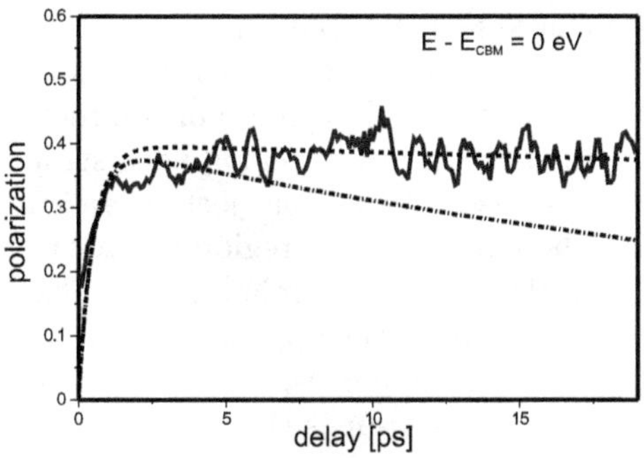

Fig. 12. Spin polarization versus delay time at conduction-band minimum. As a guide to the eye, the *dashed curve* corresponding to a spin decay on the order of 300 ps. The *dash-dotted curve* represents the bulk value of $T_1 = 60$ ps, as obtained by a time-resolved Faraday-rotation experiment

rotation experiment. Our experimental data show a lower limit for T_1 of more than 300 ps. It should be noted that for p-doped semiconductors a much faster population decay of the carriers compared to the spin decay makes the exact determination of T_1 difficult by means of a real-time experiment.

4.4 Comparison Between Experimental and Theoretical Results

For the p-doping concentration considered here, the BAP process, i.e., the electron–hole exchange interaction, is considered to be the dominating contribution to spin relaxation [18]. The reason for the experimental result is shown by our calculations: To obtain reliable spin-relaxation times one has to compute the time development of the distribution functions for spin-up and spin-down electrons including the relevant scattering mechanisms and accounting for in- and outscattering events, thus going beyond existing evaluations of the spin relaxation at the level of Fermi's Golden Rule [18, 22]. In the case of strongly p-doped GaAs these scattering mechanisms are electron–hole exchange scattering, which is comparatively slow but can flip the electron spins, and electron–hole direct Coulomb scattering, which is much faster and cannot flip spins. Because the exchange scattering alone does not lead to a thermalization between electrons and holes, the direct scattering process, which does lead to this thermalization, is needed to provide physically sensible results if the electronic distributions are tracked over several tens of picoseconds. From the time development of the electronic distribution functions one then calculates the spin relaxation and determines the relaxation time by an exponential fit. This procedure yields energy-dependent spin polarizations whose relaxation times have only a very weak energy dependence, in contrast to the rates leading to the spin relaxation.

Since the experimental results presented above were obtained using 2PPE, where the spin polarization is measured for electrons from the surface or interface region of the semiconductor, we can assess the spin relaxation in the semiconductor/metal interface region. The electrons that are detected in the interface region have been optically excited in the bulk of the semiconductor by the *red* laser pulse and have then reached the interface region with strong band bending. It has been argued that the electrons are subject to an efficient scattering process when they reach the band-bending region where they become "hot" and thus that their spin polarization should relax more quickly than in the bulk [39]. Our experimental results show the opposite effect: The spin-relaxation time is significantly longer than the corresponding value for electrons in the bulk. The explanation for this behavior is that the electrons in the interface region are effectively separated from a high hole density because they become effectively localized in the potential well formed by the downward slope of the electron band and the surface. In the surface region, on the other hand, the concentration of holes is strongly reduced because of the depletion of holes close to the interface in a Schottky barrier. This explanation is supported by our calculations: We see from Fig. 4 that the relaxation time is indeed very sensitive to the hole concentration, and that the relaxation time thus is a measure of the hole concentration in the interface region. Using the experimental value of 300 ps relaxation time, we obtain the rough extimate of a 40 % decrease in hole concentration at the surface from Fig. 4.

For the comparison between theory and experiment presented above, we have taken the theoretical bulk spin-relaxation time as a measure of the spin relaxation at the interface and we have allowed the hole concentration to vary over several orders of magnitude. For lower doping concentrations, it is expected that a contribution from the DP process will also come into play and some details of the scattering mechanisms can also be affected by the localization of the electrons close to the surface, which essentially restricts them to move in a two-dimensional layer. There is as yet no straightforward explanation for the reduced DP process at the interface. One may assume a preferential orientation of the conduction electron momentum in the (100) direction of the GaAs interface caused by the electric field in the band-bending region. Taking into account that there is no spin splitting of the conduction band along the (100) axis of GaAs [27], this alignment effect can reduce the spin relaxation due to the precession along the internal field. A quantitative theoretical analysis of the influence of the DP process, as well as an inclusion of the reduced dimensionality of the electron gas at the interface are left for future investigations

Acknowledgements

We thank B. Beschoten for helpful discussions and sharing his experimental time-resolved Faraday rotation results before publication. We are grateful to M. Bauer, M. Fleischhauer, W. Hübner, and S. W. Koch for helpful discussions. Financial support from the DFG and a CPU grant from the Forschungszentrum Jülich are gratefully acknowledged.

References

[1] G. A. Prinz: Device physics – magnetoelectronics, Science **282**, 1660 (1998)

[2] L. J. Sham: Semiconductor devices – closer to coherence control, Science **277**, 1258 (1997)

[3] H. Akinaga, H. Ohno: Semiconductor spintronics, IEEE Trans. Nanotechnol. **1**, 19 (2002)

[4] J. M. Kikkawa, D. D. Awschalom: Resonant spin amplification in n-type GaAs, Phys. Rev. Lett. **80**, 4313 (1998)

[5] G. Schmidt, D. Ferrand, L. W. Molenkamp, A. T. Filip, B. J. van Wees: Fundamental obstacle for electrical spin injection from a ferromagnetic metal into a diffusive semiconductor, Phys. Rev. B **62**, R4790 (2000)

[6] B. T. Jonker: Progress toward electrical injection of spin-polarized electrons into semiconductors, Proc. IEEE **91**, 727 (2003)

[7] D. D. Awschalom, D. Loss, N. Samarth (Eds.): *Semiconductor Spintronics and Quantum Computation* (Springer, Berlin, Heidelberg 2002)

[8] E. Johnston-Halperin, D. Lofgreen, R. K. Kawakami, D. K. Young, L. Coldren, A. C. Gossard, D. D. Awschalom: Spin-polarized Zener tunneling in (Ga,Mn)As, Phys. Rev. B **65**, 041306 (2002)

[9] S. A. Crooker, D. D. Awschalom, J. J. Baumberg, F. Flack, N. Samarth: Optical spin resonance and transverse spin relaxation in magnetic semiconductor quantum wells, Phys. Rev. B **56**, 7574 (1997)

[10] S. Hallstein, J. D. Berger, M. Hilpert, H. C. Schneider, W. W. Rühle, F. Jahnke, S. W. Koch, H. M. Gibbs, G. Khitrova, M. Oestreich: Manifestation of coherent spin precession in stimulated semiconductor emission dynamics, Phys. Rev. B **56**, R7076 (1997)

[11] P. Y. Yu, M. Cardona: *Fundamentals of Semiconductors*, 3rd ed. (Springer, Berlin, Heidelberg 2001)

[12] G. L. Bir, G. E. Pikus: *Symmetry and Strain-Induced Effects in Semiconductors* (Wiley, New York 1974)
[13] E. L. Ivchenko, G. E. Pikus: *Superlattices and Other Heterostructures*, vol. 110, 2nd ed., Springer Series in Solid-State Sciences (Springer, Berlin, Heidelberg 1997)
[14] R. Winkler: *Spin-Orbit Coupling Effects in Two-Dimensional Electron and Hole Systems*, vol. 191, Springer Tracts in Modern Physics (Springer, Berlin, Heidelberg 2003)
[15] G. E. Pikus, A. N. Titkov: Spin relaxation under optical orientation in semiconductors, in F. Meier, B. P. Zakharchenya (Eds.): *Optical Orientation*, vol. 8, Modern Problems in Condensed Matter Sciences (North-Holland, Amsterdam 1984) pp. 73–131
[16] C. Kittel: *Quantum Theory of Solids* (Wiley 1987)
[17] B. A. Sanborn, B. Y. K. Hu, S. Das Sarma: Correction to the decay rate of nonequilibrium carrier distributions due to scattering-in processes, Phys. Rev. B **49**, 7767 (1994)
[18] P. H. Song, K. W. Kim: Spin relaxation of conduction electrons in bulk III–V semiconductors, Phys. Rev. B **66**, 035207 (2002)
[19] M. I. Dyakonov, V. I. Perel: Spin orientation of electrons associated with interband absorption of light in semiconductors, Sov. Phys. JETP **33**, 1053 (1971)
[20] G. L. Bir, A. G. Aronov, G. E. Pikus: Spin relaxation of electrons due to scattering by holes, Sov. Phys. JETP **42**, 705 (1976)
[21] G. E. Pikus, G. L. Bir: Exchange interaction in excitons in semiconductors, Sov. Phys. JETP **33**, 208 (1971)
[22] M. Z. Maialle: Spin relaxation of electrons in p-doped quantum wells via the electron–hole exchange interaction, Phys. Rev. B **54**, 1967 (1996)
[23] L. Reichl: *A Modern Course in Statistical Physics* (University of Texas Press, Austin 1980)
[24] I. Žutić, J. Fabian, S. Das Sarma: Spintronics: Fundamentals and applications, Rev. Mod. Phys. **76**, 323 (2004)
[25] L. J. Sham: Theory of spin coherence in semiconductor heterostructures, J. Magn. Magn. Mater. **200**, 219 (1999)
[26] M. W. Wu, H. Metiu: Kinetics of spin coherence of electrons in an undoped semiconductor quantum well, Phys. Rev. B **61**, 2945 (2000)
[27] F. Meier, B. P. Zakharchenya (Eds.): *Optical Orientation* (North-Holland, Amsterdam 1984)
[28] M. M. Glazov, E. L. Ivchenko: Precession spin relaxation mechanism caused by frequent electron–electron collisions, JETP Lett. **75**, 403 (2002)
[29] M. Q. Weng, M. W. Wu: Spin-dephasing in n-type GaAs quantum wells, Phys. Rev. B **68**, 075312 (2003)
[30] C. Lechner, U. Rössler: An extension of the optical bloch equations: A microscopic approach including spin and carrier–phonon scattering (2004), cond-mat/0412370
[31] W. Schäfer, M. Wegener: *Semiconductor Optics and Transport Phenomena* (Springer, Berlin, Heidelberg 2002)
[32] R. Binder, S. W. Koch: Nonequilibrium semiconductor dynamics, Prog. Quant. Electron. **19**, 307 (1995)

[33] M. Kira, F. Jahnke, W. Hoyer, S. W. Koch: Quantum theory of spontaneous emission and coherent effects in semiconductor microstructures, Prog. Quant. Electron. **23**, 189 (1999)

[34] F. Rossi, T. Kuhn: Theory of ultrafast phenomena in photoexcited semiconductors, Rev. Mod. Phys. **74**, 895 (2002)

[35] H. Haug, S. W. Koch: *Quantum Theory of the Optical and Electronic Properties of Semiconductors*, 4th ed. (World Scientific, Singapore 2004)

[36] B. Beschoten, E. Johnston-Halperin, D. K. Young, M. Poggio, J. E. Grimaldi, S. Keller, S. P. DenBaars, U. K. Mishra, E. L. Hu, D. D. Awschalom: Spin coherence and dephasing in GaN, Phys. Rev. B **6312**, 121202 (2001)

[37] A. Sutton, R. Balluffi: *Interfaces in Crystalline Materials*, vol. 51, Monographs on the Physics and Chemistry of Materials (Oxford University Press, Oxford 1995)

[38] W. Mönch: *Semiconductor Surfaces and Interfaces*, 3rd ed., Springer Series in Surface Sciences (Springer, Berlin, Heidelberg 2001)

[39] D. T. Pierce, F. Meier: Photoemission of spin-polarized electrons from GaAs, Phys. Rev. B **13**, 5484 (1976)

[40] M. Aeschlimann, M. Bauer, S. Pawlik, W. Weber, R. Burgermeister, D. Oberli, H. C. Siegmann: Ultrafast spin-dependent electron dynamics in fcc Co, Phys. Rev. Lett. **79**, 5158 (1997)

[41] J. Kirschner: Sources and detectors for polarized electrons, in R. Feder (Ed.): *Polarized Electrons in Surface Physics* (World Scientific, Singapore 1985)

[42] F. Bentivegna, P. Yvernault, A. V. Petukhov, T. Rasing: Time-resolved photo-induced nonlinear magneto-optical Kerr effect for the study of spin dynamics at a GaAs (001) surface, Appl. Phys. B **68**, 519 (1999)

Index

Index

Topics in Applied Physics

85 **Solid–Liquid Interfaces**
Macroscopic Phenomena – Microscopic Understanding
By K. Wandelt and S. Thurgate (Eds.) 2003, 228 Figs. XVIII, 444 pages

86 **Infrared Holography for Optical Communications**
Techniques, Materials, and Devices
By P. Boffi, D. Piccinin, M. C. Ubaldi (Eds.) 2003, 90 Figs. XII, 182 pages

87 **Spin Dynamics in Confined Magnetic Structures II**
By B. Hillebrands and K. Ounadjela (Eds.) 2003, 179 Figs. XVI, 321 pages

88 **Optical Nanotechnologies**
The Manipulation of Surface and Local Plasmons
By J. Tominaga and D. P. Tsai (Eds.) 2003, 168 Figs. XII, 212 pages

89 **Solid-State Mid-Infrared Laser Sources**
By I. T. Sorokina and K. L. Vodopyanov (Eds.) 2003, 263 Figs. XVI, 557 pages

90 **Single Quantum Dots**
Fundamentals, Applications, and New Concepts
By P. Michler (Ed.) 2003, 181 Figs. XII, 352 pages

91 **Vortex Electronis and SQUIDs**
By T. Kobayashi, H. Hayakawa, M. Tonouchi (Eds.) 2003, 259 Figs. XII, 302 pages

92 **Ultrafast Dynamical Processes in Semiconductors**
By K.-T. Tsen (Ed.) 2004, 190 Figs. XI, 400 pages

93 **Ferroelectric Random Access Memories**
Fundamentals and Applications
By H. Ishiwara, M. Okuyama, Y. Arimoto (Eds.) 2004, Approx. 200 Figs. XIV, 288 pages

94 **Silicon Photonics**
By L. Pavesi, D.J. Lockwood (Eds.) 2004, 262 Figs. XVI, 397 pages

95 **Few-Cycle Laser Pulse Generation and Its Applications**
By Franz X. Kärtner (Ed.) 2004, 209 Figs. XIV, 448 pages

96 **Femtsosecond Technology for Technical and Medical Applications**
By F. Dausinger, F. Lichtner, H. Lubatschowski (Eds.) 2004, 224 Figs. XIII 326 pages

97 **Terahertz Optoelectronics**
By K. Sakai (Ed.) 2005, 270 Figs. XIII, 387 pages

98 **Ferroelectric Thin Films**
Basic Properies and Device Physics for Memory Applications
By M. Okuyama, Y. Ishibashi (Eds.) 2005, 172 Figs. XIII, 244 pages

99 **Cryogenic Particle Detection**
By Ch. Enss (Ed.) 2005, 238 Figs. XVI, 509 pages

100 **Carbon**
The Future Material for Advanced Technology Applications
By G. Messina, S. Santangelo (Eds.) 2006, 245 Figs. XXII, 529 pages

101 **Spin Dynamics in Confined Magnetic Structures III**
By B. Hillebrands, A. Thiaville (Eds.) 2006, 164 Figs. XIV, 345 pages

102 **Quantum Computation and Information**
From Theory to Experiment
By H. Imai, M. Hayashi (Eds.) 2006, 49 Figs. XV, 281 pages

103 **Surface-Enhanced Raman Scattering**
Physics and Applications
By K. Kneipp, M. Moskovits, H. Kneipp (Eds.) 2006, 221 Figs. XVIII, 464 pages

104 **Theory of Defects in Semiconductors**
By D. A. Drabold, S. K. Estreicher (Eds.) 2006, approx. 60 Figs. XIII, 297 pages

www.ingramcontent.com/pod-product-compliance
Lightning Source LLC
Chambersburg PA
CBHW051736180126
38435CB00003B/91
9783540201083